Genetic Nature/Culture

Genetic Nature/Culture

Anthropology and Science beyond the Two-Culture Divide

EDITED BY

Alan H. Goodman, Deborah Heath, and M. Susan Lindee

UNIVERSITY OF CALIFORNIA PRESS

Berkeley Los Angeles London

University of California Press
Berkeley and Los Angeles, California

University of California Press, Ltd.
London, England

© 2003 by the Regents of the University of California

Library of Congress Cataloging-in-Publication Data

Genetic nature/culture : anthropology and science beyond the two-culture divide / edited by Alan H. Goodman, Deborah Heath, and M. Susan Lindee.
7 p. cm.
Papers presented at a Wenner-Gren Foundation international symposium, held June 11–19, 1999, in Teresópolis, Brazil.
Includes bibliographical references and index.
ISBN 0–520–23792–7 (alk. paper)—ISBN 0–520–23793–5 (pbk. : alk. paper)
1. Human population genetics—Congresses. 2. Human genetics—Research—Congresses. 3. Human genetics—Moral and ethical aspects—Congresses. 4. Anthropological ethics—Congresses. I. Goodman, Alan H. II. Heath, Deborah, 1952– III. Lindee, M. Susan.

GN289 .G455 2003
599.93'5—dc21 2002152222

Manufactured in the United States of America
13 12 11 10 09 08 07 06 05 04
10 9 8 7 6 5 4 3 2 1

The paper used in this publication is both acid-free and totally chlorine-free (TCF). It meets the minimum requirements of ANSI/NISO Z39.48–1992 (R 1997) *(Permanence of Paper).*

CONTENTS

ILLUSTRATIONS

FIGURES

TABLES

FOREWORD

Sydel Silverman

In the last decade of the twentieth century, anthropology, like many other disciplines, was deeply affected by the revolution in genetic science. Both as a set of methodological tools and as an object of study in its own right, genetics assumed an increasingly important place in anthropological research and practice, presenting new opportunities and new challenges. At the same time, public discourse around genetics intensified, touching on long-held concerns of anthropologists; yet the anthropological voice was not often heard, even when it was sorely needed. This confluence of developments led to the idea for a conference on anthropology and the new genetics. It came to fruition as a Wenner-Gren Foundation's international symposium, "Anthropology in the Age of Genetics: Practice, Discourse, Critique," which took place in June 1999, in Teresópolis, Brazil. This volume is a product of that conference.

I had become aware of the reverberations of the new genetics in anthropology primarily from reading the nearly one thousand grant proposals submitted to Wenner-Gren each year. This perspective afforded a significant—albeit only partial—window on the discipline. From this window I could see enormous potential for research in all areas of anthropology but also some danger signs. For each of the subfields, the developments in genetics opened up new problems for study and new approaches to old problems, but they also brought new difficulties.

The anthropological study of living nonhuman primates was profoundly affected by the advent of new genetic methods. For some time in this field, the predominant goal had been to identify the evolutionary significance of behaviors and social patterns. A key question, of course, was whether genes actually did get replicated in accordance with the predictions; but until recently, this question could be addressed only by inference. The invention of the polymerase chain reaction (PCR) allowed for the amplification of small amounts of DNA sufficient for the kind of analysis that could determine paternity. With the possibility of making that determination directly, more and more research designs tested hypotheses about the selective advantage of food getting, mating, infant care, and other social behaviors. Almost every project on some aspect of primate social behavior now included DNA analysis to establish the "relatedness" of individuals whose interactions were observed. Some of the early results were surprising, and they called into question prevailing theories concerning mate selection, aggression, coalition formation, and other patterns of primate sociality. This powerful tool had a downside, however, to the extent that it tilted research toward a search for genetic explanation.

A major danger in this new primatology stemmed from its very success with genetics: the misconstrual of implications for understanding human behavior. All too often, grant applications for projects to demonstrate the evolutionary significance (selective advantage) of certain behaviors in monkeys or apes (a goal now more attainable with the new genetic technology) would conclude with the promise that this would shed light on "comparable" behavior in modern humans. But infanticide in langurs or chimpanzees is not the same thing as child abuse; dominance patterns in baboons do not equate with sexual harassment in the workplace; "demonic male" behavior in great apes does not explain proclivities to war.

This problem relates to the "98% issue," discussed by Jonathan Marks (see chapter 7, this volume), the supposed genetic commonality between chimpanzees and humans. Commonality, of course, invites comparison. The pitfall comes from using a method of comparison that takes two end points and connects them directly to a common origin. What comparative analysis of human and nonhuman primates requires is a grasp of the trajectories of human cultural evolution and historical change that account for the diversity of patterns known through the archaeological and ethnographic records. Trends in anthropology that separate primatology from archaeology and cultural anthropology can only encourage misuse of primate studies.

Signs of a rapprochement between paleoanthropology and the new genetics came first to Wenner-Gren when a few young biological anthropologists expressed interest in mitochondrial DNA (mtDNA) analysis. This interest grew rapidly in the field, opening up new research areas and proposing new, often controversial, answers to old questions. Both nuclear and mtDNA methods soon established themselves not only as powerful adjuncts to the time-honored morphological study of fossils but sometimes also as direct challenges to it. While most researchers asserted that the two approaches were complementary, the problem of bringing them together was not easily solved.

Three major landmarks in paleoanthropology resulting from the genetics revolution stand out. The first, drawing on an earlier idea of a non-Darwinian molecular clock, was the acceptance of a drastic shortening of the time period since the chimpanzee-hominid divergence, to around 5 million years. The second was the establishment of mitochondrial DNA methods of chronology to propound the "Eve" and "Out of Africa" hypotheses, initiating a new phase in an older debate over single-lineage versus multiregional models of human origins. Both breakthroughs were based on methods of inference from extant populations. The third landmark was the successful extraction of mitochondrial DNA from a Neanderthal specimen, which bolstered the argument for a species difference between Neanderthals and modern humans and an early separation between them. That issue is far from resolved, but the significance of the event lies in the potential for obtaining DNA directly from ever older remains.

The mitochondrial DNA methodology was immediately applied to the study of human population history. One of the early Wenner-Gren projects was Mark Stoneking's analysis of blood samples from populations on six Indonesian islands, designed to trace prehistoric migrations through the Pacific. Other proposals for the study of DNA in diverse populations followed, all aspiring to uncover group relationships and ultimately to reconstruct population movements and adaptations.

From my perspective as a champion of four-field anthropology (including biological anthropology, archaeology, cultural anthropology, and linguistics), this development held the promise of integrating data from all the fields to address issues of population history and relationships. Increasingly, however, what we saw at Wenner-Gren were applications claiming to reconstruct population history from DNA alone, without recourse to independent evidence from prehistory or other sources and with little questioning as to what DNA can actually reveal. There was also an unfortunate use of nonbiological concepts, such as "ethnic group" or language group (race being studiously avoided, for the most part), with the assumption that such entities can be identified directly from the DNA.

The Human Genome Diversity Project (HGDP) was a particular product of the interest in genetic relationships of populations. It was born out of the messianic vision of the geneticist Luca Cavalli-Sforza, who was engaged in reconstructing world population history and who argued for the need to bring an appreciation of human genetic diversity to the effort to map the genome. Many biological anthropologists embraced the HGDP in the early 1990s in the hope that it would yield an invaluable data bank of population genetics that could be applied to a wide range of old and new anthropological problems. A surge of criticism followed, however, not only from members and advocates of the potential "target" groups (those whose blood would be collected) but also from anthropologists who saw theoretical, methodological, and ethical difficulties in the project. The criticisms were valid; so also were the hopes for an invigorated biological anthropology equipped to take on important research problems.

In the field of bioarchaeology, DNA extraction from skeletal remains extended both the time range for which certain research questions could be asked and the kinds of data that could be obtained. An early application titled "DNA Extraction in Mummies" elicited skepticism from the reviewers, who nevertheless recommended support because, as one said, "if this can really be done it will be momentous." It could indeed be done and has been done ever more frequently. Here as in other fields, however, the danger is that genetics may forge ahead in the interpretive process (perhaps pronouncing the discovery of population origins or relationships) while leaving aside the archaeological, historical, and cultural evidence.

One trend that particularly worried me was the use of genetics to infer social patterns. For instance, the claim was sometimes made that kinship patterns could be reconstructed from DNA. Yet one of the great discoveries of anthropology has been the distinction between biological relatedness and kinship systems. Thus, genetic studies of people buried together might reveal resemblances in their mitochondrial DNA, but this cannot be taken to mean matrilineal relationships, matrilocality, or matrifocality (terms that tend to be used interchangeably, although they mean quite different things). A similar problem arises with the uncritical extension of genetic relationships in a sample to the identification of a group or, worse, with an inference of group identity. Cultural anthropologists know identity to be an extremely malleable phenomenon and a slippery concept. Group self-identification may follow a trajectory very different from a history suggested by genetic, linguistic, or cultural evidence.

Consider the example (explored by Frederika Kaestle, chapter 14, this volume) of an adult male buried with many women and children. What were their relationships in life? Was this group a noble with sacrificial slaves or captives, a polygynous kinship group, or something else? Each of these terms corresponds to a variety of possible arrangements known in the ethnographic and historical record, and each term carries assumptions that can be grossly misleading. DNA analysis can offer relevant data, but it cannot stand alone. We need an updated version of ethnographic analogy that takes account of the range of possibilities known from the ethnographic record and is sensitive to the way social systems actually work—including how people apply rules flexibly, adapt to circumstances, and invent rationales.

The genetics revolution entered cultural anthropology in several ways, reflecting, in part, the divisions in that field. Some cultural anthropologists, already committed to neo-Darwinian approaches (such as evolutionary ecologists and evolutionary psychologists) embraced it; their language of evolutionary processes shifted from metaphorical uses of the term *genes* to explicit invocations of genetics. Probably the majority of cultural anthropologists were skeptical of genetic determinism, but many were at the same time interested in the social impact of genetics. That interest surfaced first in studies of the new reproductive technologies, which struck anthropological chords because of their implications for kinship theory. New fields and new ways of thinking about old concepts emerged: the new kinship studies, the politics of reproduction, and challenges to accustomed ways of looking at gender, property, and identity.

A second kind of interest focused on the study of knowledge production, including the production of genetic knowledge. We saw a convergence with science studies in ethnographies of laboratories and cultural analyses of genetic science. There followed the beginnings of research on ways in which knowledge of genetic processes outside the human body was being applied in social and cultural contexts. The topics engaged included biodiversity, conservation, and organic-species alteration, which were joined to current concerns in cultural anthropology with transnationalism and social movements. (An example is Chaia Heller and Arturo Escobar's essay, chapter 8 in this volume.)

A third arena was medical anthropology, including research on institutional settings (e.g., in diagnosis and counseling), where genetics and diseases known or assumed to have genetic bases are at issue. In the latter category, some anthropologists took the designation "disease population" uncritically, while others focused on how the disease was culturally constructed. Still others took as their subject the social groups constituted around genetically based diseases (see the essay by Karen-Sue Taussig, Rayna Rapp, and Deborah Heath, chapter 3, this volume).

The fact that most of the work of cultural anthropologists has been critical of biological explanation has left it open to countercharges of scientific naïveté. The critical perspective is probably the major contribution that anthropology can make to understanding the social construction and impact of genetic science and practice. If this perspective is to be taken seriously, however, cultural anthropologists must show themselves to be fully competent in the biological component of their subject matter.

Many of the dangers I have alluded to derive from the growing separation of the subfields of (American) anthropology during the last decades of the century. Ironically, at the same time that the genetics revolution revealed the complementarity of the different fields, we saw institutional and intellectual barriers raised among them. This situation was a central concern for me throughout my presidency of Wenner-Gren (1987–99); I did not agree with many in the discipline who saw the barriers as inevitable.

During a symposium in March 1998, "New Directions in Kinship Study," the discussions frequently turned to the new genetics, invariably engaging both cultural and biological issues. In a coffee-break conversation I had with

Sarah Franklin and Jonathan Marks, the idea came up for a conference that would continue these discussions. I saw in that idea not only an opportunity to assess important developments in anthropology but also the possibility of showing that the increasingly divergent subdisciplines actually had a great deal to say to one another. This topic was particularly apt because it brought together specialists from the polar ends of the spectrum of subfields: on one side, the most "scientific" of the biological anthropologists, and on the other, the social-cultural anthropologists doing cultural studies of science, who favored interpretive approaches and worked in nontraditional sites. It seemed to me that if these two poles could find common ground, it would speak directly to the potential of anthropology as an integrated discipline.

To turn the idea into a symposium, I recruited as organizers Alan Goodman, a biological anthropologist who had long been an advocate of biocultural synthesis, and Deborah Heath, who had done field research on genetic practices in laboratories, in clinics, and with advocacy groups. The two had never met before; it was an arranged marriage, and it proved a success. The three of us worked together on the conference plan and program. In selecting the participants, we sought a balance of about one-third each from biological anthropology, cultural anthropology, and related disciplines (evolutionary biology, human genetics, sociology, history of science, and science studies). We also looked for individuals who were open to unfamiliar material and perspectives.

In six days of intensive meetings (and many more hours of equally intensive informal conversation), the conferees—who came from diverse specialties and viewpoints—melded into a unified group, not dissolving the differences among them but engaging one another around newly discovered common interests and commitments. Every paper, written in advance of the conference, was changed in the process. This volume presents the revised versions of the papers and reflects what transpired during our time together at Teresópolis. It offers not only a unique appraisal of issues and problems of the age of genetics and geneticization but also a testimony to the possibility of building bridges across disciplinary divides.

PREFACE AND ACKNOWLEDGMENTS

On April 17, 2002, Dr. J. Craig Venter, the scientific entrepreneur who headed Celera Genomics' commercially funded effort to sequence the human genome, revealed on the television program *Sixty Minutes II* that Celera's genome sequence data was based largely on Venter's own DNA. Up to this point, both Celera and the federally funded U.S. genome project had said publicly that they based their sequence data on anonymous donor DNA, which they described as representing a cross section of different ethnic or racial groups. Is Venter's disclosure scientifically significant to the paid subscribers to Celera's genome database? Probably not, especially since both the public and private human genome initiatives have produced to date what amount to rough drafts in which individual idiosyncrasies may not matter. Does the disclosure affect the public credibility of the genomic enterprise? Probably, given the global tensions surrounding the ownership and use of human DNA. Scientific organizations that mislead the public about whose DNA is being sequenced can expect to exacerbate international concerns about the control of genomic knowledge and materials.

Whose genome is it anyway? Venter, who lost his position as the head of Celera in January 2002, now plans to write a book about his own genome. He will be the author of the story of his own DNA. Meanwhile, just a few weeks before Venter's announcement, a media-savvy Yanomámi group met in São Paolo to demand the return of Yanomámi blood samples collected by biological anthropologists in the 1960s. The Pro-Yanomámi Commission has begun an international public campaign to bring the samples back, not for research but, as Yanomámi Davi Kopenawa put it, to be spilled into the Orinoco River. This group argues that the geneticist James V. Neel and the anthropologist Napoleon Chagnon, both of the University of Michigan at that time, violated the 1947 Nuremberg Code when they collected blood and other tissue in exchange for trade goods as part of their fieldwork in the 1960s. Now frozen in laboratories at several institutions in the United States, the samples continue to be used in biological research. "Genetic heritage" thus engages with the personal and public lives of molecular biotechnologists, Yanomámi activists, anthropologists in the field, and everyday medical consumers facing questions about genetic testing or fetal diagnostics. Genes say many things for many different people. They are worked into the nexus of desire, identity, colonialism, indigenism, parental love, and global commerce. One is left to wonder, perhaps productively, for whom the genome speaks.

In this volume, we present a nested set of complexities that refuse easy resolution. We are invested in complexity, not mystification, and committed to the possibility that the genome might speak for us all. Uncovering biocultural complexities—for example, of heritability or race—interrogates objects as processes and turns nouns into verbs. We hope that these essays illuminate the processes at risk of being obscured or made mysterious by dominant genetic discourses that reduce biologies and cultures to mechanistic metaphors and models.

Richard Lewontin, in his book *The Triple Helix,* explores the constant coproductivity of organisms, genes, and environments through the development of the organism. As scholars who study human activity, we add to this list culture, a particularly messy, meaning-making fourth helical strand. A testament to the productive dialogue underlying this collection of essays, *Genetic Nature/Culture* represents a collective commitment to a relational, dialectical perspective on genetics and its cultural-material complexities.

The chapters in this collection, representing a rich mix of perspectives from biological and cultural anthropologists, geneticists, sociologists, and historians, examine genetics at the intersection between nature and culture. The contributors share the conviction that genetic practice and discourses about genetics are fertile material–symbolic terrain for considering key questions such as the relationships between science and society and, within the academy, between the sciences and the humanities.

The book has its origin in papers presented at the Wenner-Gren International Symposium "Anthropology in the Age of Genetics: Practice, Discourse, Critique," held near the end of the second millennium, June 11–19, 1999, in Teresópolis, Brazil. During our week together, participants engaged one another, almost always patiently, exploring and clarifying divergent perspectives while focused on better understanding the work and expertise of the others. This inspirational moment of border crossing between C. P. Snow's "two cultures"—the sciences and the humanities—made this book possible.

The symposium took shape in the shadow of the so-called science wars, in which science studies was drawn into a public and rancorous manifestation of the ideological division between interpretive and scientific perspectives. We saw the symposium as an opportunity to transcend the polarities, both topical and theoretical, that separate subfields within anthropology and that continually reproduce caricatures of both scientists and humanists. Our aims were, first, to disrupt these ostensible boundaries by locating the common ground between contemporary studies of genetics and uses of genetic technique, and, second, to provide a laboratory to determine more precisely what these different realms of research and practice might have to say to each other. Although biological-scientific anthropology and cultural-interpretive anthropology increasingly are developing separate worldviews, vocabularies, and domains of practice, we saw potential for intellectual alliance through our shared interest in situating genetic knowledge within organisms, environments, histories, and cultures. It was and remains our conviction that pursuing these issues in dialogue with one another will make our various approaches to genetics more fully anthropological. A central goal of this book is to open conversations about both the growing impact of genetics on anthropological practice and the ethnographic investigation of genetic worlds inside and outside the laboratory.

We are indebted beyond words to the Wenner-Gren Foundation for Anthropological Research for both supporting the conference and providing a book publication grant. Laurie Obbink and Mary Beth Moss at the foundation made innumerable, invaluable contributions to the planning stages and to the conference itself. Cochairs Alan Goodman and Deborah Heath bow in gratitude for Obbink's insights as a veteran of many previous Wenner-Gren symposia. At the Hotel Rosas dos Ventos, the conference's hosts made all the right moves behind the scenes and provided us with a magical context. Erin Koch, the conference monitor, brought her intellectual insights and good cheer to the table each day, in addition to serving as the symposium's indefatigable scribe. We also owe special thanks to Rayna Rapp for providing key ideas and network connections in the initial stages of conference planning. Stan Holwitz, editor at the University of California Press, believed in the project from the start and expertly guided us from conference papers to an integrated book. Laurie Smith of Hampshire College made order of disordered page numbers, endnotes, and references, and then with no displeasure passed along these and other tasks to the cordial professionals Laura Pasquale and Marian Olivas at the University of California Press. Deborah Heath and Alan Goodman extend our love and appreciation to our coeditor, the science historian M. Susan Lindee, for her wit, acumen, and editorial sharpshooting and for agreeing to join our editorial adventures following the symposium in Teresópolis. We also thank Joan Barrett, Hampshire College, for proofreading the book and the University of California Press, in particular Erika Büky and Bonita Hurd, for exceptional editorial work. Finally, we offer our heartfelt appreciation to Sydel Silverman, former president of the Wenner-Gren Foundation, whose commitment to an integrated anthropology gave birth to the symposium, and whose vision guided and inspired us throughout.

Introduction

Anthropology in an Age of Genetics

Practice, Discourse, and Critique

M. Susan Lindee, Alan Goodman, and Deborah Heath

On June 26, 2000, the rival scientific factions vying to complete the DNA sequencing of the human genome declared a truce. The race that might have been won by a single victor was set aside, and credit for completing a working draft of the sequence was to be shared by the Human Genome Project's international, publicly funded consortium and by Celera Genomics, a private company. At the press conference where this laying down of arms was announced, President Bill Clinton stood flanked by Craig Venter, the head of Celera, and Francis Collins, director of the National Institutes of Health's Human Genome Project (HGP) in the United States. The sequence was front-page news; the top banner of the *New York Times* declared, "Genetic Code of Human Life Is Cracked by Scientists" (June 27, 2000).

This very public and reluctant coalition of a government-sponsored, transnational scientific program and a biotechnology industry heavyweight is just one node in a wide-ranging, heterogeneous network of human and nonhuman actors that constitutes genetics-in-action (*pace* Latour 1987; cf. Flower and Heath 1993; Heath 1998a,b). The knowable, manipulable human genome also belongs to health advocates living with particular heritable diseases, who raise research funding and run on-line forums (Heath et al. 1999; Taussig, Rapp, and Heath, chapter 3, this volume). It belongs to scientists in Japan, China, the United Kingdom, France, and Germany, as well as to DNA "donors" (voluntary or not) from Iceland and the Amazon. And it is the province of essential nonhuman players, from centralized sequence databases and their search engines to genetically modified organisms (GMOs). Genomes, human and other, are dynamic, emergent entities still under negotiation as territory, property, soul, medical resource, and national prize. Meanwhile, narratives of both technoscientific expertise and everyday life have come to be scripted in a genetic idiom deployed by laypeople and experts alike.

In the decade and a half since the Human Genome Project was launched, new technologies, institutions, practices, and ideologies built around genes have constituted a technocultural revolution. The age of genetics is also an era of what Abby Lippman calls geneticization (1991, 1992) and what Paul Rabinow (1996) calls biosociality. Lippman's geneticization describes a widely dispersed network of genetic resources, power relations, and ideas elaborating the meanings of the gene. Rabinow playfully transposes the terms of sociobiology and the credo that biological forces (genes) explain behavior and sociality. Drawing on Foucault's notion of biopower, he underscores the coconstitution of nature and culture and all their familial iterations. Both concepts aptly map the genetic borderland that this volume explores, as we present the fruits of a dialogue on genetics that brings together cultural studies of genetic knowledge production and natural-scientific studies that foreground cultural-historical context.

For anthropologists, genetics, as both technoscientific and technocultural practice, has provided a fertile medium for cultural and biological studies. Biological anthropologists who study human genome evolution and diversity have benefited immensely from the transfer of technologies like polymerase chain reaction and bioinformatics that have been integral to the HGP. Ethnographers, in turn, have found a rich array of new field sites in and beyond the lab. Sometimes they have brought to their own research firsthand participant-observers' knowledge of those aforementioned technologies so central to the work of contemporary biological inquiry. At the very moment when some have trumpeted their intentions to cleave the divisions between science and not-science more deeply, genetics has provided anthropologists from both sides with opportunities for constructive, intellectual engagement. The potential for these and broader engagements was chief among the optimistic aspirations that launched this volume.

The essays collected here began as contributions to the Wenner-Gren International Symposium "Anthropology in the Age of Genetics: Discourse, Practice, Critique." Our symposium was a social experiment informed by scholarship in science studies, in which the technical, the cultural, and the ideological are inextricably bound together. The mix of participants was carefully constructed as a test of the premise that world-making takes place in an interactive web or network, and that pulling together different bits of the network brings the silences of any particular position into sharp relief. Having come from diverse fields and stood in different places, we learned theories, practices, ideas, and perspectives from each other. And sometimes we listened but remained puzzled. In our juxtaposition and framing of the essays in this volume, we have tried to mark both the synergies of this experience and the questions that remain to be answered.

Among the most striking synergies was a deep, shared interest in the multiple meanings and consequences of "opening the veins" of indigenous people in Brazil, the Icelanders, the Amish, Africans and African Americans, Little People, Native Americans, at-risk populations, and even man's best friend. Our discussions returned again and again to the many threads running through these acts of collecting biological samples: blood, cheek swabs, bone, hair. While there may be no particular intellectual privilege in any given microcosm, this highly charged moment was clearly a point of entry to compelling concerns about love, power, and knowledge. The narrative, we concluded, can be more painful than the blood stick. In thinking about the disembodied sample and the database that can never be the product of a "clean birth," we found a shared concern with the cultural-historical contexts that link power relations and the politics of difference to the production of knowledge, with systems built around biologicals. By what standards can genetic data be made to speak about population differences, colonialism, global capitalism, human suffering, and social order?

The investigation of complexity, or complex relationalities, also emerged in our discussions as a salient concern for all participants. One participant stated flatly that s/he had a "stake in complexity," not to obscure the issues but to deepen the perspective. Complexity is important to both cultures. This insight has been reinforced since the inception of the Human Genome Project, which institutionalizes intense reductionism by its fixation on a static map, as well as increasingly facilitates the scientific study of complexity—of interaction, expression, development, and context, an era of proteomics.[1] With this in mind, one might say that genetics is taking an *anthropological turn.* We hope that this volume can begin to map the overlapping networks that bind a sheep named Dolly to the Yanomámi of South America, and the African diaspora to the genome of the daffodil.

Two stories from our conference are illustrative. One evening in Teresópolis, a group of locals, primarily employees of our hotel and surrounding horse ranch, staged a traditional Brazilian harvest festival around a bonfire in an open meadow. The actors were wildly attired and included men dressed as women and both men and women with painted black faces or long blonde wigs or both. Presented in Portuguese and therefore incomprehensible to most of the attending scholars, the skit seemed to involve a minister, a marriage, and jokes about sex, religion, and drunkenness. It produced laughter in some members of the audience, which included local residents, and bewilderment in most of us. Some of us found the skit and the costumes offensive and left. Others, unaware of their colleagues' departures, joined the dancing around the fire at the end of the show. Coincidentally, we were scheduled to discuss race, genetics, and anthropology the next day.

The following morning, the skit and varying responses to it became a way to explore the specificity of racialized meanings and experiences. Brazilian racial politics made interpreting the blackface difficult. The dancers themselves were people of color, at least by European and North American standards. They were also lower-level employees in the service economy of a less than affluent region. And their burlesque could be seen to be racist as well as sexist and classist. The carnivalesque elements in the skit suggested the overthrow of accepted hierarchies of power (the mocking trickster), while the costumes and sexualized joking seemed to replicate the long history of Western oppression of marked bodies. In some ways the skit was a perfect lesson, an intersection of power, culture, history, and biology that refused all categories. When we discussed it the next day, nearly every participant had a different perspective.

The same week, a controversy erupted in Brazil over genetically modified soybean seeds, illegal in Brazil but apparently being smuggled in and used without deference to the proprietary rules devised by Monsanto, which produces both the transgenic seeds and the powerful pesticide Roundup that the seeds can tolerate. Farmers buying the modified soybean seeds have to agree not to save seeds for the following years and to permit Monsanto investigators, known as gene police, to walk their fields and take samples to ensure compliance if they stop buying the seeds. But farmers in Brazil apparently were acquiring the seeds on a GMO black market and reusing them without approval from their corporate overseers (DePalma and Romero 2000). During our meeting, several of us were interviewed by Brazilian television journalists about GMOs and the soybean trade.

We thus participated in Brazil's complex history of racial politics and in the complex local and global politics of GMOs. These two incidents capture a central concern of the essays to follow: the tangled politics, and coconstitution, of nature and culture.

PROVOCATIONS

Anthropology has been in some ways ground zero in the latest elaboration of what C. P. Snow construed in 1959 as the "two cultures"[2]—the apparently incompatible humanistic and scientific ways of understanding the world. Anthropology as a discipline has been deeply affected by the imperfect fit between technical and cultural explanations. It is a field that takes seriously both nature and culture, and both scientific and humanistic analyses. And the techniques and practices of the new genetics, as they have come into wider use in anthropology, have become a source of contention (see Sydel Silverman, foreword to this volume).

Paul Rabinow has proposed that the new genetics represents the apotheosis of modern rationality in that the object to be known "will be known in such a way that it can be changed" (1996: 93). And this power to produce change, including technical change mediated through laboratory or industrialized manipulation of biological materials, will also produce a new nature "remodeled on culture." Nature, he suggests, will become overtly artificial just as culture becomes natural. The technical-discursive achievements of modernity will lead to the collapse of the distinctions out of which that modernity emerged. Biosociality describes what we are calling nature/culture, or the labyrinthine intermingling of realms that calls into question both categories.

In an attempt at productive provocation, we have organized chapters under these categories—nature and culture—as we simultaneously interrogate and destabilize them. In part 1, which we are calling "Nature/Culture," we turn our attention to the sites of the critical cultural project of constructing and defining boundaries between populations and between species. In other words, we consider the technocultural domain of making differences and making nature. These are places where the age of geneticization plays out in extraordinary ways. In some cases, they are places deeply imbricated in the history of anthropology, such as the study of indigenous populations and the identification of a "pure line" in human groups. In other cases they are novel sites reflecting shifts in the landscape of the field, including the materiality of the "bodies that matter" (Butler 1993). These corporeal encounters involve Little People or the Amish, Icelanders or indigenous groups in Brazil, all of whom confront the interventions of geneticists. They also involve the genomes of the dog, the cloned sheep, and the chimpanzee, and the many ways that other species are implicated in contemporary genomics. We are interested in the stories told about such sites, and in the storytelling art in all its manifestations.

In part 2, titled "Culture/Nature," we consider the intersections of biosociality, complexity, and reductionism. Transnational processes and national identities are increasingly bound up in genetic history and genetic debates, about GMOs and their national meaning, the new eugenics, sovereignty, ethnic or racial identity, and the biological or cultural differences between groups. "Culture/Nature" includes the future of Japanese genomics, and of Japan, as imagined through the genome; the politics and complex historicities of genetic inquiry in South Africa; and the historical events and present-day identity politics embedded in ancient DNA. It includes fears and hopes about the future expressed in the responses of French farmers to GMOs, and the fears and hopes expressed in the enduring scientific effort to make sense of that chameleon-like categorizing idea, race. As our playfully serious couplings indicate, all the essays in this volume engage in resistance to simple determinisms.

Certainly, for both anthropology and genomics, this is a period of growing attention to complexity and new questions about the reductionism that has served so amiably as a self-evident justification of the ascendance of molecular genetics. In this light, we consider how critical theory can swerve anthropology and genetics in ways that respond to these issues. Genetics itself has become a focus of anthropological research; in a sort of feedback loop, critical cultural studies of genetics are raising questions relevant even to the most unrepentant reductionist. This is part of our project: we want to suggest how the productivity and potential of genetic explanations can be effectively integrated with other ways of understanding words, blood, and history. How can the burgeoning, and increasingly well-institutionalized, genetic narratives so characteristic of this era become a resource for justice and equity? How can both genetics and anthropology work in ways that recognize the tight bonds linking the techniques and practices of molecular genetics to the systematic exercise of power?

NATURE/CULTURE

The Human Genome Diversity Project (HGDP) as first proposed by Luca Cavalli-Sforza and colleagues (1991) strongly resonated with salvage anthropology, though in this case what was to be salvaged was DNA rather than culture and people (Goodman 1995; Marks 1995).[3] Blood samples from isolated or specialized populations of anthropological interest from around the world would be stored indefinitely, immortalized so to speak, in a public archive that could have many possible uses.[4] Cavalli-Sforza was a strong promoter of the historical relevance of DNA. He believed that the HGDP could help answer questions about ancient human population shifts such as the spread of agriculture, the peopling of Africa, and other events that were undocumented in any written record. DNA also appeared to be material that could be acquired without any particular attention to culture. Proponents, in their meetings and appeals for public funding in 1994 and 1995, seem to have assumed that taking blood was a simple technical act. Their plans became the focus of intense criticism by not only the indigenous groups targeted and their supporters, including the Rural Advancement Foundation International, but also anthropologists concerned about research ethics, power relationships, and scientific soundness (Goodman 1995; Marks 1995). The original plans for the HGDP combined technical sophistication with inattention to the political or cultural implications of opening the veins of people around the world.

The controversy may have killed the HGDP as a global project, but it did not stop the continued collection of biological samples and analysis of genetic variation. That larger project continues to be funded not only by the anthropology program at the National Science Foundation but also by the National Institutes of Health, where changes in focus are taking place. The goal of the HGDP has shifted from understanding "the" genome to exploring variations in genomes.[5]

The HGDP was a collision between postcolonial theory and geneticization. By the 1990s, the blood samples that could have been collected without controversy by earlier generations (who would not have been able to use them so effectively) were seen as deeply embedded in power relations and subject to all the constraints of informed consent, ethical disclosure, and sensitivity to cultural context—and this at the very moment when their utility as scientific objects of interest was highest. Interestingly, the power of the Internet, the motor and icon of informational capitalism, allowed indigenous groups to communicate and thus form a more powerful coalition to resist the HGDP (Lock 1994). In a sense this illustrates the power of the technical to undermine its own authority (Rabinow 1996), or what might be called the self-sabotage of the technical.

As the HGDP controversy suggests, those whose bodies are necessary participants in the networks of the new genetics can no longer be construed as invisible or silent. The postcolonial critique, human rights movements, changing standards for human subjects research, and the rise of the institutional review board have all affected field research in human genetics and biological anthropology. In the wake of recent controversies over the work of Napoleon Chagnon and James V. Neel with the Yanomámi in Venezuela, such questions have taken on a new, highly public urgency (see Ricardo Santos, chapter 1, this volume). How can anthropologists construct their work in ways that benefit vulnerable populations? Human subjects have long been important to biomedical knowledge, but this importance is now underlined by their institutional and organizational power to shape the research in ways that reflect their perceived advantage.

Human Populations/Genetic Resources

Some groups have become active and effective participants in genetic science. Four essays here explore populations that have been remade as genetic resources, examining how these scientific subjects have participated in the construction of new knowledge.

Ricardo Santos begins by considering the fieldwork of the geneticist James V. Neel, of the University of Michigan, who became the focus of a dramatic international controversy in the fall of 2000. Though Santos wrote this essay before accusations appeared claiming that Neel's use of a particular measles vaccine caused an epidemic among the Yanomámi in Venezuela in 1968, his text provides critical perspective on a scientist whose work has provoked intense debate. Exploring Neel's construction of the indigenous populations in Brazil as one of the last representatives of "primitive man," Santos compares Neel's work in the 1960s with the HGDP and with other research involving indigenous populations in the 1990s, much of it conducted by Neel's former students. Subjects seen as untouched by Western history became resources in various biological projects, including the Human Adaptability Project of the International Biological Programme, and Santos suggests that the concerns driving biological research among indigenous peoples since the 1960s have been relatively consistent, even if the population response has not.

Considering field studies of a very different population in the 1960s, the Pennsylvania Amish, M. Susan Lindee explores the intense social work built into producing the pedigree, as this textual record of a family line was transformed into a molecular resource by Victor McKusick, a contemporary and competitor of Neel. McKusick's work with the Pennsylvania Amish was an effort to track the biological—in this case, the gene for Ellis–van Creveld syndrome—through the disciplined deployment of the social, including birth and death records, the culture of the Amish, social networks, and specialized texts such as notations in Bibles. His Amish subjects were often cooperative, though some contested their status as objects of scientific curiosity, and McKusick was able to track a rare form of dwarfism through community history and through state records in Harrisburg, Pennsylvania. Lindee's study suggests that the pedigrees built on the exhaustive field studies carried out by many investigators interested in human genetics in the 1960s became molecular records and laboratory objects precisely because of their detailed social embeddedness.

Karen-Sue Taussig, Rayna Rapp and Deborah Heath explore the complex stakes made manifest in the contemporary phenomenon they call flexible eugenics as it plays out in the technical and social cultures built around dwarfism. The practices and discourses of the Little People of America, and of the scientists and physicians they engage, reflect a new convergence of genetic normalization and biotechnological individualism. As these authors demonstrate, the Little People of America's coalition with technical people, machines, and processes facilitated both a productive resistance to prejudice or exclusion, and a sociotechnical normalization that is in tension with that resistance. The "obligation to be free," they suggest, is a social practice shaped by technical interventions ranging from the molecular or genetic intervention of the prenatal test or the genetic diagnosis, to the older, if increasingly baroque, interventions of surgery and pharmacology.

Hilary Rose explores still another population that has been the focus of intense genetic interest, the people of Iceland, who sold their genome to deCode Genetics in 1998 in what seemed at first to be a bizarre and unprecedented act of national commercialization. The Icelandic genome and its commodification provide Rose with an opportunity to explore the rise of pharmacogenomics, in which the joint interests of the state and of venture capital remade a seemingly isolated population into a commercial and public health resource. As she demonstrates, the Icelandic case must be understood as part of a much wider program of supposedly cost-effective preventive medicine and genetic pharmacology. Rose excavates the concerns of those who have chosen not to participate, considering particularly how women expressed distrust of the database and questioned the confidentiality of information collected. Finally, she suggests that the database is a manifestation of expert-driven technological innovation common in the old welfare states, and an example of long-standing traditions badly in need of reform.

Animal Species/Genetic Resources

As human populations have provided data and ideological support for cultural hierarchies and corporate value to the emerging biotechnology industry, animals have been an equally exploited genetic resource. The negotiations between nature and culture are in some ways easier to see when they focus on companion animals, experimental organisms, genetically engineered mice, or cloned sheep.

Drawing on the technical frames of feminist theory, kinship theory, and molecular genetics, Sarah Franklin explores the notion of viable offspring when viability is biological, economic, strategic, and corporate. Dolly, Franklin proposes, is viable not only in the sense that she is capable of living outside the womb but also in the sense that she demonstrates a viable technique, a viable merger between corporate sponsorship and academic science, a viable investment driving up the value of the stock of the company that financed her creation, and a nuclear transfer technology producing a reliable natural-technical product. With Dolly, not only life itself but also the means of its production can be owned. She is therefore an unnatural kind, in an uneasy relationship to existing ideas of species, breed, property, gender, and sex.

Donna Haraway offers a "low-resolution linkage map" of the complex cross-species world of canine genetics. Presenting us with an "apparatus of naturalcultural production," Haraway shows us how the dog genome serves as the catalyst and central node in a network of human and nonhuman actors who engage one another through an interwoven array of practices and narratives, both popular and scientific. Offering a historical perspective on the genetic concerns of the present era, the article considers paleoarcheological portraits of canine agency, with the descendents of wolves successfully enlisting humans as purveyors of garbage dumps—perhaps before companionship—and puppy tenders. If an earlier epoch gave us the Birth of the Clinic, what can we learn from the elaborate technologies of canine care engendered by what Haraway would call the Birth of the Kennel?

Animals function as boundaries and can come threateningly, or alluringly, close to humanity. Jonathan Marks's quarry is a single factoid: chimpanzees and humans are commonly described as sharing a significant proportion of their genes—between 97 percent and 99 percent. Yet what does this number mean? That humans are hardly more than chimps, genetically, or that genetics is irrelevant because humans are obviously very different from chimpanzees? Proposing that the genetic claim of great likeness is often deployed to suggest that human beings and chimps share unsavory qualities, Marks goes on to play with the numbers himself, taking quantification to absurd lengths. By exploring a particular fact and its cultural moorings, Marks demonstrates the stakes involved in cross-species comparisons. The meanings of relatedness—between individuals, groups, nations, regions, and species, past and present—are always contested and contextual. Making relationships solid is a high priority in many different disciplinary and institutional settings; getting the world to hold still is one of the great Western projects. For many observers, including geneticists and anthropologists, genetics has promised to provide a particularly compelling way of defining relationships of all kinds, producing solidity and stability. At the same time, new genetic technologies such as cloning undermine the notion that genes can or should define both naturalness and relatedness in some straightforward way. Similarly, the technical invocation of DNA as the site at which race can be obliterated because we are genetically alike must confront the social reality that race has been literally written onto and into the body by history and social practice (see Alan R. Templeton and Troy Duster, chapters 12 and 13, this volume). When biology is a product of social organization, what is biological?

CULTURE/NATURE

Anthropologists historically have played a critical role in conceptualizing and studying human variation and identity. Race, ethnicity, and nationality are salient identity signifiers regardless of whether they are biologically legitimate categories. Sovereignty has sometimes functioned as a biological resource, a form of power that reinforced claims about the body and its value. And racial science—the science that validated the legitimacy of racial categories and that provided stories about racial difference which conformed to prevailing power relations—has been a sovereign resource deployed in law and nation building. In this group of essays, contributors explore the deep linkages binding state, race, and genome.

Political and Cultural Identity

We first present three essays that explore nature as an explicit cultural and political resource. While anthropology has begun to problematize the geneticization of medical domains such as disease-gene mapping and screening, it also must address the cultural reverberations that emerge as genetic science moves into the world of plant biology and agriculture. Indeed, as agricultural and pharmaceutical production are absorbed into the global biotechnology industry, novel sets of actors, including small farmers and local community activists, are emerging to contest an industry that is encroaching on cultural understandings and practices of food, land, and nature. Chaia Heller and Arturo Escobar explore two social movements, one in Colombia, the other in France, that represent early and formative case studies in what has since continued to become a global and potent movement in which activists around the world are contesting biotechnology.

Anthropology is well suited to exploring the novel intersection of genetic knowledges and globalization. For Heller and Escobar, this intersection results in the emergence of powerful networks that both produce and are produced by novel discourses of biodiversity and genetically modified organisms. While these networks are the site of science, capital, and government bodies, they are also the site of new social movements in which actors resist a perceived commodification of nature and a loss of cultural autonomy linked to agricultural and other land practices.

Joan Fujimura here explores views of genomics promoted by two prominent Japanese scientists, each of whom is engaged in imagining the future consequences of genomics as a social system and as a technological enterprise. She proposes first that imagination is a critical social practice through which global futures are designed, emphasizing the practical, fundamental importance of the discourses deployed around biotechnological change. She also points out that the Japanese tradition of translating foreign technology in ways that make the foreign "native" plays out around genomics in novel ways. The pseudonymous genomics promoter Suhara, for example, constructs the findings of genomics as a spiritual problem for the Christian West, which, in his interpretation, resists the embeddedness of human beings in nature. The Japanese, in contrast, he proposes, can readily accept the biological truths that genomics will reveal, including the truth about "what man is." Culture, therefore, in his account, encourages genomics in Japan but retards it in the West, a play that deftly severs science from "the West" and locates the problems of science not in technoscientific rationality but in the problematic orientations to life expressed through Christianity.

As Fujimura suggests, genomic scientists are building maps of genomics, of national and transnational identities, and of culture, and new institutions that encode structural visions of new futures. National identities linked to genomic science are not second-order effects, she proposes, but are instead inseparable from the first-order effects of gene maps and databases, cloned organisms, and pharmacogenetic commodities.

Africa is a hot spot of anthropological genetics. The continent was a focal point of the HGDP, and the interrelationships of African populations have long puzzled scientists. For example, Linnaeus thought that the San people of southern Africa were a different species, and it has been said that, up to the 1950s, some scientists even questioned whether the San could reproduce with Europeans. Himla Soodyall here explores how those outmoded scientific perspectives intersect with her own field research. Officials of the new South Africa embrace genetics to show the goodness of Africa, just as others once embraced genetics to show its backwardness. Yet how much can the technoscientific network be reformulated as an African resource? Soodyall relates her first venture out of the laboratory to take samples from conscripted San soldiers, and her realization that others had sampled the same group of individuals. How different is the drawing of blood for racist reasons from the same act undertaken for libratory reasons? Does it matter if the blood samples are sent to U.S. laboratories or held at a local lab in South Africa?

Race and Human Variation

The idea that technical expertise can be libratory, despite its historical relationships, threads through the next four essays, which explore race and human variation. The authors elaborate on the plastic and contested qualities of racial and ethnic variation by considering race and difference as historical problems accessible through the politics of processing and making sense of ancient DNA, as mathematical problems of gene frequencies, and as medical problems of phenotypic diagnosis and effective intervention.

Racial privilege and the injustice it has produced have a precise technical dimension in Rick Kittles and Charmaine Royal's exploration of an excavated burial ground in New York City. The authors draw on results from mitochondrial DNA studies both to illuminate the ethnicity of African Americans brought to North America enslaved and, thereby, to understand the ethnicity of contemporary African Americans. They studied mitochondrial DNA extracted from the bones of individuals who were buried in the 1700s at the New York African Burial Ground in lower Manhattan. Kittles and Royal hope that the DNA preserved in bones of eighteenth-century slaves will serve as a historical resource for populations whose history has been effectively obliterated (or almost so) by the slave trade.

While acknowledging the tangled history of biomedical research and practice on African Americans, particularly the history of medical racism and barriers to care, Kittles and Royal strongly support genetic studies of African Americans. Like Soodyall, they propose that technical knowledge can become a cultural resource even for those who historically have been oppressed by it. Alan Templeton implicitly adopts a related perspective in his examination of gene frequencies.

Since Richard Lewontin's famous study of the apportionment of human genetic diversity (1972), it has been shown repeatedly that most variation occurs within populations and races rather than among them. Populations can be defined as races, but they can also be defined in other ways, for example, strictly in geographical terms. Lewontin's conclusions called into question the biological reality of race; the genetic study that an earlier generation expected to demonstrate that races were biologically distinct (Boyd 1950) instead suggested that race had no biological meaning at all. Alan Templeton goes a step further in the formal disproof of race. He applies Wright's F_{st}, a measure of diversity within and among groups, to show that humans did not evolve as separate lineages (races). Templeton also provides an alternative explanation for human genetic variation: geographic distance. He argues strongly that applying different standards to human populations is scientifically indefensible. If race is to be considered biologically valid, then it must meet the standard scientific criteria for subspecies: Genetic diversity is genetic diversity, no matter the species. The science that helped to reify race, now buries it.

From a different perspective, Troy Duster explores the fluidity of the scientific concept of race by following the feedback loops linking biological research to culture and to practices of social stratification. While many anthropologists have sought to declare that the scientific concept of race is meaningless, Duster suggests that "purging science of race" is not practicable, possible, or even desirable. Scientific communities, legitimately troubled by commonsense interpretations of race as a biological justification for inequality, have oversimplified the issues.

Race, Duster asserts, is a stratifying practice of profound importance, and while the socially decontextualized concept of race as biological taxonomy is clearly groundless, the stratifying practice is a complex interactive feedback loop directly relevant to science and health care. Racial and ethnic classifications are in practice critical resources for the routine collection and analysis of medical data. Duster proposes that, when race is used as a stratifying practice, there is a reciprocal interplay of outcomes in which it is impossible to completely disentangle the biological from the social. Race is always, he suggests, a complex interplay of the social and the biological. It is neither meaningless nor trivial, and science cannot be purged of a category that has had such dramatic consequences for social organization. Ignoring race, Duster argues, also ignores or denies racial privilege. The paradox is that, as long as race plays a role in stratifying practices, it cannot be ignored.

The final chapter, by Frederika Kaestle, provides a site-specific window onto the technical, moral, and political worlds built around a found object, the remains of a human being. The Kennewick Man, the nearly complete skeletal remains of a man found in Washington State and dated to about 8,500 years before the present, is subject to a complex web of legal and historical frames. The man was first interpreted as Caucasoid from the historic period, but an archaic spear embedded in his hip suggests an older origin. If he were modern, the case would fall under the jurisdiction of the coroner.

If he were historic and non–Native American, then his disposition would fall under the jurisdiction of the U.S. Archeological Resource Protection Act. And if he were ancient and Native American, then the remains would be subject to the Native American Graves Protection and Repatriation Act. The body would have to be given to a Native American group—but which one? When one congressional representative proposed that human remains should not be returned to particular tribes unless "we can be reasonably confident that the remains are affiliated with that particular tribe," the National Congress of American Indians and the Clinton administration opposed the plan. Anthropologists sued to continue their studies, suggesting that scientific evidence drawn from DNA could be interpreted to contradict the creation myths of the tribes living in the region.

The Kennewick Man saga illustrates many of the cultural, ethical, and scientific issues that increasingly collide in the study of ancient DNA. Biological materials drawn from ancient remains may belong (in some senses) to indigenous groups in which there is profound mistrust and even outright rejection of Western science. Reflecting the genuine injustices of the past two centuries of racially driven research with Africans, African-Americans, South American groups, and Asian groups, such skepticism has a dramatic effect on contemporary research. Scientists and anthropologists working with such groups face complicated ethical dilemmas and biological problems. So, for example, the Native American Graves Protection and Repatriation Act requires proof of cultural (which often means biological) connection to a recognized Native American group in order for repatriation to occur; but populations are not closed systems, and many remains have contingent links to many groups, depending on how evidence is organized and interpreted. The linear connections over millennia that such legislation demands are neither realistic nor easily traceable.

Race, ethnicity, nationalism, and global capitalism increasingly play out in technoscientific debates that draw on cultural identities and laboratory techniques. Genes are resources for many different groups, deployed to resolve long-standing disputes about race, negotiate international trade, explain historical events inaccessible in any other way, and contest oppression and racism. Genetics in practice is plastic and contingent, embedded deeply in culture, time, and place.

CONCLUSION

The cover story of the September 13, 1999, issue of *Time* focused on the IQ gene purported to have been found in a strain of mice. The same issue included a report on the acts of resistance of the French farmers of Confédération Paysanne to genetically modified organisms, including the farmers' recent trashing of fields growing GMOs and of McDonald's restaurants. What are the links among IQ genes, the farmers' resistance to GMOs, the global hegemony of McDonald's, and the intelligence of laboratory-manipulated mice, which were among the first standardized animals and among the first patented experimental organisms? How does the network of complex meanings operate?

Bruno Latour, in a survey of a single daily newspaper, suggests that reports of computers, ecological disasters, pharmaceutical regulation, AIDS, and forest fires bring together "heads of state, chemists, biologists, desperate patients and industrialists" in a single story. The "imbroglios of science, politics, economy, law, religion, technology, fiction" produce a world in which "all of culture and all of nature get churned up again every day" (1993: 2). Meanwhile, the biologist Scott Gilbert has recently suggested that the grand narratives of the biological sciences are taking the place of the grand narratives of Western civilization. The "Western Civ" course, with its political origins in a "War Issues" course developed during World War I, has faded from the curriculum at most institutions. But introductory biology remains a vibrant core course, and biological narratives now provide what once came from Greek mythology, Dante, Shakespeare, Rousseau, and Goethe. The stories that are said to define our culture increasingly involve DNA, cells, organs, animals, plants, and ecosystems, Gilbert has suggested.

As though to validate Gilbert's claim, *Newsweek*'s first issue of the new millennium featured a striking image of a young man, bare-chested, longhaired, cradling in his hands a glowing strand of DNA. He looks down at the double helix while a serpent whispers in one ear and a dove in the other. In this obvious iconography, the young man is Adam, or perhaps the new American Adam, the contemporary molecular geneticist. The serpent is a devious character we all recognize, and the dove is the Holy Ghost, the voice of God, presumably offering good advice about what to do with the powers symbolized by a molecule whose existence and properties the majority of readers must take on faith.[6] A few weeks earlier, the cover of *Nature* featured an amended reproduction of the familiar detail from Michelangelo's Sistine Chapel. The hands of God and Adam, stretched toward each other, were connected by the sequence of chromosome 22, the first human chromosome to be fully sequenced.[7] The spark of life passing from the divine to the human was not the soul but the DNA sequence. Such images suggest the cultural significance attached to DNA, and this significance, as it plays out in multiple sites, poses the central problem of this volume.

One of the great ironies of the celebration of reductionism that produced the Human Genome Project is that the genome-in-practice has proven to be a bit more like the coyote than the architectural blueprint, the dictionary, or the machine. As the mapping proceeds, a Harry Potter world of unexpected doorways, secret passwords, and strange monsters has emerged. The early comparisons to the Bible begin to seem cogent in new ways, for like the Bible the genome is full of contradictions, inexplicable passages, historical errors, and ambiguity.

In the early years, when it was necessary to convince Congress that the genome should be mapped, James Watson and others prophesied a complete text that would explain "who we are."[8] Yet the genome, as Watson and other leading genomics scientists recognized, is in practice exceedingly complex, and any explanations it can provide of who we are will be equally complicated. While the *New York Times* of June 27, 2000, featured the cracking of the genetic code on its front page, the headline of the "Science Times" section was more somber: "Now the Hard Part: Putting the Genome to Work."

Perhaps genetic science is entering an era in which complexity and context are more important, both internally and externally, than reductionistic causal models. Perhaps genetics and anthropology have the potential to provide a sort of fusion in which questions about how facts become obvious and how categories silence questions are relevant to all sides. Perhaps the age of genetics will allow "geneticists to remake themselves as anthropologists."[9] And if the language of the gene is not well suited to anthropological questions, is the language of anthropology well suited to genetic questions?

Genetics at the beginning of the new millennium is a corporate, personal, medical, ideological, emotional, and bodily conglomerate stretching across and through many institutions and many layers of society. It is a way of thinking about the body and about the state, a way of organizing social expectations and making decisions about what questions are worth answering. Haraway has proposed that there is no innocent place to stand in this network. The common life and future imagined through genomics and all its corollaries imposes on us all, and the "sticky threads of DNA wind into the frayed planetary fibers of human and nonhuman naturalcultural diversity" (chapter 6, this volume). We are both bound to all other living things through DNA and separated from them by DNA, which defines both similarity and difference.

For anthropologists, genetics increasingly defines new questions and new methods, sharpening tensions within the field, attracting public notice, and raising new ethical quandaries. The new genetics has entered an older landscape in anthropology with a range of revolutionary or apocalyptic claims. Blood rewritten as genes provides powerful frames for kinship and identity, race and culture, history and the human future. What stories do genes tell? And what stories do we tell about genes and, in so doing, about others and ourselves, science and society, and nature and culture?

Anthropologists have long been critical players in constructing the narratives that define culture. Making the world, building narratives, is a craft, and we need to become skilled at that craft. We must learn to notice the networks of systems that sustain geneticization and identify some of the conceptual barriers that have made these networks so difficult to trace. The following chapters explore some problems posed by the intersections of words, blood, and history and show how those intersections reflect inequities, shape social policy, and privilege particular frames of meaning.

NOTES

1. As the HGP's era of DNA sequencing nears completion, there are those who project an impending era of proteomics, marked by increased efforts to achieve rapid progress in studying the complex structure and function of the proteins encoded by DNA sequences.

2. D. G. Burnett (1999) demonstrates the continuing power of what was in retrospect a relatively pedestrian analysis presented in a 1959 Rede Lecture at Cambridge University. The positing of "two cultures" provoked a spirited response and became a way of talking about many crises in the 1960s.

3. The idea of rapid loss of valuable data frequently has been used to justify "salvage anthropology." Much credit for this insight goes to Jonathan Marks.

4. In fact, a point of the scientific critique was the dubious utility of the data. Cavalli-Sforza first seemed to be interested only in using the data for historical reconstruction. When this purpose was deemed insufficient by many, not least the objects of the study, other reasons for the study, such as showing race to be a biological myth or using the data for genetic epidemiological purposes, were forwarded. The scientific design, however, is insufficient for genetic epidemiology, and we already know that race is a myth (Goodman 1995, 1996).

5. The future course of the HGDP is uncertain. The project is related to a much broader research program in genetic diversity, which can be expected to continue whether or not a formal HGDP program gears up. Soon after the announcement of a plan for global collection of human genetic data, biological anthropologists became involved; the Biological Anthropology Program at the National Science Foundation helped fund an HGDP conference in 1992 and held an HGDP grant competition in 1996. In 2001 no projects explicitly investigating human genome diversity were supported, but genetic diversity research continues to be funded. Anthropological studies of diversity are now overshadowed by genetic epidemiological studies, particularly of single nucleotide polymorphisms and their potential as risk factors for diseases.

6. See *Newsweek* (1 January 2000): 75. We are grateful to Scott Gilbert for calling this image to our attention.

7. *Nature* 2 (December 1999): cover, "The first human chromosome sequence." Thanks again to Scott Gilbert.

8. For a discussion of the early negotiations over the Human Genome Project in the United States, see Cook-Deegan 1994, especially pp. 148–85.

9. This was a comment by the biological anthropologist Frederika Kaestle on the first day of our meeting at Teresópolis.

REFERENCES

Boyd, W. 1950. *Genetics and the races of man.* Boston: Little, Brown.

Burnett, D. G. 1999. A view from the bridge: The two cultures debate, its legacy, and the history of science. *Daedalus* 128, no. 2 (spring): 193–218.

Butler, J. 1993. *Bodies that matter: On the discursive limits of "sex."* New York: Routledge.

Cavalli-Sforza, L., A. C. Wilson, C. R. Canton, R. M. Cook-Deegan, and M. C. King. 1991. Call for a worldwide survey of human genetics diversity: A vanishing opportunity for the Human Genome Project. *Genomics* 11:490–91.

Cook-Deegan, R. M. 1994. *Gene wars: Science, politics, and the Human Genome Project.* New York: W. W. Norton.

DePalma, A., and S. Romero. 2000. Crop genetics on the line in Brazil: A rule on seeds may have global impact. *New York Times,* May 16.

Flower, M. J., and D. Heath. 1993. Micro-anatomo politics: Mapping the Human Genome Project. In *Biopolitics: The anthropology of genetics and immunology.* Special issue of *Culture, Medicine, and Psychiatry,* ed. D. Heath and P. Rabinow, 17:27–41.

Foucault, M. 1980. *The history of sexuality.* New York: Vintage.

Gilbert, S. 1997. Bodies of knowledge: Biology and the intercultural university. In *Changing life: Genomes, ecologies, bodies, commodities,* ed. P. J. Taylor, S. E. Halfon, and P. N. Edwards, 36–55. Minneapolis: University of Minnesota Press.

Goodman, A. 1995. The problematics of "race" in contemporary biological anthropology. In *Biological anthropology: The state of the science,* ed. N. T. Boaz and L. D. Wolfe, 215–39. Bend, Oregon: International Institute for Human Evolutionary Research.

———. 1996. Glorification of the genes: Genetic determinism and racism in science. In *The life industry,* ed. M. Baumann, J. Bell, F. Koechlin, and M. Pimhert, 149–60. London: Intermediate Technology Publications.

Heath, D. 1997. Science studies: Beyond the war zone. *American Anthropologist* 99, no. 1:144–80.

———. 1998a. Locating genetic knowledge: Picturing Marfan syndrome and its traveling constituencies. *Science, Technology, and Human Values* 23, no. 1:71–97.

———. 1998b. Bodies, antibodies, and modest interventions: Works of art in the age of cyborgian reproduction. In *Cyborgs and citadels,* ed. G. Downey and J. Dumit, 67–83. Santa Fe, N.M.: School of American Research.

Heath, D., E. Koch, B. Ley, and M. Montoya. 1999. Nodes and queries: Linking locations in networked fields of inquiry. *American Behavioral Scientist* 43, no. 3:450–63.

Latour, B. 1987. *Science in action: How to follow scientists and engineers through society.* Cambridge: Harvard University Press.

———. 1993. *We have never been modern.* Trans. Catherine Porter. Cambridge: Harvard University Press.

Lewontin, R. 1972. The apportionment of human diversity. *Evolutionary Biology* 6:381–98.

Lippman, A. 1991. Prenatal genetic testing and screening: Constructing needs and reinforcing inequities. *American Journal of Law and Medicine* 17, nos. 1–2:15–50.

———. 1992. Led (astray) by genetic maps: The cartography of the human genome and health care. *Social Science and Medicine* 35, no. 12:1469–76.

Lock, M. 1994. Interrogating the Human Genome Diversity Project. *Social Science and Medicine* 39:603–6.

Marks, J. 1995. *Human biodiversity: Genes, race, and history.* New York: Aldine de Gruyter.

Nelkin, D., and S. Lindee. 1995. *The DNA mystique.* New York: W. H. Freeman.

Rabinow, P. 1996. Artificiality and enlightenment: From sociobiology to biosociality. In *Essays on the anthropology of reason.* Princeton: Princeton University Press.

PART 1

Nature/Culture

SECTION A

Human Populations/Genetic Resources

Chapter 1

Indigenous Peoples, Changing Social and Political Landscapes, and Human Genetics in Amazonia

Ricardo Ventura Santos

In his memoirs, the North American geneticist James V. Neel described a particular evening in July 1962, when he was carrying out field research among the Xavànte Indians in central Brazil:

> Beneath a fantastic canopy of stars . . . I listened uncomprehendingly, as the mature males, gathered in a group, discussed . . . the day's events, and planned for the next day. In the background, the young males began to discharge their nightly function of chanting before each house. Suddenly the thought came to me that I was witness to a scene which, in one variation or another, had characterized our ancestors for the past several million years. . . . Here was the basic unit of human evolution—the band or village—considering its interaction with other similar units and the environment. We were as close as modern man can come to the circumstances under which our species had evolved, under which our present attributes had arisen. (Neel 1994: 129)

This research marked the beginning of Neel's involvement with a research program in human biology in South America that, with a strong focus on population genetics, would produce hundreds of publications in the following decades, mainly in the 1960s and 1970s. In the opinion of one influential analyst, this program would become one of the "main developments in recent years in the physical anthropology of modern human populations" (Harrison 1982: 469).

Let us consider another setting. On a Sunday morning in August 1996, the Rio de Janeiro newspaper *Jornal do Brasil* published a front-page headline on the genetics of indigenous peoples. The tone was that of an exposé: biological samples from two Amazonian indigenous peoples (Karitiána and Suruí) were being marketed by a foreign company through the Internet. The issue cropped up again in the *Folha de São Paulo* several months later, in a special section on biopiracy (*Caderno Mais*, June 1, 1997). The news was also published in papers from several other Brazilian states. Seconding the written press, Rede Globo, Brazil's largest TV network, devoted part of its weekly science program to the case (*Programa Globo Repórter*, October 6, 1997). In fact the Brazilian samples of DNA and cell lines were stored in the early 1990s at the Coriell Institute for Medical Research in Camden, New Jersey, a part of the Coriell Cell Repositories' Human Variation Collection, or Human Diversity Collection.[1] As of December 1999, there were samples from eighteen populations from several parts of the world, most of them indigenous peoples from Asia, Africa, and Latin America.

Nongovernmental organizations backing the cause of indigenous peoples mobilized. The leader of one of the indigenous communities (the Karitiána) filed a complaint with the regional office of the Federal Attorney General in Rondônia demanding an investigation to identify the responsible parties. The Brazilian House of Representatives, located in Brasília, held hearings on the use of genetic resources and intellectual property rights in 1998, during which the Karitiána and Suruí issues were addressed. The issue reverberated widely in Brazil for several reasons, including the public confusion about how and why the samples were sent to Camden and public concern about the use of material from indigenous groups.[2]

The two events—the one from the 1960s and the other from the 1990s—involved different groups of scientists and, to a large extent, distinct methodological approaches and research goals. Yet they suggest trends in human biological research in Amazonia in the latter half of the twentieth century. Beginning in the 1960s, the vision of indigenous groups as models to investigate the biological history of humankind was a critical element in one of the largest genetic research programs on indigenous peoples ever to be carried out. The more recent Brazilian response to the biological material stored in Camden reflects the "participation" of these same groups in late-twentieth-century human genome research.

In this essay I compare research in human biology (and genetics in particular) in Amazonia in the 1960s and 1970s with later genomic programs (see also Santos 1999, 2002). As human subjects standards have shifted and indigenous groups have become more actively involved in making decisions about participation and resistance, such research has become a site of profound conflict and tension. Many indigenous peoples and activists have come to strongly oppose genomic variation research, or at least to see it with increased suspicion, as in the Brazilian case. Scientific research on indigenous groups has had to be renegotiated. The trends in research in Amazonia during recent decades reflect important changes and tensions in the relationship between science and society, particularly in the fields of human population biology and biological anthropology.

PHYSICIAN TO THE GENE POOL

James Neel was unquestionably one of the most prolific and influential researchers working in Amazonia in the second half of the twentieth century. What were the questions that motivated his investigations in Amazonia? Why would a scientist previously involved in research on the genetic effects of atomic explosions turn his attention to South America? Why would populations associated with specific attributes (i.e., local, particular, native, autochthonous, etc.) attract the interest of a scientist whose previous work adopted the perspective of "big science" (global, universal, generalizing, transnational)?

In the preface to his autobiography, Neel attempts to situate himself in the broader history of twentieth-century genetics, a field he joined in the 1930s:

> I am of the school that believes that along with . . . support and freedom [provided by society] there comes an implicit social contract. That contract requires the geneticist to be ever sensitive to the societal implications of his new knowledge. The contract involves a two-way exchange: the scientist informs society as to what genetics has to offer, and society in turn must decide along which of the many possible avenues of progress it will proceed. (Neel 1994: vii)

Neel was one of the main figures in the establishment of human genetics after 1945 (Kevles 1995). Like so many other geneticists who were trained in the first half of the twentieth century, he began his career working in experimental genetics with fruit flies *(Drosophila),* on which he wrote his Ph.D. dissertation in 1939. Neel recalls that, having concluded this phase of his scientific training, he was still asking himself to "what kind of genetics" he would devote his work. There was human genetics, which appealed to him in a way, but which, in his view, involved enormous methodological difficulties ("humans, as viewed through the eyes of the *Drosophila* geneticists, were not a favorable object of genetic study") and the burden of its recent history ("badly stigmatized . . . by the many incredibly sloppy and biased studies of the 1920s and 1930s that had provided the 'scientific' justification for much of the American eugenics movement" [Neel 1994: 9]).

In 1946 Neel became involved in research on the effects of the nuclear explosions in Japan. After finishing his graduate training at the University of Rochester, he spent a few years working at the University of Michigan. Motivated to enter the field of human genetics, he returned to medical school in Rochester. During the war years, his old Ph.D. supervisor (the geneticist Curt Stern) participated in secret research sponsored by the Manhattan Engineering District, the project in charge of building the atomic bomb. Through contacts with military physicians linked to the Rochester Manhattan District Unit, Neel went to Japan as a junior medical officer on a team whose task was "making recommendations to the National Academy of Sciences concerning the feasibility of appropriate follow-up studies in Hiroshima and Nagasaki" (Neel 1994: 57). Neel was later appointed by the academy to the position of director of human genetics research to be conducted by the Atomic Bomb Casualty Commission. Neel participated actively in the research program in Japan throughout the following decade, and summarized his results in a monograph published in 1956, titled *The Effect of Exposure to the Atomic Bombs on Pregnancy Termination in Hiroshima and Nagasaki* (Neel and Schull 1956; see also Lindee 1994).

In addition to studying the effects of the bomb, in the 1950s Neel was involved in establishing a genetic counseling service at the University of Michigan Medical Center and in various studies on genetic epidemiology (Neel 1994: 21–55). His interest in hemoglobinopathies, and particularly sickle-cell anemia, took him in the 1950s to Africa, where he collaborated in an investigation on the relationship between the geographical distribution of hemoglobin variants and malaria epidemiology. At this point his career turned increasingly to studies on the functioning of genetic systems at the population level (45–55). During the 1960s and part of the 1970s, Neel's research focused mostly on the genetics of Amazonian indigenous peoples.[3]

THE UNIVERSAL "PRIMITIVE"

> The study of these savages does not reveal a utopian state of nature; nor does it make us aware of a perfect society hidden deep in the forests. It helps us to construct a theoretical model of a society that corresponds to none that can be observed in reality, but will help us to disentangle "what in the present nature of man is original, and what is artificial." (Lévi-Strauss 1968 [1955]: 391)

> To understand ourselves, and how the conditions regulating survival and reproduction had changed, we must understand the biology of precivilized man much better. I realized we would probably never assemble from studies of existing tribal populations the numbers of observations necessary to relate specific genes to specific selective advantages, but at least we could take steps to define the range of population structures within which the evolutionary forces shaping humans had to operate. (Neel 1994: 118–19)

The lines of argument from the two quotations above are notably similar: to understand the past, present, and future of humankind, one might start with "savage" or "precivilized" peoples. Both are also autobiographical reflections, stemming from the experiences of two intellectuals from the same generation (the first was born in 1908 and the second in 1915) and whose interests were focused on indigenous societies in South America. Although one excerpt comes from the reflections of an ethnologist and the other from a physician-biologist, the point is virtually the same, expressed with different words: "to disentangle 'what in the present nature of man is original, and what is artificial' " or "to understand ourselves." The first excerpt is from the famous travel report *Tristes Tropiques* by ethnologist Claude Lévi-Strauss, originally published in 1955; the second is from *Physician to the Gene Pool,* the autobiography by James Neel.

In comparing Lévi-Strauss and Neel, two scholars who have produced vast, complex intellectual work, I am less concerned with comparing theoretical details (difficult to pinpoint because of their distinct interests and different disciplinary locations) than with calling attention to a specific point: the search for the Other through Amazonian studies in human biology in the second half of the twentieth century had motivations similar to those that had, for a long time, inspired anthropological and philosophical reflections. By seeking the so-called exotic, native, primitive, savage, or precivilized, anthropological inquiries historically have been aimed at understanding the roots, possibilities, and limits of human beings; at understanding where they came from and determining where they were going—in a word, at understanding what it means to be minimally and essentially human (see Diamond 1993). Lévi-Strauss, through his analyses of the myths of "savages," hoped to discover universal forms of thought and morality. According to Neel, what was "primitive" could provide "insights into problems of human evolution and variability" (Neel 1970: 815). As stated by Adam Kuper in *The Invention of Primitive Society,* throughout the history of anthropology, the "primitives" have served as a counterpoint in analyses that, in one way or another, have been intended to shed light on the societies to which the anthropologists themselves belonged: "They [the anthropologists] had particular ideas about modern society and constructed a directly contrary account of primitive society. Primitive society was the mirror image of modern society—or, rather, primitive society as they imagined it inverted the characteristics of modern society as they saw it" (1988: 240).

What information was Neel (and by extension a large portion of the research program in human biology in Amazonia) interested in obtaining? Fundamental to answering this question is his 1958 article "The Study of Natural Selection in Primitive and Civilized Human Populations," published four years before he carried out his first fieldwork in Brazil. Neel begins by stating that "the principle of natural selection as a guiding factor in human evolution is today universally accepted" (43). Nevertheless, he claims, even though sophisticated mathematical models had been developed (by F. A. Fisher, S. Wright, and J. B. Haldane) demonstrating the behavior of genetic frequencies in theoretical populations, the way natural selection would function under "real" conditions remained virtually unknown. In Neel's opinion, little progress had been made beyond what had been anticipated by authors like Wallace in the nineteenth century: "Our knowledge of the actual working of natural selection in human populations is almost nil" (43). Understanding the functioning of human population genetic dynamics was crucial and transcended "mere" theoretical relevance: "At this moment one of the most actively discussed topics in human biology is the genetic risk of the increased amounts of ionizing radiation to which human populations all over the world are being subjected" (43). In Neel's view, the relationship between nuclear technology and its potential mutagenic effects was just one example of the value of understanding the functioning of the human genetic pool under "natural" conditions and the way it had played out during most of evolutionary history. Information on the balance between the occurrence of mutation and selection in populations not exposed to radiation and other mutagenic agents and living under other demographic patterns (patterns of fertility, mortality, etc.) could be important in evaluating how far the human species had distanced itself from the selection circumstances under which it had evolved. In "The Study of Natural Selection in Primitive and Civilized Human Populations," Neel defined a research agenda whose seed he attempted to sow in subsequent years.

An opportunity came in 1962, soon after his first research expedition to study the Xavánte Indians of central Brazil, when Neel chaired a meeting of experts held by the World Health Organization. This meeting resulted in the technical report *Research in Population Genetics of Primitive Groups* (WHO 1964). The document presents recommendations for specific issues and methodological procedures (biological and demographic data to be collected, populations considered suitable for investigation, research design, etc.) in genetic surveys of indigenous peoples. The participants included researchers who in subsequent years became leading figures in the field of human biology and anthropological genetics, like F. M. Salzano, D. C. Gadjusek, R. L. Kirk, W. S. Laughlin, and J. S. Weiner, among others. Besides indicating what were considered the relevant issues for human population genetics in the 1960s, the WHO report forecast the areas that would receive greatest attention in the Amazon studies: genetic components in mortality and fertility differentials, biological consequences of inbreeding, disease patterns, biological relationships among populations, and so on.

THE FOCUS OF RESEARCH IN HUMAN BIOLOGY IN AMAZONIA

How did the research in human biology in Amazonia situate the subjects under investigation in the broader scenario of world history? Two statements help us answer these questions, one from Neel's first study in South America, among the Xavánte, and one from an article published in *Science* several years later summing up the results of the initial years of research, which also included the Yanomámi and other groups:

> For perhaps 99 per cent of its biological history, the human species has lived in small aggregates whose livelihood came primarily from hunting and gathering. The time factor in evolution being what it is, there can be little doubt that many—most—of the genetic attributes of civilized man have been determined by the selective pressures and breeding structures of these primitive communities. If we would understand modern man, we must study such of these primitive groups as still remain in a way in which they have rarely if ever been investigated to date. So rapidly are the remaining primitive communities disappearing, the matter of these investigations has an urgency not common in scientific problems. (Neel et al. 1964: 52)

> The general thesis behind the program was that, on the assumption that these people represented the best approximation available to the conditions under which human variability arose, a systems type of analysis oriented toward a number of specific questions might provide valuable insights into problems of human evolution and variability. We recognize, of course, that the groups under study depart in many ways from the strict hunter-gatherer way of life that obtained during much of human evolution. . . . We assume that the groups under study are certainly much closer in breeding structure to hunter-gatherers than to modern man; thus they permit cautious inferences about human breeding structure prior to large-scale and complex agriculture. (Neel 1970: 815)

By focusing their attention on South American indigenous peoples, Neel and several other researchers who worked in Amazonia in the 1960s hoped to find populations whose genetic and demographic dynamics were close to those that had characterized "perhaps 99 per cent" of human biological history. Understanding reproductive dynamics was essential to unveiling how genetic variability had emerged and spread. In order for the indigenous peoples to represent "the best approximation available to the conditions under which human variability arose," it was necessary above all that their fertility and mortality patterns correspond to those that supposedly obtained throughout most of human biological history.

In large part, the focus of the research program in human biology in Amazonia was the isolated, the pristine, but it also recognized that those societies were no longer untouched by outside influences. In 1967, Neel and Salzano, who often expressed in their writings a deeply humanistic concern for the indigenous peoples they had worked with, noted that "there is no Indian group completely untouched by the discovery of America and subsequent contacts, direct or indirect, with the Western world" (246). Even though no group could be considered completely unchanged, the researchers still sought subjects as isolated as possible. Neel justified beginning research with the Yanomámi on the basis that they were more culturally intact than the Xavánte, which he believed made them more appropriate for research on internal tribal dynamics (1994: 134).

The search for human communities living in isolation and biologically representative of a world on the verge of disappearing (i.e., subjects perceived as untouched by Western history and thus having generalizable attributes) was an approach that not only oriented the research in Amazonia but also served as the framework for much research in human biology during the 1960s. For example, it was this logic that permeated the human adaptability component of the International Biological Programme (HA-IBP), whose activities lasted from the early 1960s until the mid-1970s and which included part of the genetic research carried out by Neel and collaborators in Amazonia (Neel 1968, 1994: 130; Neel and Salzano 1967).

The IBP was intended to facilitate a "comprehensive global understanding of the processes and forces responsible for the properties of our complex planetary shell" (Collins and Weiner 1977: 1). It had a strong ecological focus and was grounded in the idea that populations living in natural conditions could provide objective lessons about human adaptability. In the view of its participants, the world was experiencing a historical moment that "represented probably a last chance of making a concerted study of the still remaining communities of hunters and gatherers and simple agriculturalists" (3–4). The history of research in human biology during the 1960s and 1970s is closely associated with the trajectory of the HA-IBP. It was an undertaking on an international scale that at one point encompassed some three hundred projects from forty countries, covering a broad thematic range in the areas of human physiological, developmental, morphological, and genetic adaptability (13).

THE 1990S AND THE HUMAN GENOME DIVERSITY PROJECT

If the HA-IBP was the leading research endeavor in the field of human population biology in the 1960s and 1970s, in the 1990s this role was to be played by the Human Genome Diversity Project (HGDP). Important differences, and also some important parallels, are discernible when one compares the two projects. In the HA-IBP, human genetics was not the central issue but one of six themes; the others were human growth and development, physique and body composition, physical fitness, climatic tolerance, and nutritional status (Collins and Weiner 1977: 13). That is, during the 1960s there was a great emphasis on the analysis of ecological relations and phenotypical manifestations. In the HGDP, the central concern is with the genome. Among the parallels between the projects, two seem particularly significant. Like the HA-IBP, the HGDP (at least as originally conceived) aimed to be a worldwide program carried out with support from international organizations like UNESCO and WHO (see Cavalli-Sforza et al. 1991: 491). And in both cases, the justification for the research alluded to notions of scarcity and disappearance. The world was on the verge of losing the information that could elucidate "the interaction of nature and nurture on the physiological, morphological, and developmental characters of human populations [living under 'natural' conditions] on a world scale" (Collins and Weiner 1977: 3). Moreover, it was the genetic material necessary "to illuminate variation, selection, population structure, migration, mutation frequency, mechanisms of mutation, and other genetic events of our past" (Cavalli-Sforza et al. 1991: 490) that was about to disappear.

The proposal for the HGDP was launched in an article published in the early 1990s and titled "Call for a Worldwide Survey of Human Genetic Diversity: A Vanishing Opportunity for the Human Genome Project" (Cavalli-Sforza et al. 1991).[4] By emphasizing human genomic diversity, the HGDP would complement the Human Genome Project, whose goal of sequencing did not involve analyses of interpopulation variability. The HGDP, in the words of its proponents, is a "concerted effort to obtain and store [blood] samples from diverse populations in order to understand human variation" (490–91; see also Kidd et al. 1993; Weiss et al. 1992). The Internet site of the HGDP North American Regional Committee states:

> The HGD Project is an effort by anthropologists, geneticists, doctors, linguists, and other scholars from around the world to document the genetic variation of the human species worldwide. This scientific endeavor is designed to collect information on human genome variation to help us understand the genetic makeup of all of humanity and not just some of its parts. The information will also be used to learn about human biological history, the biological relationships among different human groups, and may be useful in understanding the causes of and determining the treatment of particular human diseases.[5]

The HGDP's reception by organizations representing indigenous peoples, backed by nongovernmental organizations working on human rights and environmental issues, took its organizers completely by surprise (Butler 1995; H. Cunningham 1998; Dickson 1996; Friedlaender 1996; Gutin 1994; Kahn 1994). The argument voiced by the HGDP—that it was important to carry out the project as soon as possible because the populations that could provide more information to elucidate human evolutionary history were in danger of dying out or being assimilated—was perceived by many as outrageous (Kahn 1994: 720). Particularly decisive in building up opposition against the HGDP was the disclosure in the mid-1990s that the U.S. Patent and Trademark Office had granted a patent to the U.S. Department of Health and Human Services on a human T-lymphotropic virus derived from the Hagahai, a native population from Papua New Guinea.[6] For the critics, this very case, called to public attention by the Ottawa-based Rural Advancement Foundation International, was predictive of what would happen to blood samples of indigenous peoples collected for the HGDP: they would be turned into profitable commodities, though not for the benefit of indigenous groups. The HGDP was perceived as a new form of biocolonialism, the exploitation of nonhuman genetic resources (plants, for example) that had been going on at least since the 1930s (A. Cunningham 1991; Dickson 1996; Kahn 1994). The Rural Advancement Foundation International became one of the most vocal critics of the HGDP, meanwhile continuing its campaign against Western companies' exploitation of plant genetic resources from Third World countries.

Henry Greely, who chaired the ethics committee of the North American branch of the HGDP, stated in the early 1990s, when the project was just receiving the first waves of criticism: "I don't think it [initially] crossed anyone's mind that [the project] would be controversial, although it should have" (*apud* Kahn 1994: 720). The HGDP was launched in the midst of long-term intensive political struggles over the utilization of nonhuman genetic resources from developing countries. The overall atmosphere of the early 1990s, observers have suggested, was not suitable for the launching of a project with the characteristics of the HGDP (Butler 1995; H. Cunningham 1998; Dickson 1996; Friedlaender 1996; Gutin 1994; Kahn 1994).

The reasoning above helps explain the reactions elicited when the HGDP was presented. However, it does not address another relevant issue: why did those planning the HGDP, as one reads in "Call for a Worldwide Survey of Human Genetic Diversity: A Vanishing Opportunity for the Human Genome Project" (Cavalli-Sforza et al. 1991), present the project the way they did? The thinking underlying the HGDP had long resonated in biological circles, and the new project was folded into the logic of world-scale biological endeavors that preceded it. Yet projects that did not elicit major public concern in the 1960s—like the HA-IBP, which attracted wide support—had become impossible by the 1990s. Changes in expectations about relationships between researchers and subjects, science and the international community, and the developed and developing worlds all undermined the HGDP. The public and political climate was so different—and the proposed scientific plan so familiar to human biologists—that both sides were talking across a culture gap that heightened the emotional nature of the debate.

Under the social and political conditions prevailing in the 1990s, the time of the HGDP, adherence to previous modes of justification, without reframing and incorporating new elements, would prove no longer adaptive. This point becomes evident when comparing the contents of two key documents: *Research in Population Genetics of Primitive Groups* and "Call for a Worldwide Survey of Human Genetic Diversity." The first expresses the perspective of population genetics of indigenous peoples (and to a great extent that of the HA-IBP) in the 1960s; the second, that of the HGDP in the 1990s.[7] Since both documents are proposed guidelines for carrying out human genetic research, it is obvious that they generally address the same topics (research goals, methodology, populations to be studied, and so forth). In the comparisons that follow, I am less concerned with the specifics of content than with how these contents are expressed.

While there are (expected) differences in the focus of analysis (blood groups, saliva, anthropometry, etc., compared to mitochondrial and nuclear DNA), both documents emphasize the importance of investigating the genetics of "primitive groups" or "vanishing" populations, seen as important repositories of the "genetic endowment of modern man" or as "informative genetic records." In both cases it is emphasized that research should be undertaken as quickly as possible, before the complete "cultural disintegration and . . . loss of physical identity" or "closing" of "the gate" take place. In addition to the impossibility of putting off the research, in both the 1960s and 1990s it is argued that the point finally has been reached when "the appropriate techniques" are available or "tools for understanding our species" have been created. Both documents express ethical concerns. Both also present lists of the populations regarded as most suitable for study, and both comment on the logistical difficulties to be dealt with (that is, in research in the Third World, as one can read between the lines). In the face of imminent "loss of physical identity" or the closure of the gate, preserving samples is seen in the two pieces as fundamental: in one, the suggested strategy is deep freezing; in the other, immortalization through cell cultures.[8]

There is no doubt that the world changed over the three decades separating *Research in Population Genetics of Primitive Groups* (1964) and "Call for a Worldwide Survey of Human Genetic Diversity" (1991). Over the course of these turbulent twenty-seven years, with the end of the cold war, there was a decrease in the political antagonism between East and West and even greater growth in the chasm separating North and South. New theoretical perspectives for interpreting power relations took hold, numerous colonies became nation-states, and ethnic affirmation movements emerged and gained strength, including the political activism of indigenous peoples. The relationship between science and society, including the supposed neutrality of scientific practice, was questioned even more intensely than in previous periods. Finally, molecular genetics evolved from a basic science discipline into an economically promising field of investigation.

With the privilege of hindsight in interpreting this process, it is difficult to imagine that the HGDP, as originally presented, could have had any chance of universal acceptance. The project was backed by the technical discourse and the most advanced technology in molecular biology available at the end of the twentieth century, but with a line of argument concerning obtaining data for scientific purposes that did not recast previous modes of justification and explanation in order to take into consideration the social and political conditions prevailing in the 1990s.[9] Back in the 1960s, not only was the possibility of commercial uses of human genetic resources still far ahead, but the social, cultural, and political implications of biological research among indigenous peoples in Amazonia and elsewhere were sharply different, to the extent that the overall tone of a document like *Research in Population Genetics of Primitive Groups* was never disputed in its time with the intensity that the HGDP has been.[10]

CONCLUSION

My aim in this essay has not been to provide a detailed account of the historical trajectory of HA-IBP, much less the HGDP, but instead to reveal resonances between the two. To date, little research has been done on the history of the HA-IBP, despite the fact that it was extremely influential in shaping the field of human biology from the 1960s onward. As for the HGDP, it had a rather complex trajectory in the 1990s, during which time an intense debate and a sizable body of literature were generated. Over the years, some scholars affiliated with the HGDP wrote papers addressing fundamental issues related to the justification and collection of biological specimens, in some instances reframing early positions and formulations put forward by HGDP proponents in the beginning of the 1990s, which were the ones I examined more closely in this essay.[11] One way or another, the HGDP was mired in controversy throughout the 1990s and did not take off as planned originally.

Taking as a point of reference the previous and highly influential research agenda of the HA-IBP, a major focus of this essay was the inception of the HGDP. That is why in the second half of this paper I mostly emphasize "Call for a Worldwide Survey of Human Genetic Diversity," the memoir through which the HGDP was first presented to the world at large. Few would dispute that the debates which arose in the early 1990s have had a major impact on the long-term trajectory of the HGDP.

Examining how indigenous peoples were pulled into human biological research in two different moments of the latter half of the twentieth century has required an exercise in comparative contextualization, with attention to the social, economic, and political conditions related to the process of knowledge production. The relative success of research agendas such as the HA-IBP and the HGDP can be understood only in relation to the rapidly changing social and political landscapes within which they emerge. In contrast with the widespread acceptance of the HA-IBP, the controversy surrounding the HGDP signals a dialectical disjuncture between the framing of the research agenda—in terms drawn from an earlier era—and historical changes in human rights discourses, means of inclusion, patterns of justification, timing related to the assembling of collections of human biological materials, and technological developments, among other factors.[12]

NOTES

I wrote the first version of this paper in 1998 and 1999, when I was a visiting scholar at the Department of Anthropology, University of Massachusetts at Amherst, and at the Program in Science, Technology, and Society, Massachusetts Institute of Technology (STS/MIT). During this period I was supported by a postdoctoral fellowship from the Brazilian Ministry of Education through CAPES. I thank Deborah Heath and Alan Goodman for inviting me to the 1999 Wenner-Gren International Symposium. I also thank M. Susan Lindee, who provided helpful comments on the manuscript.

1. See http://locus.umdnj.edu/nigms/cells/humdiv.html (accessed on 27 December 1999).

2. The Human Genetic Cell Repository at Coriell is sponsored by the National Institute of General Medical Sciences. One reads in the web site of Coriell that its goal is "to provide essential research reagents to the scientific community by establishing, maintaining, and distributing cell cultures and DNA derived from cell cultures." The collection of the Amazonian samples became a widely discussed topic in Brazil because Brazilians questioned whether the federal agency Fundação Nacional do Índio (the Brazilian Indian Agency) had granted the researchers permission that allowed for the samples to be placed in a repository like Coriell, where the samples would be available to investigators worldwide. It was considered troubling that blood and DNA samples could be stored, transformed into cell lines, and made widely available without explicit individual and community consent. There was considerable concern that, while consent may have been given for a particular project, this long-term storage would make it possible to use samples in ways not originally described or intended. The potential uses of the samples were certainly much broader than the specific, approved research projects for which the samples were collected. In addition, the fact that the samples were priced, even considering that these funds might be intended to maintain the cell lines and DNA samples, also raised serious concerns. The Brazilian samples are part of the so-called Yale-Stanford Collection and were deposited at Coriell by a team of researchers from Yale University in the early 1990s (see Kidd et al. 1991; see also Kidd et al. 1993; and Weiss et al. 1992). According to Kidd and colleagues (1991), the Brazilian samples have been utilized in studies on the origins of, and genome diversity among, world populations, aiming at providing insights into the recent evolutionary history of human diversity.

3. The journalist Patrick Tierney, in his explosive account of anthropological research on the Yanomámi, *Darkness in Eldorado,* constructed James Neel as a sort of evil mastermind embedded in the global machinations of the Atomic Energy Commission. Several authors have already disputed Tierney's allegation that Neel intentionally spread the measles virus in order to cause epidemics and, subsequently, to study genetic and biomedical outcomes based on eugenic premises (see, for instance, contributions by several authors in *Current Anthropology* 42, no. 2 [2001]; *Science* 292, no. 5523 [2001]; and *Interciencia,* 26, no. 1 [2001]). It is not my goal here to discuss

Tierney's book in detail, but I consider Tierney's interpretations of genetic and human biological research in indigenous peoples in Amazonia rather biased, superficial, and lacking historical depth and context.

4. The two Amazonian samples, from the Karitiána and the Suruí, were collected and placed at Coriell in the early 1990s (see Kidd et al. 1991), prior to the first call for an integrated project to investigate human genome diversity (Cavalli-Sforza et al. 1991). Nevertheless, the Human Diversity Collection at Coriell seems to be related to the HGDP in several ways. For example, the Amazonian samples are managed by a genetics laboratory at Yale that is closely associated with the HGDP (see Kidd et al. 1993; Weiss et al. 1992). In papers related to the HGDP, Coriell is listed as one of the repositories where samples would be stored (Cavalli-Sforza et al. 1991: 490). In addition, approaches for preserving and supplying DNA and cell lines in association with collections such as the Human Diversity Collection have been presented as strategies for the study of human genome diversity (Bowcock and Cavalli-Sforza 1991: 495–96). Perhaps even more significant is the fact that the moral, social, and ethical concerns that have been expressed in relation to the Brazilian samples (nature of consent, minority participation, etc.) mirror important discussions that have been carried out in association with the HGDP.

5. See http://www.stanford.edu/group/morrinst/hgdp/faq.html (accessed on December 27, 1999).

6. For details, see Friedlaender 1996; and H. Cunningham 1998: 208–10.

7. This point should be well clarified, since *Research in Population Genetics of Primitive Groups* resulted from a meeting organized by WHO, and not by HA-IBP. Notwithstanding, the approach outlined in this document, which closely reflects Neel's views and interests (he was the chair of the committee that prepared the report), had a lasting effect on the population genetics projects conducted under the HA-IBP. In addition, influential geneticists who carried out genetic research under IBP participated as members of the committee. As pointed out by Collins and Weiner (1977: 9 and 25), Neel was one of the two "theme consultants" in human genetics (the other was A. E. Mourant) for HA-IBP, and his laboratory in Michigan was a major center for advanced training for those interested in pursuing human genetic research under HA-IBP.

In the same issue of *Genomics* in which "Call for a Worldwide Survey of Human Genetic Diversity" was published, Bowcock and Cavalli-Sforza's 1991 paper titled "The Study of Variation in the Human Genome" also appeared. It can be said that, even more than "Call for a Worldwide Survey of Human Genetic Diversity" alone, the combination of the two pieces fits the role of founding memoir of the HGDP. I use the latter article as the piece representative of the HGDP despite the fact that, to some extent, in response to the many criticisms, its proponents had to reformulate it in the course of the 1990s (see, for instance, text on the home page of the North American component of the project [http://www.stanford.edu/group/morrinst/hgdp.html, accessed on December 27, 1999]; see also Kahn 1994: 722). In this essay, however, I am less interested in the "mutations" experienced by the HGDP than in how it was originally conceived. For comments on some of the directions taken by the project in the 1990s, see Butler 1995, Friedlaender 1996, Macilwain 1996, Pennisi 1997, and Weiss 1998, among others. Henry Greely, a member of the North American Regional

Committee of the HGDP and professor of law at Stanford University, published a number of papers in the 1990s in which he addressed important social and legal issues related to human genomic research, including the HGDP (see Greely 1998a for an overview). In one of them, he states, "The people whose genetic and clinical data will be essential for the next phase of human genomics research need to be treated not merely as 'subjects' but more as (somewhat limited) partners. . . . The goal of this approach is not to prevent research but to prevent research subjects from feeling cheated, powerless, misled, or betrayed" (1998b: 625).

8. See Santos 2002 for a detailed comparison of the two documents.

9. Cavalli-Sforza's research on the Pygmies in the 1960s and 1970s was carried out as part of HA-IBP (Cavalli-Sforza 1986; Collins and Weiner 1977: 151–53). The project, titled "Genetics of Primitive Human Populations: The Babinga Pygmies," aimed in part "to obtain a picture of population structure of the Babinga Pygmies which may serve as an example of a population living in conditions very nearly as primitive or at least very little different from those that must have prevailed for perhaps hundreds of thousands of years" (Collins and Weiner 1977: 151).

10. The first reaction against the HGDP came in 1993 after the nongovernmental organization the Rural Advancement Foundation International made public a list that had been prepared during an HGDP-sponsored workshop in 1992 and that named the populations considered most suitable for study. At this time the project was still a fresh idea that had just come out of the minds of a few geneticists and anthropologists (see Kahn 1994; Roberts 1992). According to Kahn (1994: 720–21), some groups representing indigenous peoples were outraged, not only because some populations were being targeted without consultation but also because of the emphasis on carrying out the research before these populations disappeared. By contrast, interestingly enough, *Research in Population Genetics of Primitive Groups,* in addition to presenting a list of populations to be studied and explicitly stating that they should be investigated before loss of physical identity, was prepared under the auspices of an international and official institution (the World Health Organization).

11. See "Appendix: The Human Genome Diversity Project" in Weiss 1998 (295–98) and documents on proposed ethical guidelines available at the HGDP Internet site (http://www.stanford.edu/group/morrinst/hgdp; accessed on December 27, 1999).

12. Issues surrounding the assembling of collections of human biological materials have much broader implications for anthropology, and for biological anthropology in particular. To what extent should the controversies concerning collections of DNA samples and cell lines in the context of the HGDP be regarded as isolated events? Today, bones and blood, more traditional physical anthropology and high-tech human genetics, museums of natural history, and laboratories of anthropological genetics are all converging through unexpected means, one of them being the controversies surrounding the assembling (and disassembling) of collections of human biological materials. A case in point concerns the Native American Graves Protection and Repatriation Act, which has been in force in the United States since 1990. Under this act, museums and other institutions receiving federal funding are required to prepare inventories of human remains and artifacts of Native American origin present in their collections. Legislation along similar lines has been passed in

other countries, such as Australia and New Zealand. These policies have been interpreted as the result of a long process of questioning the ways in which Western science has dealt with indigenous peoples over past centuries—as signs that power relations and past practices of collecting human biological data are being recast or at least deeply questioned (see Ferguson 1996; Martin 1998; Rose et al. 1996; Simpson 1996: 173–89, 223–42). The HGDP launched its large-scale proposal to mount collections of certain types of human biological materials (cell lines and DNA samples), to be stored in molecular biology labs and cell culture banks, at the very moment when one of the most traditional branches of biological anthropology (skeletal biology and related areas of investigation) is watching the drawers and shelves of natural history museums being emptied of their contents of bones and other human remains. This simultaneity in disassembling and assembling of collections of human biological materials (in several instances obtained from roughly the same sources—past and present living indigenous populations from various parts of the world) has had, and certainly will have in the future, consequences for the practice of human biological research that are yet far from being fully recognized, analyzed, and understood.

REFERENCES

Bowcock, A., and L. Cavalli-Sforza. 1991. The study of variation in the human genome. *Genomics* 11:491–98.

Butler, D. 1995. Genetic diversity proposal fails to impress international ethics panel. *Nature* 377:373.

Cavalli-Sforza, L. L., ed. 1986. *African pygmies.* Orlando: Academic Press.

Cavalli-Sforza, L. L., A. C. Wilson, C. R. Cantor, R. M. Cook-Deegan, and M. C. King. 1991. Call for a worldwide survey of human genetic diversity: A vanishing opportunity for the Human Genome Project. *Genomics* 11:490–91.

Collins, K. J., and J. S. Weiner. 1977. *Human adaptability: A history and compendium of research in the international biological programme.* London: Taylor and Francis.

Cunningham, A. B. 1991. Indigenous knowledge and biodiversity. *Cultural Survival Quarterly* (summer): 4–8.

Cunningham, H. 1998. Colonial encounters in postcolonial contexts: Patenting indigenous DNA and the Human Genome Diversity Project. *Critique of Anthropology* 18:205–33.

Diamond, S. 1993. *In search of the primitive: A critique of civilization.* New Brunswick, N.J.: Transaction Publishers.

Dickson, D. 1996. Whose genes are they anyway? *Nature* 381:11–14.

Ferguson, T. J. 1996. Native Americans and the practice of archaeology. *Annual Review of Anthropology* 25:63–79.

Friedlaender, J. 1996. Genes, people, and property: Furor erupts over genetic research on indigenous peoples. *Cultural Survival Quarterly* 20, no. 2:22–25.

Greely, H. T. 1998a. Legal, ethical, and social issues in human genome research. *Annual Review of Anthropology* 27:473–502.

———. 1998b. Genomics research and human subjects. *Science* 282:625.

Gutin, J. 1994. End of the rainbow. *Discover* 15:70–74.

Harrison, G. A. 1982. The past fifty years of human population biology in North

America: An outsider's view. In *A History of American Physical Anthropology,* ed. F. Spencer, 467–72. New York: Academic Press.

Kahn, P. 1994. Genetic diversity project tries again. *Science* 266:720–22.

Kevles, D. 1995. *In the name of eugenics: Genetics and the uses of human heredity.* Cambridge: Harvard University Press.

Kidd, J. R., F. L. Black, K. M. Weiss, I. Balazs, and K. K. Kidd. 1991. Studies of three Amerindian populations using nuclear DNA polymorphisms. *Human Biology* 63:775–94.

Kidd, J. R., K. K. Kidd, and K. M. Weiss. 1993. Human genome diversity initiative. *Human Biology* 65:1–6.

Kuper, A. 1988. *The invention of primitive society: Transformation of an illusion.* London: Routledge.

Lévi-Strauss, C. 1968 [1955]. *Tristes tropiques.* New York: Atheneum.

Lindee, M. S. 1994. *Suffering made real: American science and the survivors at Hiroshima.* Chicago: Chicago University Press.

Macilwain, C. 1996. Tribal groups attack ethics of genome diversity project. *Nature* 383:208.

Martin, D. 1998. Owning the sins of the past: Historical trends, missed opportunities, and new directions in the study of human remains. In *Building a new biocultural synthesis: Political-economic perspectives on human biology,* ed. A. H. Goodman and T. L. Leatherman, 171–90. Ann Arbor: University of Michigan Press.

Neel, J. V. 1958. The study of natural selection in primitive and civilized human populations. *Human Biology* 30:43–72.

———. 1968. The American Indian in the International Biological Program. In *Biomedical challenges presented by the American Indian,* ed. Pan American Health Organization. Scientific Publication no. 165. Washington, D.C.: Pan American Health Organization.

———. 1970. Lessons from a "primitive" people. *Science* 170:815–822.

———. 1994. *Physician to the gene pool: Genetic lessons and other stories.* New York: John Wiley and Sons.

Neel, J. V., and F. M. Salzano. 1967. A prospectus for genetic studies on the American Indians. In *The biology of human adaptability,* ed. P. T. Baker and J. S. Weiner, 245–74. Oxford: Clarendon Press.

Neel, J. V., F. M. Salzano, P. C. Junqueira, F. Keiter, and D. Maybury-Lewis. 1964. Studies on the Xavánte Indians of the Brazilian Mato Grosso. *American Journal of Human Genetics* 16:52–140.

Neel, J. V., and W. J. Schull. 1956. *The effect of exposure to the atomic bombs on pregnancy termination in Hiroshima and Nagasaki.* National Research Council Publication no. 461. Washington, D.C.: National Academy of Sciences.

Pennisi, E. 1997. NRC OKs long-delayed survey of human genome diversity. *Science* 278:568.

Roberts, L. 1992. Anthropologists climb (gingerly) on board. *Science* 258:1300–1301.

Rose, J. C., T. J. Green, and V. D. Green. 1996. NAGPRA is forever: Osteology and the repatriation of skeletons. *Annual Review of Anthropology* 25:81–103.

Santos, R. V. 1999. Bioética, antropología biológica y poblaciones indígenas amazónicas. *Estudios de Antropología Biológica* (México) 9:13–26.

———. 2002. Indigenous peoples, postcolonial contexts, and genomic research in

the late twentieth century: A view from Amazonia (1960–2000). *Critique of Anthropology* 22:81–104.
Simpson, M. G. 1996. *Making representations: Museums in the post-colonial era.* London: Routledge.
Tierney, P. 2000. *Darkness in Eldorado: How scientists and journalists devastated the Amazon.* New York: W. W. Norton.
Weiss, K. M. 1998. Coming to terms with human variation. *Annual Review of Anthropology* 27:273–300.
Weiss, K. M., K. K. Kidd, and J. Kidd. 1992. Human Genome Diversity Project. *Evolutionary Anthropology* 1:80–82.
World Health Organization (WHO). 1964. Research in population genetics of primitive groups. WHO Technical Report Series, no. 279. Geneva: World Health Organization.

Chapter 2

Provenance and the Pedigree

Victor McKusick's Fieldwork with the Old Order Amish

M. Susan Lindee

Provenance is defined in the *Oxford English Dictionary* as the record of the "ultimate derivation and passage of an item through its various owners." The term is most commonly used to describe the history or pedigree of a painting—who has owned it, its value at various stages—but it also has a meaning in silviculture, in which it refers explicitly to genetic stock. Provenance, for forestry professionals, is the record of where a seed was taken and of the character of the "mother trees." In this essay I explore provenance in both senses, as a textual record of the origins of a given object (in this case a blood or tissue sample) and as a record of genetic stock. I focus on fieldwork, which creates a record of origins that can certify the authenticity and reliability of a particular pedigree, which then can acquire status as a form of scientific evidence.

In the 1950s and 1960s, human geneticists undertook wide-ranging field studies of human populations around the globe. They tracked visible anomalies, such as Ellis–van Creveld syndrome in the Pennsylvania Amish and albinism in the Hopi of Arizona. They also tracked geographical anomalies, such as the presence in the Pacific Rim of small populations that appeared to be African. Identifying suitable populations, assessing their genetic status, learning their reproductive histories, and extracting from them blood, tissue, and pedigrees were important activities in postwar human genetics.

The medical geneticist Victor McKusick, of Johns Hopkins University, was among the most prominent practitioners of this genetic fieldwork in the 1960s. This essay focuses on McKusick's field practices in the early 1960s with the Old Order Amish in Lancaster County, Pennsylvania. He was tracking a rare form of hereditary disease, Ellis–van Creveld syndrome, a dwarfing condition, and within a few years he had identified as many cases of this syndrome in the Pennsylvania Amish alone as had previously been reported in the entire medical literature (McKusick 1978: 104).

As he continued to work with the Amish, McKusick found many other recessive conditions in this inbred population, and he published many papers on genetic disease in Amish populations. Though I deal here only with his early work with the Lancaster County Amish, and with his efforts to understand one rare genetic disease, this case illuminates more generally the labor involved in genetic fieldwork in this period and, by extension, in the study of genetic disease as it became the focus of scientific and medical interest after 1955. I look at McKusick's methods, his recording systems, and his data collection network, attending particularly to the ways he enrolled different social actors in his project and deployed different kinds of knowledge. McKusick drew on field methods used in anthropology, sociology, and history in order to bring human pedigrees, notoriously complicated social documents, into the laboratory. His labor turned the Amish into a medical and scientific resource. Gossip, X rays, feelings, blood tests, and social consensus were resources for the construction of the pedigree, and any data point might have more than one axis running through it, from notes in Bibles to state public health records to reports from the local undertaker. Knowledge of heredity and disease was craft knowledge, dependent on a wide range of diagnostic and social skills, documents, and practices.

McKusick was acutely sensitive to questions of legitimacy and authority—he was himself a skilled clinician and, therefore, not quite a scientist in the eyes of some of his peers—and he kept scrupulous records of his field activities. I do not claim that he invented the field methods I examine here, but rather that his field methods exemplify the ways in which human geneticists began to remake human genealogy as a scientific resource. The tabulated lists of ancestors had long been suspicious in the eyes of some geneticists, and there had been various calls over the years for human genetics to find a way to go "beyond" the pedigree (see, for example, Haldane 1942). The pedigree was burdened by its transparently social nature, its dependence on the words of the subjects describing their parents or grandparents, and perhaps even by its connection to the project of eugenics and to the questionable data collection practices of the American eugenicist Charles Davenport. Davenport organized dozens of field studies before 1924 of albinos in Massachusetts, of juvenile delinquents in Chicago, and even of the Amish in Pennsylvania, but his workers' field methods were casual and the resulting pedigrees later were considered to be of relatively little scientific value (see Kevles 1985: 55, 199–200). In the molecular era, and with the explosion of new work in human population genetics, physical anthropology, human cytogenetics, cancer genetics, and related fields, the pedigree was being remade into a resource for laboratory science.

As Yoshio Nukaga and Alberto Cambrosio point out in their ethnographic study of the pedigree in contemporary genetic counseling, pedigrees "still constitute the basic investigative tool" in human genetics. The stories people tell about their families move from a "web of oral narratives to a sequence of visual inscriptions which, in turn, become part of larger inscriptions connecting medical pedigrees to the visual display of, say, cytogenetic or molecular biological test results" (1997). Even the most technical, machine-driven inscriptions of molecular genetics are grounded in the social complexity of the pedigree, which is nature-culture, and which represents a signal case of the employment of cultural resources to achieve erasure—of the cultural.

McKUSICK AND THE AMISH

In the fall of 1962, McKusick, then head of the Division of Medical Genetics at Johns Hopkins School of Medicine, happened to read a profile of a country doctor in Lancaster, Pennsylvania. This doctor, David Krusen, suggested that achondroplasia was frequent among the Amish. McKusick had been working with Marfan's syndrome patients and had an interest in disorders of connective tissue. Because most dwarfing conditions involve a defect in connective tissue, he was interested in achondroplasia in the Amish. He thought, however, that the rates the doctor described seemed much too high, and suspected that the Amish had some other condition.[1]

A few months later, John Hostetler, a Penn State sociology professor, submitted a book proposal to the Johns Hopkins University Press. The book was to focus on the medical, social, and cultural beliefs of the Old Order Amish; McKusick read the proposal for the press and was intrigued. He invited Hostetler to give a talk on the Amish to his research group (see Hostetler 1963, 1963–4). In his invitation, McKusick noted, "We have had occasion to become much interested in blood group and other physical anthropological characteristics of the Old Order Amish in Mifflin County"; the interest derived from "an observation of an unusual type of hereditary disorder which seems to occur with relatively high frequency in this Amish group." McKusick had not yet been out in the field but explained that he was interested in "making some arrangement to get blood samples on a representative group of individuals." He realized that acquiring these blood samples would be facilitated by a knowledge of "family structure, the attitude of the group toward illness and conventional medicine, etc., etc."[2] It is perhaps noteworthy that McKusick here explicitly construed the blood as historical and social, a material embedded in local narratives and Amish culture, and a material whose acquisition would require knowledge not of genetics but of social organization, history, and medical belief.

Just as knowledge of genetics required knowledge of social organization, so social organization and practice could produce biological qualities, bringing genetic disease out into the open through a series of cultural and reproductive choices. McKusick's later list of the qualities that made the Amish good research subjects included "great interest in illness," "clannishness," and a high rate of cousin marriage.[3] The cultural produced the biological. The Amish made genetic disease socially visible and easier to track as a result of their acceptance of cousin marriage, their closed breeding population, their meticulous genealogical records (records that had a religious significance but then became scientific resources), and their practice of publishing reports of diseases of all kinds in local newspapers. Indeed, the practices of this population seemed to be almost tailored to the priorities of field research in human genetics. In addition, the Amish, despite their isolation from mainstream life in Pennsylvania, were subject to the standard collection of vital statistics that applied to all residents of the state. Their births and deaths were recorded in Harrisburg. Their death certificates, with the names of attending physicians, were filed in state records, and these records, too, became a part of McKusick's information network.

McKusick's first foray into the field included two local guides: Hostetler, whose scholarly work on Amish culture he found so useful, and who had been born into an Amish family; and Krusen, the country doctor who thought he was seeing a specific genetic disease in the Amish, and who brought the Baltimore physician to meet the families in which it was present. Characteristically, McKusick pulled together various forms of local knowledge and created allies that could help him build an information-gathering network. Finding one's way through the social and cultural system, and through the country roads of Lancaster and Mifflin Counties, required many informants.[4]

Over the next year, McKusick, Hostetler, and a Yale University Ph.D., Janice Egeland, who recently had written her dissertation on the medical sociology of the Amish, conducted a formal survey of five hundred physicians who worked with Amish patients in Pennsylvania, Ohio, Indiana, and Ontario. Their first publication on the Amish as subjects of genetic study appeared in the *Bulletin of the Johns Hopkins Hospital* in 1964. This paper, "Genetic Studies of the Amish: Background and Potentialities," proposed that "the interest of many simple peoples in genealogy is a matter of note."[5] Geneticists, like "simple peoples," were interested in genealogies, of course, and their specific needs intersected with those of their subjects. McKusick, Hostetler, and Egeland stated, "In some primitive people, such as the Navajo Indians, descent, kinship, and clan identification are important in connection with decisions on whom to marry." Such details were therefore accessible and well-known to the populations under study. For the Amish, the model for genealogical record-keeping was the Bible, and "most Amish can trace their complete ancestry back to the immigrants from Europe." One family genealogy, that of the Fisher family, contained data on thirty-six hundred families and had some relation to almost all living Amish in Lancaster County.

As subjects of genetic studies, the Amish were a closed, defined population, the authors pointed out. No one could join, and there could be "no question of who is presently Amish." The Amish were also producing large families. Seven to nine children were common, and parents did not stop having children after the birth of an ill or abnormal child. Furthermore, mentally retarded or disabled children were kept at home, which meant that they could be readily studied "in relation to the rest of the family." Populations with similar characteristics had been studied in Switzerland, Sweden, and other areas of Europe, but the authors pointed out that the most informative such population in the United States was the Utah Mormons, who, like the Amish, were genealogically inclined, relatively immobile, clannish, and closed to outsiders (McKusick et al. 1964).

An entire range of cultural and religious choices made such isolates scientific resources. But these choices could also interfere with the fieldwork. The difficulties in "realizing the full potential of the Amish for genetic studies" related to Amish suspicion of outsiders, and reactions to some aspects of medical science. Amish families, for example, were in general reluctant to agree to autopsy. Of the thirty-six deceased persons with Ellis–van Creveld syndrome reported in a separate paper in the same issue of the journal, only one was autopsied (McKusick et al. 1964).

McKusick and his coauthors reprinted a letter from a thirty-year-old Amish man who had Ellis–van Creveld syndrome and who refused to be examined by the Hopkins researchers. The man stated, "I feel I am exactly the way the Good Lord intended for me to be, even before I was born. So I feel no human hands or brains can do a thing about me or anyone like me, if it is the way the Lord wants it, no matter how highly educated anyone is. I am happy, have work, friends and can support myself. So what more do such people want?" (McKusick et al. 1964). Another prospective participant in the study sent a postcard declining: "I am not interested in going in the hospital, so don't come around for me because I am not going in. And you don't have to stop by to see me either. I am allright[,] there isn't anything wrong with me and I don't think much of those x-rays you want. So don't stop in to see me. I am not interested in your stopping by."[6] In both cases respondents were contesting their status as objects of medical interest. The first was satisfied that his condition was God's will; the second that there was nothing wrong with him. They would not participate in the medical research, nor in the construction of Ellis–van Creveld syndrome—of their short stature and extra fingers and toes—as a genetic disease, and they were resistant to the technologies that McKusick's work would require.

Over the next two years, McKusick built a sieve that could lift a specific, visible form of genetic disease out of a social network. He was trying to find all cases of short stature and extra fingers. He used public records, Amish genealogy books, birth reports, newspapers, health professionals, and Amish contacts. He began to subscribe to *The Budget,* the Lancaster County newspaper that published reports about why someone missed church (a knee injury), tonsillectomies, birthday parties, and bad backs. He kept in touch with the undertaker who handled most Amish deaths. He surveyed the records of hospitals. He wrote to school officials to ask them to excuse students who missed school to be examined in his clinic.[7] He wrote personal letters to teachers, nurses, parents, and physicians. He was looking for extra fingers (polydactyly), which were signs that left a record in the state capital in Harrisburg years after a neonatal death in Lancaster, or that could be remembered years later by a midwife, a grandmother, a sibling. The disease that ended a life of only a few hours could be seen despite the poor resolution of the record and the temporal and cultural distance from a midwife on an Amish farm to the head of the Division of Clinical Genetics at the Johns Hopkins University years later. Every baby mattered, including those who lived only twenty minutes. "Did this baby have extra fingers?" McKusick asked a nurse present at a birth in September 1969.[8] The entire population needed to pass through the sieve. (See figure 2.1.)

Records of field trips from Baltimore to Lancaster and Mifflin Counties in McKusick's papers suggest the many kinds of information McKusick brought back. His notes describe families returning from funerals, conversations at the vegetable market, a dog bite he endured at one home, a frightened young child who "sobbed throughout our time there." He visited one family to learn that "they were no longer Old Order Amish" and another in which a teenager with Ellis–van Creveld was a "very fine boy," a junior in high school who had a job at a hardware store. "All the children work hard on the farm taking care of 27 milk cows," noted McKusick. Another boy in this family had been run over by a "wagonful of stones" but now was doing all right: "He sings a great deal and takes voice lessons."[9] On another field trip, the group "stopped by Kaufman's Orchard to get some apples and other things" and ran into one of the families they were going to visit. They also stopped by the home of a "very attractive young couple" who were "prepared for having church at their house the following day. They expected 70 or 80 people. She had made 32 pies the day before!" When the group visited an Amish school they learned that there recently had been a "discipline problem" at the school, where some of the boys had refused to wear their hats on the playground. At another home they found a blind father, a daughter with serious medical problems, and a "competent" mother: "She has to run a farm and raise a family with a blind husband." The group ended up "having a family style meal at the Harvest Drive Restaurant."[10] I mention these details to capture something of the tone and feel of this fieldwork. McKusick and his assistants were collecting many kinds of information about Amish culture and Amish people and made many kinds of observations in these field notes.

Amish participation in McKusick's study seems to have been enthusiastic. While there were a few resistant reactions, most Amish queried were willing

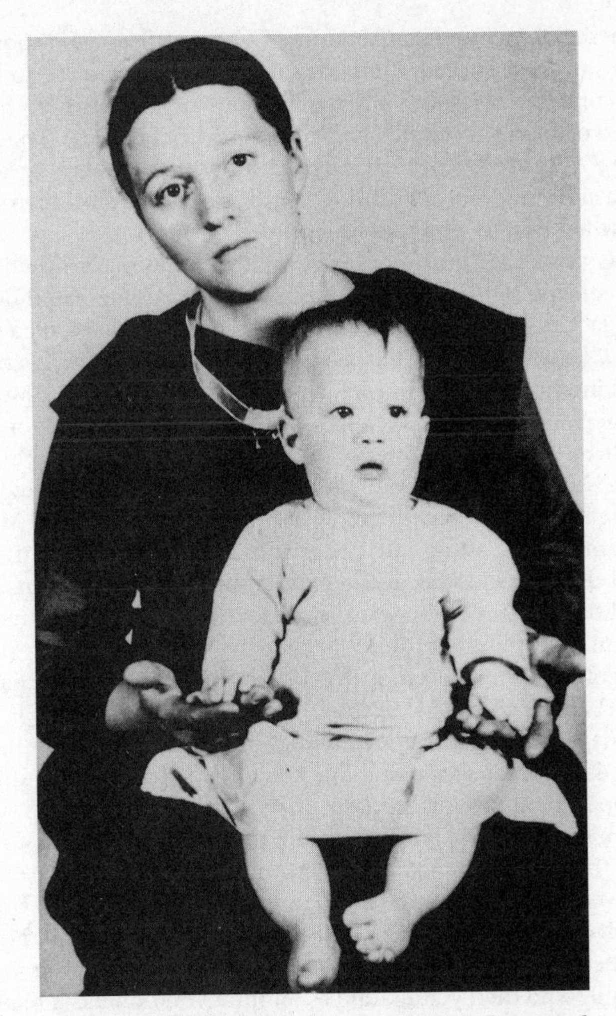

Figure 2.1 The "Amish Madonna." This was the second child in sibship 14 affected by polydactyly. Courtesy of Dr. Robert Weilbaecher.

to participate, and some served as informants and field-workers for his project. In one of the early papers, the authors hint at female skills in the management of awkward social approaches, noting, "Several unmarried Amish women collected information both of medical and genealogic nature and provided introductions to affected families who in many instances were relatives" (McKusick et al. 1964). Amish provided McKusick with clues, leads, and suggestions for tracking other cases. They wrote to McKusick for advice about whether to marry distant cousins. They spontaneously reported abnormal births in their extended families, and many cases came to McKusick's attention through the Amish. His families were organized in sibships, groups composed of all the offspring of a particular pair of parents. The "mother in sibship 23" reported two other cases that became sibships 25 and 26; the "father in sibship 3" was the "informant" for a case in sibship 10. McKusick was recording exactly who told him what.[11]

Parents provided him, moreover, with descriptions of their children's bodies, recalling the morphologies of stillbirths and neonatal deaths years after the fact, so that those pathological forms could become a part of the pedigree. "Mother states no extra fingers. Four children living and well," he recorded in one case of a four-week-old infant who died; or "Mother states extra fingers were present but apparently not of type in EvC" in the case of a thirty-three-year-old daughter who died in 1959, before McKusick began his study.[12] Maternal descriptions could rule out Ellis–van Creveld syndrome or certify its presence. Amish family members could also lead McKusick to other communities, as did one Pennsylvania father who said his siblings in Ohio had children with the same condition that affected his own. This father theorized that the condition was hereditary. "Now this may sound strange," he wrote to McKusick, "but five years ago my sister had a baby girl with the same trouble as ours. And a month or so later my brother's wife gave birth to a baby girl also with this same thing. Could it be hereditary?"[13] There was what might be called a folk epidemiology in the community itself, a network of knowledge and interpretation that could help McKusick identify relevant families and relevant bodily forms.

The Amish genealogy books encoded this folk epidemiology, and McKusick used them to construct pedigrees. In the fall of 1963, for example, he encountered an Amish family in which there were three adult siblings with polydactyly and achondroplasia, Ellis–van Creveld syndrome. He found in reading the genealogy book of the family that there were two other siblings in the family who died young, one as an infant and one as a teenager. Did these children have the same traits? He made his inquiry to the family physician, but the family physician asked the father and the father reported that the infant probably did, for it was a "short, chunky little baby," but the teenager certainly did not and had died of pneumonia.[14] A genealogical text, prepared for religious reasons, helped make the family a scientific resource, and a physician consulted for his specialized knowledge simply asked the father, who diagnosed both infant and teenager. There were many kinds of knowledge in this reconstructed pedigree.

Local physicians were of course an important resource, and tracking down a new case usually began with an appeal to the attending physician. In September 1963, for example, McKusick queried a Pennsylvania physician about a child who had died at Lancaster General Hospital in the spring of 1962, and who reportedly was born with signs of achondroplasia and heart problems. McKusick had probably learned about this death from the Amish newspaper, which reported the details of neonatal deaths, including the presence of abnormalities. He asked the attending physician questions about the size of the family, the presence of extra fingers, and the health of the parents, noting, "I have been much interested in the last year in hereditary disorders among the Amish and have been making a particular study of dwarfism. I would appreciate any information you can give me on Amish dwarfs." He enclosed a list of cases and families he already knew of and asked the physician if he knew any others, closing with a proposal that he would drop by the physician's office on his next field trip. This particular physician cheerfully answered all the questions and invited McKusick to stop by his office.[15]

As McKusick's database grew, he recorded where and how he had learned of the existence and status of each diseased person. The provenance of any given case included how it was ascertained initially, on what basis Ellis–van Creveld was diagnosed, what the health status of the affected individuals was, and how the pedigree was constructed in relationship to other pedigrees such as the Fisher genealogy. McKusick recognized fully the importance of this documentation, and he even included it as an appendix to his 1978 paper "Dwarfism in the Amish," stating that the "frequency of EvC as determined in this study is so unusually high [,] and such a large proportion of the cases had died before the study was performed [,] that it is deemed essential to outline briefly the features of each sibship, and to indicate the mode of ascertainment and basis for diagnosis in each case" (McKusick 1978: 119). The appendix listed twenty-nine sibships, and a typical listing included a description of the affected child and some indication of the reliability of the information ("polydactyly is . . . absolutely certain in the minds of the Amish informants" [McKusick 1978: 121]). In sibship 11, the affected family member was reported to "work hard with horses on farm" (suggesting perhaps that he was relatively healthy), and in sibship 3 there was a strange coincidence that led to ascertainment, a coincidence reported in the appendix. For most of his cases, and certainly for his most important pedigrees, he had multiple sources, and these were all recorded. A case might be documented in birth and death certificates in Harrisburg, a family Bible, hospital records, phone calls from a physician, letters from family members, personal observation on a particular field trip, reports from a descendant, or by one of his field-workers. As McKusick followed these signs, he collapsed distinctions between sources of information, accepting as equivalent the reports from midwives about recent births, from mothers about infants born a decade earlier, from physicians and nurses, and from a great-grandson reporting on the health and stature of a long-dead great-grandfather. The pedigree was like a patchwork quilt, pulled together from multiple fabrics into a pattern that rationalized the heterogeneity of the sources.

His papers from this period contain hand-sketched maps and directions telling field-workers where particular houses were and where families lived. At the same time, field-workers were mapping relationships between families on similarly scribbled pieces of paper, which were also tucked away in archived files. Both forms of maps—those depicting roads and landmarks, general stores and silos and red barns, across many miles; and those using darkened and clear circles and squares to depict complex familial lines across generations—were ways of organizing the Amish.

In Edward R. Tufte's explorations of envisioning information, he distinguishes between pictures of nouns and pictures of verbs. Maps and aerial displays, he says, "consist of a great many nouns lying on the ground," while pictures of verbs involve "the representation of mechanism and motion, of process and dynamic, of cause and effect." The directional maps of Lancaster County were "nouns lying on the ground," while McKusick's tentative pedigrees, hand-sketched and tucked in folders and notebooks as the work progressed, were pictures of verbs, arguments about cause and effect, and stories about history and heredity. They depicted a flow chart that made its case using the "smallest effective difference" between diseased and not diseased, male and female, alive and dead (Tufte 1997: 73–78, 121–27). The genetic pedigree, a standardized genre by the 1960s with rules about circles, squares, shading, and arrangement, is a record of field labor. In this case the labor engaged an entire community. Both the Amish and McKusick were proficient collectors of genetic disease.

Despite their participation in his fieldwork, McKusick did not particularly want the Amish to see the scientific papers he published. When he sent a reprint in 1965 to a physician in Strasbourg, Pennsylvania, who was treating one of the Ellis–van Creveld patients, he noted, "I of course do not want it to get into the hands of our Amish neighbors."[16] He was clearly assuming that the Amish would not read the scientific journal in which the paper was published. The social and intellectual gulf between his own world and that of the Amish seemed large enough to prevent any chance encounter between his subjects and his published work. It may be that he was concerned about the photographs, which featured recognizable Amish people, both adults and children, who had Ellis–van Creveld syndrome. Their names were not included in the text, but they presumably would have been known in the community. But I wonder too if he was concerned about his translation of their own genealogies, from religious texts to secular texts, from records of relation to records of pathology.

Another individual case was pictured in one of the early papers, in the form of a reproduction of a seventeenth-century Dutch drawing of an elegantly posed human skeleton (see figure 2.2). The skeleton was that of a newborn infant, drawn standing, skull tipped quizzically, arms slightly raised, in a formal posture that would have been impossible for the living child. "Thrown into the river at birth," according to the accompanying text, the newborn had been retrieved for its scientific interest. If not the focus of maternal love, it could at least be the focus of the love and desire that informed natural philosophy. It had seven digits on each hand, eight and nine digits on its feet, and some had theorized that it was an Ellis–van Creveld case. The Amish, with all their historical specificity, could be linked to a seventeenth-century Dutch newborn. Their disease bound them to a distant place and time.[17] The Dutch newborn was perhaps a record of the provenance of Ellis–van Creveld syndrome, a record of the "derivation and passage of an item through its various owners." The gene moved through human populations, leaving traces, signs, clues in the standard systems of accounting for people and recording their medical status. The pedigree highlighted these clues, brought them together, and situated them in a narrative that could make them scientific resources.

THE NATURE OF THE PEDIGREE

Finding any gene requires extracting words and blood from people, convincing them to contribute some portion of their bodies and some portion of their personal histories to science. The blood and the narrative are embedded in a larger narrative, a pedigree documenting family history, a causal model documenting the nature of a genetic defect based on inheritance patterns revealed in the pedigree, an origin story about the source of the mutation based on its population distribution, or a map upon which a particular disease can be placed in relation to all other genetic diseases. The potential inscriptions of fieldwork are multiple, complex, cumulative. But the basic inscription, the first point of translation, is the pedigree, and producing a pedigree is unquestionably social work.

Recent work in the history and sociology of science has explored the properties of the field sciences with special attention to the "chronic issues of status and credibility that derive from the social and methodological tension between laboratory and field standards of evidence and reasoning." Fieldwork involves phenomena that are "multivariate, historically produced, often fleeting and dauntingly complex and uncontrollable," as Henrika Kuklick and Robert Kohler have noted, and the field seems to be almost unsuited to the production of scientific knowledge. "It may seem astonishing that any robust knowledge comes out of fieldwork. Yet it does, abundantly and regularly." They also point out that in the field, unlike in the private space of the laboratory, scientific work is shaped by the social interactions of profession-

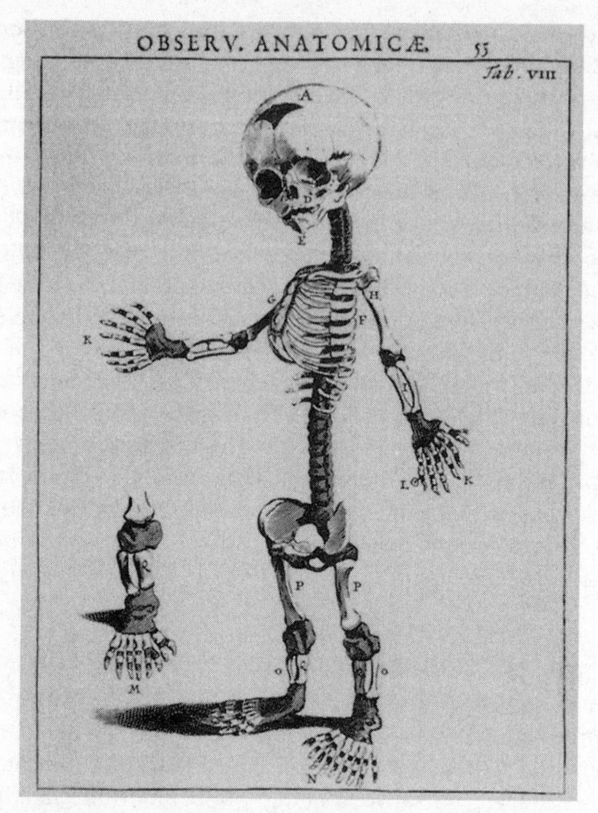

Figure 2.2 A polydactylous dwarf. Pictured in a 1670 Dutch publication, it was identified in 1940 as a possible Ellis–Van Creveld case.

als, amateurs, and local residents whose cooperation is both necessary and rarely acknowledged (Kuklick and Kohler 1996).

In a scribbled, undated set of notes in McKusick's papers, he outlined his concerns about pedigree methods. First, he said, genetic disease involves a "dynamic process": "At the time you study them you can't give the last word." The disease might develop as a person ages, and human subjects, McKusick pointed out, live as long as geneticists do. They were therefore outside the normal time frame for scientific study. Second, he said, there was a "lack of specificity" in clinical manifestations. The trait was "far from the gene." Reading down to the hereditary material from the clinical sign was difficult, complicated, and sometimes not possible. McKusick fully appreciated the complexity of moving from the body to the genotype. Finally, there was the question of familiar seeing. He knew himself that the more you looked at anyone the more you saw. The clinician in a standard examination might simply miss something that was relevant to understanding the disease and, therefore, the pedigree: "The more [an] individual personally studied and studied from a specific point of view, the greater the reliability."[18]

Yet the pedigree, despite its difficulties, was the "cornerstone" of medical genetics.[19] The preparation of pedigrees and the listing of new genetic diseases did not amount to mere "stamp collecting," he argued. "The catalogs of simply inherited genetic traits in man are like photographic negatives from which a positive picture of the normal genetic constitution can be constructed" (McKusick et al. 1964). McKusick suggested that cataloguing genetic traits, birth defects, and diseases in human populations could transform medical practice and patient care. This idea—I have called it the cataloguing imperative—has seen its most recent formulation in the promotion of the Human Genome Project; but it predates the project itself, and it is possible to argue that the project is a consequence of this expectation, which was so important to the rise of a scientifically legitimate human genetics. The listing of traits (diseases, chromosomal anomalies, birth defects) became central to the promotion of medical genetics. The catalogue seemed to be a crucial resource, a database that could demonstrate the relevance of genetics to medical practice.

In a 1977 grant proposal, McKusick said his project would permit him to document "new recessive disorders among the Old Order Amish." He could also gain insights into "incompletely characterized Mendelian entities" by studying this population. There were at least sixty cases of cartilage-hair hypoplasia, for example, and there were some new reported cases of chondrodysplasia with severe combined immune deficiency. There were also a "presumably autosomal recessive form of osteogenesis imperfecta," many cases of cleft hand and foot, and six sibships with Kaufman syndrome.[20] The Amish were a rich treasure trove of genetic disease, and McKusick's field trips provided access to these diseases, which were called to his attention in casual conversations at the market and in trips to the local school. The fragments of detail—a child born dead, a sister who was mentally retarded—became part of grant proposals. The "medical tourist" brought back enough information to secure funding for another trip.[21]

A grant reviewer, when considering this proposal for the National Institutes of Health, wrote that "the type of research is more descriptive than innovative, but an enormous amount of data which makes a significant contribution to clinical genetics results from astute clinical observation and intelligent interpretation of the results. It is in this field that Dr. McKusick is a world leader."[22] As this assessment suggests, McKusick's own relationship to both clinical medicine and high science mirrored the complicated interplay of clinic and laboratory in the rise of medical genetics. McKusick occupied a critical borderland between science and medicine. When he was nominated for membership in the National Academy of Sciences, one skeptic asked, "But is he a scientist?"[23] And was human genetics a science?

In the 1970s, McKusick kept a dwarfed miniature poodle, named Vanilla, as a pet. In March 1973 Vanilla was bred back to her father, Can.Ch.Andeches Ready to Go. The pregnancy was expected to be risky, with a potential need for a cesarean section, and McKusick made provisions for Vanilla to be kept and closely watched at the Division of Laboratory Animal Medicine at the Johns Hopkins University School of Medicine, under the care of Edward C. Melby, professor and director of this division, during the final stages of her pregnancy.[24] I could not find any documentation in McKusick's papers describing the outcome of this high-risk pregnancy. I mention the pregnant poodle because she was an animal model—and a companion animal—manifesting a condition related to those McKusick was studying in human populations. McKusick could control the breeding and pregnancy of Vanilla and could produce, through inbreeding, a biological state that mimicked Ellis–van Creveld. He was an overseer of dwarfing conditions, a manager of bodies and genomes and catalogues, and a perceptive student of the pedigree, both canine and human, in all its biosocial complexity.

Like other human geneticists of his generation, he was struggling to find a way to study human populations and human genetic traits that mitigated the chaotic conditions of the (human) field. Genetic fieldwork was the study of phenomena that were deeply disordered and uncontrollable. Managing this disorder required a keen attention to provenance, to the histories of people, blood, families, and cultures as they intersected in a single phenotype. It required following a malformation of the limbs and an abundance of fingers through the many texts, stories, and memories where they were documented, and translating these signs into an authentic record of heredity that could presume to hold its own in the laboratory. Creating the impeccable pedigree was one of the great achievements of postwar human genetics. Making its contingency functionally invisible—or irrelevant to its usefulness as a guide to the genome—was a social project of enormous complexity, as McKusick's fieldwork makes clear.

NOTES

This paper is part of a larger study of postwar genetic disease supported by the Burroughs-Wellcome Fund Fortieth Anniversary Award, and I am grateful to the BWF for its support. Special thanks are due to archivists Nancy McCall, for guidance and suggestions, and Gerard Schorb, for his patient and efficient assistance in my efforts to negotiate the papers of Victor McKusick at the Alan Mason Chesney Medical Archives at John's Hopkins University (hereafter Papers of VAM). This collection is certainly one of the finest records extant of postwar human genetics. For our ongoing dialogues, and for critical comments and suggestions, I am grateful to Dorothy Nelkin, Rayna Rapp, Barton Childs, Charles Rosenberg, and Henrika Kuklick and to my coeditors Alan Goodman and Deborah Heath. Following the practices of the sociologist or anthropologist, I do not use the names of Amish subjects here because the records I have explored are basically confidential medical records, but I recognize that constructing McKusick, physicians, and other researchers as the only named people in this narrative represents a form of asymmetry that I would prefer to avoid.

1. "Since true achondroplasia is a dominant with markedly reduced reproductive fitness, one would not expect it to have a high frequency in an inbred reproductive group." McKusick, "Medical Genetic Studies of the Amish," grant application to National Foundation–March of Dimes, for 1979–80, Box R109C16, "National Foundation 1979–80," Papers of VAM.

2. McKusick's invitation to Hostetler to speak is in McKusick to Hostetler, 7 May 1963, Box R115F1–2, "The Amish Population," Papers of VAM. He offered Hostetler a fifty-dollar honorarium, and as was often his custom with invited speakers, an invitation to stay at McKusick's home. See McKusick to Hostetler, 7 May 1963.

3. McKusick discusses these advantages in many sources both published and unpublished, but a specific list, "Characteristics of Amish Society Favorable for Genetic Studies" is in Ellis–van Creveld 8005, Box R115 F1–2, Papers of VAM.

4. This first trip is discussed in David Brown, interview with Victor McKusick, p. 35, Papers of VAM.

5. McKusick, Hostetler, and Egeland 1964, reprinted in McKusick 1978.

6. January 1964, postcard to Victor McKusick, Ellis–van Creveld 8005, Box R115F1–2, Papers of VAM.

7. 13 September 1966, Ellis–van Creveld Syndrome File, Box R115F1–2, Papers of VAM.

8. Letter in Ellis–van Creveld 8005, Box R115F1–2, Papers of VAM.

9. Andrew Gale, E. A. Murphy, V. W. McKusick, and V. A. McKusick, "Trip to Lancaster County," 17 February 1975, Lancaster File, Box R110D8, Papers of VAM.

10. Notes on a trip to Lancaster with Clair Francomano and Guadalupe Gonzalez-Rivera (known as Lupita), 4/6/82, Red Binder, Box R109I18, Papers of VAM.

11. Twenty-one pages of handwritten notes describing these sibships and this informant network, probably written in July or August of 1963, are found in Ellis–van Creveld 8005, Box R115F1–2, Papers of VAM.

12. Elis–van Creveld, Box R115F1–2, Papers of VAM.

13. Letter in Amish, Box R110D8, Papers of VAM.

14. McKusick to Robert Baur, 26 November 1963, with attached handwritten notes in response, in P8005, Ellis–van Creveld, Box R115F1–2, Papers of VAM. Other correspondence with physicians who treated the Amish is also in this file.

15. McKusick to Clinton Lawrence, 12 September 1963, Ellis–van Creveld 8005, Box R115F1–2, Papers of VAM.

16. McKusick to Henry S. Wentz, 10 March 1965, Ellis–van Creveld 8005, Box R115F1–2, Papers of VAM.

17. This illustration appeared originally in Theodor Kerckring, 1670, *Spicilegium anatomicum* (Amsterdam: n.p.). It was reprinted in McKusick 1978.

18. McKusick (his handwriting), "Pedigree methods," in the Pedigree Method in Medical Genetics file, Box R109C8, Papers of VAM.

19. McKusick used this word in a grant application in 1957 to the National Institutes of Health for support for a study that could "define, in as much detail as possible, the mode of inheritance, mechanisms of clinical manifestations . . . [and] factors affecting penetrance and expressivity" in cardiovascular disease. The pedigree method, he said, would be "the cornerstone of the program," and the "machinery" for ascertainment of relevant pedigrees was already in place in Baltimore. He would need a nurse for "pedigree tracing, checking hospital records, assistance in clinical testing," part-time secretarial help, and a technician for chemical tests. He also asked for $150 for the drawing of pedigree charts. Application for research grant: "A study of the genetic factor in cardiovascular diseases," 20 February 1957 (not in folder), Box R109C8, Papers of VAM.

20. See Amish Grant, Box R110D8, Papers of VAM.

21. This term was used by a physician currently treating the Amish for genetic diseases to refer to medical researchers of an earlier era who were "less interested in the health-care needs of the community than they were in the diseases themselves." This is presumably a reference to McKusick and his field-workers. D. Holmes Morton, quoted in Stranahan 1997.

22. The summary referee report is attached to a letter. Paul A. Deming of the National Institute of General Medical Sciences to McKusick, 28 December 1977, for application 1RO1 GM 24757–01, filed with grant proposal, Amish Grant, Box R110D8, Papers of VAM.

23. McKusick was elected to the National Academy of Sciences (NAS) in 1973. See the large file on his nomination and admission, NAS, Box R109E4, Papers of VAM.

24. McKusick to Edward C. Melby, 6 April 1973; and Melby to McKusick, 7 March 1973. See correspondence and related materials in Folder M, Box R109I9, Papers of VAM.

REFERENCES

Haldane, J. B. S. 1942. *Heredity and politics.* New York: W. W. Norton.

Hostetler, J. 1963. *Amish society.* Baltimore: Johns Hopkins University Press.

———. 1963–64. Folk and scientific medicine in Amish society. *Human Organization* 22, no. 4 (winter): 269–75.

Kevles, D. J. 1985. *In the name of eugenics: Genetics and the uses of human heredity.* Berkeley: University of California Press.

Kuklick, H., and R. Kohler. 1996. Science in the field. *Osiris,* 2nd ser., 11:1–14.

McKusick, V. A., ed. 1978. *Medical genetic studies of the Amish: Selected papers assembled with commentary.* Baltimore: Johns Hopkins University Press.

McKusick, V. A., J. A. Hostetler, and Janice A. Egeland. 1964. Genetic studies of the Amish: Background and potentialities. *Bulletin of the Johns Hopkins Hospital* 115:203–22.

Nukaga, Y., and A. Cambrosio. 1997. Medical pedigrees and the visual production of family disease in Canadian and Japanese genetic counseling practice. In *The soci-*

ology of medical science and technology, ed. M. A. Elston, 29–55. Oxford: Blackwell Publishers.

Stranahan, S. Q. 1997. Clinic a lifeline to children: A doctor finds his calling among the Amish and Mennonites. *Philadelphia Inquirer,* 15 September.

Tufte, E. R. 1997. *Visual explanations: Images and quantities, evidence and narrative.* Cheshire, Conn.: Graphics Press.

Chapter 3

Flexible Eugenics

Technologies of the Self in the Age of Genetics

Karen-Sue Taussig, Rayna Rapp, and Deborah Heath

In other words, our essence is ours to choose, depending on how we direct our selves with all our baggage, DNA included.

DAVID BARASH, *"DNA and Destiny," 1998*

In 1994, John Wasmuth and his laboratory colleagues published an account of the discovery of FGFR3, the gene for achondroplasia—the most common form of heritable dwarfism—in the journal *Cell* (Shiang et al. 1994). Hailed soon after in the *Scientist* as the article most frequently cited during 1995, Wasmuth's publication revealed that 98 percent of those affected with achondroplasia have an identical mutation in the molecule FGFR3, a receptor for what is called a growth factor.[1] Among other things, the discovery opened the possibility for prenatal screening for this condition. During the many years of work that led to the publication of Wasmuth's article, molecules, scientists, and technicians were drawn into engagements not only with one another but also with patients, physicians, and genetic counselors. Genetic knowledge emerged, in this case as in others, as a coproduction of clinical diagnosis and treatment regimes as well as the molecular technologies and other research practices that constitute laboratory life. Patient populations contributed to laboratory and clinical knowledge through their tissue samples in countless experimental and diagnostic contexts, and through what the historian M. Susan Lindee (chapter 2, this volume) describes as the emotional knowledge that families living with genetically different members accumulate.

The long-term work on dwarfism and related skeletal dysplasias depended on the collection of research samples from individuals from all over the world affected by these conditions. The samples were held in a tissue registry established to bank research materials. This story, too, had its fair share of competition and collaboration, not only in the search for a "dwarfism gene" but also in the quest for the gene for Huntington's disease,[2] on which Wasmuth had previously worked, and which, as it turned out, lies on the same chromosome as FGFR3. Indeed, the successful search for the Huntington's gene figures prominently in the mobilization of scientific and popular support for initiating the Human Genome Project, but that is another story. The multilayered discovery processes we recount here are instances of science-as-usual at the beginning of the twenty-first century.

One year after Wasmuth published his article, Clair Francomano, chief of medical genetics at the National Human Genome Research Institute at the National Institutes of Health, attended the national convention of the Little People of America (LPA), the U.S. national organization for people of short stature. Dr. Francomano is a long-standing researcher and health service provider for people with heritable dwarfism and a member of the LPA Medical Advisory Board. As she tells her story, "The first thing I saw when I came to this convention last year [after the discovery of the gene was publicized] was one of the people wearing that 'Endangered Species' T-shirt.[3] It really made a very big impact on me. And I really worry about it. I worry about what we're doing and about how it's going to be used and what it means to the people here" (Francomano 1997, pers. comm.).

Dr. Francomano's response was to chair several workshops for LPA members on the Human Genome Project. There, she explained genetic technologies and programs, listening attentively to the fears and hopes of short-statured people. She also expressed her own aspirations concerning the possibilities opened up by genetic research, and her dismay that new discoveries might be eugenically deployed. Her aspirations centered on gene therapy for specific ailments—such as ear and breathing problems, back pain, and skeletal problems—associated with dwarfism. In addition, Dr. Francomano collaborated in designing a membership-wide survey for the LPA on attitudes toward prenatal testing. Like Dr. Francomano and the officers of the LPA, we also want to know what Little People—a term widely used among people with dwarfing conditions to refer to themselves—in all their biomedical and political diversity want and do not want from this emergent genetic technology. We consider such desires to be part of science-as-usual in this history of the rapidly transforming present.

In this essay, we examine forms of embodiment and subjectivity emerging from relations between biomedical experts and lay health advocates in an era when genetic explanations, and desires for genetic improvement, appear to be proliferating throughout U.S. public culture. We address both biomedical technologies, like limb-lengthening surgery and prenatal diagnosis, and social technologies, like the organization of self-help groups such as the Little People of America. Our analysis of genetic and eugenic thinking in action underscores what Foucault (1988) calls "technologies of the self," the practices by which subjects constitute themselves, and work to improve themselves, while living within institutional frameworks of power. The expansive salience of genetic narratives and practices across a broad range of social groups in the United States today shapes embodied understandings of selfhood in historically specific ways. Those living with heritable dwarfism, and the researchers associated with them, are no less subject to these social and historical processes than the general population is: increasingly, we all live inside a world saturated by genetic discourses. Yet the consequences of dwelling inside these geneticized perspectives and practices are highly differentiated.

Born with bodies that historically have been stigmatized, dwarfs were among the first people in the United States to form an organization of social solidarity based on phenotypical difference.[4] The LPA, founded in 1957, became one of the first U.S. health advocacy groups to cooperate with biomedical and, especially, genetic researchers. This biosocial coalition between those born with a stigmatized difference and researchers and medical service providers was at once a site of productive resistance to widespread social prejudice and a domain of normalization. More recently Paul Rabinow (1996) has used the term *biosociality* to describe the conscription into a new identity politics as people come to align themselves in terms of genetic narratives and practices. This is something that Little People (LPs) began experimenting with as a social form decades before recombinant technology called into play new social forms.

By elaborating the diverse strategies through which dwarfs deploy technologies of the self, or an "ethics of self-care," we are able to illustrate the types of agency by which individuals "can resist the normalizing effects of modern power" (Bevir 1999: 78). In the contemporary United States, LPA members act within a society marked by a long-standing attachment to ideologies of individualism and free choice, which are increasingly imbricated with the intensified commodification and market orientation of the recent neoliberal era. LPs, along with the rest of us, are obliged to be free and are presented with an array of technically mediated choices and with varied discourses of perfectibility: we all live within dominant ideologies of power (Althusser 1971)—in this case, the idea of both choosing and perfecting oneself.[5] There is a convergence, or constitutive tension, between genetic normalization and an individualism that increasingly engages biotechnology—*biotechnological individualism.* From this tension, what we call *flexible eugenics* arises: long-standing biases against atypical bodies meet both the perils and the possibilities that spring from genetic technologies.

We have learned about the genetics of dwarfing conditions and the advocacy of the LPA through our collaborative ethnographic project on new knowledge production in the field of genetics.[6] In order to understand how scientists, clinician-physicians, and members of lay health organizations perform their daily work, we constituted ourselves as a mobile research team. In addition to ourselves, we worked with three graduate research assistants—Erin Koch, Barbara Ley, and Michael Montoya. During the project, we lived on two coasts and were attached to five institutions; much of our communication took place over the Internet, a common enough situation among the genetic knowledge producers we were tracking, but an uncommon way for cultural anthropologists to conduct research. Our traveling methodology followed genetic stakeholders in and out of their various milieus, from national meetings of health advocacy groups to basic research laboratories, and from interviews with clinicians to encounters with families living with heritable conditions like achondroplasia. Like Dr. Francomano and the members of the LPA, our team is concerned about the ways in which molecular discoveries may reinforce eugenic thinking and practices. And like many members of the constituencies among whom we conducted fieldwork, we also recognize the complex interplay that makes it difficult to distinguish the gifts from the iatrogenic poisons of contemporary medical genetics.

A discourse of benefits and burdens, perils and possibilities, and danger and opportunity now surrounds contemporary discussions of genetic technologies and their presumed power to rock the foundations of nature (Paul 1995; Strathern 1992). The attribution of social upheaval to scientific advancement is, of course, not new: interwoven fears and hopes have long been attached to biomedical attempts to "play God" with nature, as the history of nineteenth-century surgery or twentieth-century reproductive medicine bears out. Here and throughout our collective work, we hope to tease out imbrications of the old and the new, the innovations and constraints through which public enthusiasm and dis-ease regularly collide. On this unstable terrain, other powerful cultural discourses surrounding notions of the mastery and perfectibility of nature—including human nature, biology, and molecular genetics—intersect one another with complex and often contradictory effects.

While eugenic thinking has a long and tenacious history in Western societies, we want to be attentive to the specificities of the present moment. Under the shadow of the Human Genome Project and the rise of the biotechnology industries, a heterogeneous array of actors has been drawn into a worldview in which human diversity is increasingly ascribed to genetic causality.[7] In many ways, this perspective builds on older versions of biological reductionism in which barely concealed, barely secularized Protestant notions of predestination identified a social elite by its alleged physical, mental, and social superiority. At the same time that contemporary medical geneticists powerfully distance themselves from prior notions of biological superiority and inferiority, they relocate the intervention of authority and explanations of the body to the molecular level. We have all benefited from previous forms of scientific reductionism and medicalization, as well as suffered their social consequences selectively.

Yet in the popular imagination, as Abby Lippman (1991) points out, the connection between the perceived heritability of complex social traits like intelligence or criminality and the assumed explanation at the level of individually carried DNA underlies powerful beliefs in both genetic determinism and the importance of new biotechnologies of genetic improvement. Thus we see the persistence of eugenic thinking in the United States today, where many people across a broad spectrum of social groups consider the genome to be the site at which the human future must or can be negotiated. For us, this expanding genetic worldview among all constituencies, including research scientists, clinicians, lay support groups, and more general populations, is constituted dialectically: on the one hand, an ever increasing number of actors and practices are conscripted into a world defined genetically, in which reductive determinism looms large. On the other hand, democratic possibilities open up as genetic discourses and practices come to occupy multiple locations and to conscript a wider range of actors. Some of those actors may use their new and multiple locations to contest a too-easy determinism or to develop interventions—molecular and otherwise—that they consider choice-enhancing. They may well be viewed as a vanguard in the politics of biosociality, a vanguard from which the rest of us have much to learn. Those who have a consequential stake in this story have taught us to appreciate and track this dialectic in practice, as illustrated in the following narratives drawn from our observations at the LPA national conventions in 1997, 1998, and 1999.

AGENCY, NORMALIZATION, AND CONTESTED IDENTITIES

The LPA offers not only a site for biomedical research but also a self-affirming social environment. Most members bear a diagnosis of one of the many heritable dwarfing conditions, and the organization brings them together in a well-elaborated example of biosociality. Genetic enrollment includes conscription into a new identity politics as people come to align themselves with categories increasingly refashioned through emergent genetic discourses and practices. In our era, contemporary social life is rapidly being rescripted in terms of genetic narratives and practices (Taussig et al. 1999). But some aspects of biosociality build on forms of medicalization that predate molecular genetics, providing other embodied foundations for individual recruitment to group identity.[8] Indeed, although the LPA initially was founded as a support organization for all people of short stature, and its membership requirement is based on height rather than medical diagnosis, the LPA has long been interpolated into the milieu of medical genetics. At the same time, not all those born with heritable dwarfism accept the body politics that have emerged from the LPA's hard-fought advocacy.

One site at which we witness the tensions embedded in contemporary genetic and eugenic thinking in action is the meetings of the medical advisory boards that virtually all lay health groups organize. These advisory groups help members communicate with researchers and biomedical service providers who are experts in their particular (and often rare) conditions. For example, the Medical Advisory Board of the LPA comprises both members of the organization and medical professionals who serve at the invitation of its officers. Since its inception in 1957, the LPA has maintained a strong grassroots orientation. While there is a tradition of cooperative medical research conducted within the LPA membership, the organization's leadership has asserted conscious control over researchers' permissions and protocols. Engagements between medical professionals and membership are carefully negotiated, as is the membership's access to the results of that research. When we interviewed one senior LPA member, for example, he stressed that the organization insists that researchers cooperate with each other, sharing blood and tissue samples that are already banked, avoiding oversampling. He told us a joke about LPA members who have become polka-dotted from the numerous skin biopsies they have provided to researchers over the years. Specific medical interventions are sharply debated and contested in ongoing and negotiated relations.

During our most recent visit to the Medical Advisory Board, a longtime physician member of the board reported on his recent trip to Spain. There, he had visited a surgeon who has been doing limb-lengthening surgeries on dwarfs for twenty years, a procedure that many dwarfs find controversial. The physician presented a video of a young American woman who had gained 12 inches in height through multiple surgeries. The ten-minute video documents a testimonial speech she gave at a fund-raiser for a genetics medical center. The opening image—a life-size blowup of the young woman before the surgeries, standing at 3 feet 10 inches—is followed by her dramatic appearance onstage on crutches (a result of her last operation). She gives a polished and thoughtful speech about how limb lengthening was not merely cosmetic. It gave her not only 12 inches in height but also the experience of being wheelchair-bound for two years, providing enforced tolerance for a range of disabilities. The woman tells her audience, "It's a better life and I'm happier. I'm more independent and confident. Many inner changes took place. I learned that the change was everything I ever wanted." She was fifteen when she reached her decision to have the surgery. Hers is a narrative of challenge, perfectibility, and growth. Reactions on the Medical Advisory Board immediately challenged the young woman's narrative.

During the video screening, the room buzzed with sotto voce comments; as the video ended, the room fell silent. The first to speak was a doctor who raised issues about insurance coverage and the competence of certain surgeons to perform such a complicated task. The first LP to speak asked, "Did she have any involvement with the LPA?" to which the presenting doctor answered, "Yes." The LP continued, "I find it surprising, when you can come here and see at least five hundred successful adults; most would say—if asked the question 'Would you change it'—they'd say 'No.' Here in America, accessibility is a minor issue [that is, for LPs in the United States when compared to Spain]. I like to keep an open mind, but I think it's easier to adapt the environment than to adapt the person." Doctors and LPs rapidly entered the fray. One of each deemed the video's life-size blowup offensive. As the physician put it, "The video itself presents a cardboard cutout of an LP as undesirable and unattractive, in contrast to the whole person who is a foot taller than before."

Dissent broke out among the physicians, including practicing orthopedic surgeons who do not perform limb lengthening. One said, "I have devoted my life to treating the medical symptoms [of dwarfism], and I could never bring myself to lengthen limbs, because I find it abhorrent. I cannot stretch them out for social acceptance. It's more abhorrent to me than prenatal diagnosis." His denunciation highlights the perceived continuities between orthopedic and genetic interventions in the presumed foundational and moral rectitude of "nature" and "natural" variation. It also highlights diverse responses among researchers and clinicians, many of whom express complex critiques anchored in worldviews ranging from religion to political economy and civil rights.

The presenting doctor replied that in a society which promotes breast enlargement, rhinoplasty, and liposuction, dwarfs, too, deserve their right to aesthetic free choice in the medical marketplace. Yet even as he defended the practice, he also stressed that the operation should not be done before adolescence, when the patient and not the parents (who are most often of average stature) can consent to the procedure. Furthermore, he thought most people should not have the operation. He pointed out that, over the past decade, he and his colleagues had performed only thirteen surgeries. One of the elders of the LPA responded, "This is good information to have, and . . . it would be good to make it widely available, because it counters the widespread impression that the clinicians carrying out limb lengthening had created a surgical 'production line.'" Another LP, who works in a clinical setting, offered a final word:

> It's an attitude thing. I look at this as an enhancement, not a correction. But I don't need a correction. I'm OK. Most LPs, especially in this organization, look upon this as, you're telling me something's wrong. I'll make that choice. But I worry about calls from [average-statured] parents with new [dwarf] babies. I get phone calls every day from parents who aren't worried by [serious medical conditions associated with dwarfism, e.g.,] decompression, sleep apnea, with two-week-, four-week-, two-month-old kids, but they want to know about limb lengthening.[9] We'll all benefit from bringing this information out in the open.

Despite their different subject positions, the LPA officers and the physicians all inhabit a world in which the benefits of individual access to information and tropes of free marketplace choice predominate. A controversial surgical orthopedic intervention into body morphology shows how both its supporters and detractors invoke free choice in presenting their views on variations of biotechnological individualism. Indeed, the LPs, the members of the Medical Advisory Board, and ourselves are no less citizens of what the legal historian Lawrence Friedman (1990) has so usefully labeled the "republic of choice." Limb lengthening proposes to change the individual's recognizable phenotype without intervention into the underlying genotype, a kind of aesthetic and highly technical mastery of normalization. Those who choose the surgery, demonstrating the agency of choice in the biomedical marketplace, elude the judgment of prescriptive "natural" dwarfism inherent negatively in the form of social prejudice and positively in the biosociality of the LPA. Notions of mastery and perfectibility extend well beyond the contemporary United States, of course. But they have been given an upgrade and brought into the realm of science and technology within the rubric loosely identified as modernity, in which individual embodied choices reveal an attachment to the pursuit of progress and perfectibility (Berman 1982). What C. B. Macpherson classically labeled "possessive individualism" (1962) is here linked to identities, the realm of the body, and indeed, genetics.

What we are describing here as flexible eugenics thus involves technologies of the self through choosing and improving one's biological assets.[10] The desire to choose one's self in terms of technological interventions into the individual body incorporates both old and new aspects, from the distant promise of gene therapy to low-tech or routinized technologies such as cosmetic surgery.[11] Such instances signal a shift, one that Emily Martin, inspired by Michel Foucault, identifies as a move away from the powerful external interventions that produced the "docile bodies" so essential to the success of an earlier era of capitalism. Now, with postwar neoliberalism and its expanding emphasis on commodification and marketability, we see the emergence of "flexible bodies" (Martin 1994) obliged to be free, constrained by the tyranny of choice. In this marketplace of biomedical free choice, technology and technique become objects of desire invested with diverse meanings that surely vary for producers and consumers, for research scientists, clinicians, and individual patients, all of whom may imagine their relationship to choice and perfectibility quite differently.[12]

With advances in molecular biology, through which genes are becoming alienable and the modification of specific genes and bodies imagined more and more as an individual choice, biotechnological interventions in the service of individual perfectibility become the objects of desire. Deploying both social and biomedical "technologies of the self" enables people to modify, and imagine modifying, what is seen as natural, while our collective and individual stakes in what counts as natural are continuously renegotiated (Franklin 1997; Ragoné 1994; Strathern 1992). That is, in a world increasingly marked by flexible eugenics, self-realization can become attached to genetic characteristics, increasingly understood as susceptible to improvement and choice. Thus, long-standing discourses on individualism and choice are now filtered through newer interventions that include the molecular or genetic, as well as older and constantly escalating ones provided by pharmacology and surgery, all in the service of sculpting flexible bodies. It is this flexibility of the individual body as an object of biotechnological choice and desire that then intersects innovations in eugenic thought and practice.

LOVE, DEATH, AND BIOTECHNICAL REPRODUCTION

How might the discourses of biotechnological individualism observed in action at the LPA highlight some social values while obscuring others? This question is richly woven through myriad discussions of love, marriage, family formation, and children, objects of desire prominent in many of the LPA workshops and informal conversations in which we participated. These, of course, involve relations across the generations and are therefore not only aesthetic but also eugenic in the classic sense of the term. For example, discussions concerning aspirations for and celebrations of dwarf children were common throughout the LPA. We also noted a particular emphasis on the value of dwarfs having babies. Thus an affirmation of the value of dwarf children struck us as a sign of resistant biosociality: although dwarfs conventionally have been despised and labeled as imperfect, kinship with and by dwarfs across the generations here has been given an elevated significance, an affirmation of diffuse and enduring identity in the face of the widespread discrimination LPs often face in the larger world.

In a workshop for new parents that was packed with family members of both average and short stature, for example, all participants introduced themselves by saying where they were from and what type of dwarfing condition their child had. Many new parents of average stature were seeking support as they dealt with the shock of having a dwarf child. Other average- and short-statured parents were there to lend such support. An achondroplastic dwarf introduced himself and his wife as expecting a child and said, "And we hope it's a dwarf!" The audience responded to this comment with loud applause.

Two of us attended a workshop on women's health chaired by two female high-risk obstetrician-geneticists. In an audience of twenty short-statured women and two average-statured anthropologists, the first comment offered came from Katherine,[13] who shot her hand up, saying, "I'll start. I'm four months pregnant. . . . " She was interrupted instantly with enthusiastic applause and murmurs of delight from everyone, including the two physicians in the room. Then Katherine asked about her childbirth options, and a long discussion, framed in extremely positive and supportive tones, ensued about the logistics of childbirth for women of short stature. Doctors and audience were united in viewing pregnancy and childbirth as highly desirable, both looking to biomedical technology to offer progress in obtaining safer and less complicated reproductive outcomes.

Katherine, like many pregnant American women, was concerned about how soon she would be able to hold her newborn child. Her physician had told her that because of prior adhesions in her lumbar area (a common problem associated with dwarfism), her only option for childbirth was a cesarean section under general anesthesia. Women with dwarfing conditions virtually always have cesarean sections, because the shape of the pelvis does not allow for passage of a baby's head. In the United States, cesarean sections typically are done with a spinal block rather than general anesthesia. One of the physicians explained that a spinal block was complicated in cases of people with spinal differences: it "is a really controversial issue. . . . Anesthesiologists are really afraid of . . . [spinal] abnormalities, and with good reason. It's really uncharted territory. . . . If your anesthesiologist is most comfortable doing general [anesthesia], . . . then you're going to have a good outcome; and putting a needle in your back is risky after adhesions, so I wouldn't take the risk." Katherine then asked, "What about the short stature makes it dangerous?" The response from one of the doctors highlights the fact that what are often considered routine medical procedures may be linked to conceptions of standardized bodies. She explained:

> When they're doing regional anesthetic, whether it's spinal or epidural, what they need to get is either a catheter or a needle into that little space—and what is the space like, is there a space? Sometimes in LPs there is no space. . . . They [the doctors] have to have landmarks. They're doing this blindly. . . . If they push on your back . . . they're looking for landmarks and they're saying, 'There's a landmark.' . . . If you have an alteration in your landmark, they have nowhere to start.

The conversation continued, with the physicians focusing on childbirth protocols for women of short stature, members of the audience chiming in with their own experiences, and Katherine trying to figure out how she could ensure that she would hold her child as quickly as possible after delivery.

The encounter between physicians and women with dwarfing conditions underscores three salient points. First, at this meeting the issue was not whether LP women should have children, but the physical logistics of pregnancy and delivery. We imagine that this discussion is a new one: it is unlikely that twenty years ago there even existed high-risk obstetrician-geneticists who would universally support LP women having pregnancies, and dwarfs' aspirations to reproduce were more highly stigmatized. Second, the concern about obstetrics brings to light the challenges of applying standardized medical techniques to people with nonstandard bodies. The familiar waltz between the normal and the pathological reveals the hidden costs of standardization (Canguilhem 1989 [1966]; Starr 1991). Finally, the encounter illustrates that the different subject positions of the participants shape their concerns about reproduction. The physicians are caught up in the practical matters of applying standardized medicine to people with spinal differences: their agency is best expressed through continuous enhancement of the expertise that will make pregnancies safer.[14] But Katherine, whose questions about anesthesia prompted the discussion of obstetric procedures, is caught up in issues of love and kinship. She wants to know about medical feasibility because she is concerned with maternal-infant bonding after surgery.

The complications of pregnancy and delivery, made even more difficult by the physical challenges of certain forms of dwarfism, may prove too daunting for some short-statured women. Such concerns may be part of the reason there is an active adoption network coordinated by the LPA. The national LPA newsletter, *LPA Today,* says, "The purpose of this service is to find a loving home for every dwarf child. . . . By outreaching to adoption agencies, doctors, hospitals and geneticists and others, we are able to locate available dwarf children for adoptions, and perspective *[sic]* parents who are interested in adopting. The LPA adoption service is not limited to the dwarf community. Average size parents are more than welcome."[15] At the three sessions on adoption at the LPA meetings we attended, flexible eugenics was the norm. Two sessions provided information to people seeking to adopt dwarf children, while the third presented an opportunity for people to discuss their experiences with adoption. All the sessions were attended by both short- and average-statured people interested in, or having experience with, adopting dwarf children, and all offered positive models of self-help.

In each of these sessions, questions arose about the scarcity of American dwarf children available for adoption. The coordinator of the adoption program responded to such questions by telling people they should expect to adopt foreign children. She then explained how she handles the rare American dwarf child who becomes available for adoption. Underlining the predominance of foreign children in the adoption network, the adoption page of the LPA newsletter lists children from India, Bulgaria, and Colombia as available for adoption,[16] and a long article describes a short-statured couple's trip to Russia to adopt a dwarf child there (Dagit 1998: 8).

During the several discussions of the scarcity of American children available for adoption that we witnessed, invariably someone expressed hope that parents in the United States were choosing to keep their dwarf children and not opting to terminate pregnancies after a prenatal diagnosis of a dwarfing condition. This discourse about dwarfism, adoption, and abortion after prenatal diagnosis reveals participants' awareness and imagination of the future in light of recent and expected scientific discoveries and their application in medical practice. Here, heightened consciousness of individual choice and biotechnological futurism converge.

As we have described, many short- and average-statured people we encountered at the LPA celebrate dwarf children but are well aware of the potential for eugenic practices to emerge from the discovery of the genes causing different forms of dwarfism. Although the gene for achondroplasia is known, and prenatal testing is therefore available, it is not routinely tested for today. The condition is simply too rare for widespread prenatal screening to be conducted expressly to detect it. Rather, prenatal diagnosis usually is made as a by-product of routine ultrasound testing sometime during the third trimester, making pregnancy termination illegal and, therefore, unlikely.

As both clinicians and many people affected by genetic abnormalities are aware, however, scientists have developed high throughput biochips with the potential to dramatically change prenatal diagnosis as we know it. Already on the market, these microarrays use silicon chips etched to receive multiple, minute samples of DNA, which may then be rapidly screened using automated computer technology.[17] Instead of testing for a few of the more common genetic conditions, such as Down's syndrome, or a condition specific to the family of a particular couple, as is now usual, the biochip technology will soon provide the means to offer rapid and relatively cheap diagnosis of a wide range of genetic conditions. Achondroplasia is regularly mentioned as one condition for which prenatal DNA chip screenings should and would become generally available. Once again, both the power of biotechnological individualism and the quite understandable fears of a marketplace-driven flexible eugenics are evident in LPA discussions of the chip.

The prospect of a highly efficient diagnostic chip also underscores the significance of speed in contemporary imaginings of the future (Rifkin 1987; Virilio 1986). The cage of late capitalism is a silicon cage and the tempo with which it is associated increases the velocity of industrial machinery (Weber 1958 [1904–05]) to that of the nanosecond tempo of computer technology. The changes suggested by such near-futuristic technologies are deeply unsettling. Yet we stress that, at the present moment, virtually all of us already live "inside" scientific and rapid technological innovation, culturally speaking. Many social groups, from the Catholic Church to some highly creative and respected feminist scholars (e.g., Hubbard 1990; Hubbard and Wald 1993; Rothman 1990, 1998) call for resistance to our inscription into new reproductive technologies like prenatal diagnosis, labeling these technologies perversions of nature and repressive aspects of capitalism. We contend that the issues involved are more complicated. There are enabling as well as constraining aspects to genetic knowledge and its associated technologies (Giddens 1984: 177).

This is a position we have come to appreciate ethnographically, through our work with the LPA. The point is nowhere more clear than in the fraught politics of prenatal diagnosis. Historically, and even now, LP couples may have opted for adoption because of the double dominant effect when two people with genetically caused dwarfing conditions reproduce together. Achondroplasia is dominantly inherited in a simple Mendelian fashion. Thus two people with achondroplasia have a 25 percent chance of producing a child with that condition, a 25 percent chance that the two nonachondroplastic genes will combine to produce an average-statured child, and a 25 percent chance (considered very high by genetic counseling standards) of producing a child with what is known as double dominance, an inevitably fatal condition. Prenatal testing allows LP couples to learn whether their fetus has this double dominant condition and to make a choice about whether or not to terminate the pregnancy rather than deliver a dying baby.

The issue of double dominance was raised during the LPA session on reproductive health. One woman asked whether the physicians knew the consequences of double dominance in cases of partners who are both short statured but have different dwarfing conditions. Another woman explained that her husband had achondroplasia and she had spondyloepiphyseal dysplasia (SED, another type of dwarfism); they had had five children with the double dominant condition, all of whom had died. In response to the question about the effects of hybrid double dominance, one of the physicians offered the observation that almost nothing was known. She then gestured toward the speaker saying, "We have the best evidence [of the consequences] right here." The doctor's evocation of "evidence" tempts us to imagine the complex imbrications between *Laboratory Life,*[18] where animal models are developed for rare conditions that cannot be investigated through human breeding experiments, and the data real people unexpectedly produce for scientists in the course of living their lives as bearers of both rare conditions and children. Here, too, we see flexible eugenics at work.

Physicians at the LPA session stressed the importance of having a definitive diagnosis for one's own dwarfing condition in advance of becoming pregnant. One told a story about a pregnant couple who thought they were both achondroplastic dwarfs, but

> lo and behold they weren't. One was and one wasn't, and we didn't know what the other had and no way of finding out, . . . so the pregnancy was on the

> line. . . . If you were in a situation . . . where you had SED and you were pregnant and your partner had achondroplasia, and your concern was that you would have a double dominant, then you might want to have amniocentesis for a prenatal test. Some people would choose to end the pregnancy and other people would not do that, [but] they would have to know the prenatal diagnosis early in the pregnancy. . . . But if you become pregnant [without knowing your own diagnoses], that all becomes either not possible or extremely difficult.

The idea of choice is powerfully present in this discussion. Here, physicians encourage genetic tests so their dwarf patients can make individual choices about their own reproduction.[19]

The story of double dominance illustrates how a controversial technology involving reproductive choice and eugenic abortion holds different meanings when used inside and outside a particular community. Within the LPA there are widespread fears that the general public will use testing to eliminate dwarf fetuses, not to prevent the birth of dying infants, as dwarfs themselves may choose to do. Indeed, in discussing her aspirations for gene therapy with us, Dr. Clair Francomano was very clear that she believed the only appropriate use of prenatal diagnosis was to avoid the birth of a child with the double dominant condition. The value of choice also underlies the apocryphal stories we have heard repeatedly about dwarf couples using prenatal testing to prevent the birth of children of average stature.[20]

We use the term *flexible eugenics* to underscore the sort of productive and problematic contradictions outlined above. These examples illustrate the complexities of living in a market-driven society that places a premium on individual choice and, at the same time, largely embraces the emergent standards posed by genetic normalization. But as our analysis demonstrates, the idea of a specifically eugenic relation to one's individual genes does not play out in a simple fashion. The people we have met through the LPA are highly attuned to the perils of eugenic thinking; many of them alternately resist and counterappropriate the push to perfectibility as specifically biological or biomedical. Yet like the rest of us, they may desire individual improvement or perfectibility in other ways that are deeply consonant with shared aspects of our cultural milieu.

PESSIMISM OF THE INTELLECT, OPTIMISM OF THE WILL

Genetic counseling and the kind of advice we see circulating at the LPA provide arenas in which both flexible eugenics and resistance to it may become operationalized. At the LPA meetings, one of our team who has conducted long-term fieldwork among genetic counselors met an unusual genetic counselor. As a person with osteogenesis imperfecta (brittle bone syndrome), the genetic counselor told the story of both her struggles and the support she had received in becoming a genetic counselor with great reflexivity. Some doctors did not want an obviously disabled person confined to a wheelchair to counsel pregnant women about conditions that might include her own. Others immediately defended her right as a professional to work with *all* clients, not merely the ones who could handle what they presumed to be the visual impact of her condition. Volatile mixes of paternalism, affirmative action, and eugenic and feminist thinking swirl through the personal life and professional experiences of this young woman. In response, she has resolved to specialize as a genetic counselor in reproductive issues affecting people with disabilities. She is surely well positioned to hear the aspirations, fears, and consequences that molecular genetic technologies invoke as they are played out in the lives of those whose stake in their outcome is most direct. Yet in less obvious ways, we all have a stake in this unfinished story.

Flying home from the LPA meetings in Los Angeles, we chatted with a flight attendant whose family, as it turned out, lives in the suburb where the LPA meetings had just been held. When she heard the reason for our journey, she immediately commented that her town was buzzing: her mother and her mother's friends had all noted the presence of Little People at the many malls and restaurants where tourists and locals might mingle. They found the LPs "cute" or "interesting." She, however, had gotten into a fight over the dwarfs with her best friend from high school. The friend had exclaimed, "I just saw the most disgusting thing: two dwarfs, a couple, with a baby carriage and a baby dwarf. Why would people like that want to reproduce?" The flight attendant said to us, "I told her they probably want to have babies just like you and me; everyone wants to have babies, why not them? I bet their lives aren't so bad. You've got [facial] neuralgia, I bet your life is tougher than theirs is." Our airborne informant continued for some minutes to express her shock and indignation at her friend's bad attitude.

Reframing the problem, if we engage an understanding of the impact of contemporary American genetic thinking and practices empirically, both flexible eugenic thinking and resistance to it are everywhere, permeating outward from the researchers, clinicians, and affected people to the suburban residents, service personnel, and sympathetic anthropologists who encounter them in daily life. We are all rapidly being interpolated into the world of genetic discourse, where resonances, clashes, and negotiations among interested parties occur at increasing velocity. While all historical moments are, by definition, transitional, we live in particularly fraught times insofar as an understanding of a shift in scientific and social thought surrounding genetics is concerned. At the risk of abusing a Gramscian truism, we note that a working knowledge of the political history of eugenics gives us reason for pessimism of the intellect, but an ethnographic perspective on the openness of these encounters and practices may give some cause for optimism of the will.

NOTES

1. The FGFR3 mutation is a genetic rarity in which all cases of achondroplasia are caused by the same mutation. The general rule is that different mutations within a given gene lead to the same disorder. For instance, virtually every family affected with Marfan syndrome (Heath 1998a,b), also a focus of our ongoing research, has a distinctive mutation in the gene for the connective tissue molecule fibrillin.

2. A dominantly inherited, fatal neurological disorder that has played an important role in the development of the Human Genome Project and in recent discoveries in molecular biology.

3. At the annual LPA meetings, a number of T-shirts are available for purchase at the expo. One such T-shirt in 1998 was a takeoff on the Tommy Hilfiger logo with the words "Tommy Dwarfiger." Another looked like a university T-shirt, with the text "Dwarf U." One of the more popular T-shirts in the last few years has been one with the text "Dwarf, Endangered Species" on the front.

4. Representations of dwarfs wishing ill to people of average stature resonate with a discriminatory apparatus that dwarfs face which is deeply rooted in popular culture and folklore and evident in stories like "Rumplestiltskin" and in movies like *Freaks* and, more recently, the Austin Powers movies (for the masses) and the *Red Dwarf* (for cognoscenti). Literature abounds with dwarf protagonists: *Mendel's Dwarf* (Mawer 1999), *The Tin Drum* (Grass 1959), *Stones from the River* (Heigi 1994), and *The Dwarf* (1967), by the Nobel prize-winning author Pär Lagerkvist.

5. In part, Americans operationalize the push to perfectibility by relying on an ideology of exercising individual choice. Discussions of individualism have a long history in American studies, one that can be traced back to de Tocqueville, who identified individualism as a distinctively American characteristic (1835). C. B. Macpherson (1962) examined a more broadly Western notion of individualism in political theory. Linking individualism and capitalist accumulation, Macpherson describes a concept of "possessive individualism."

6. The field research on which this essay draws was supported by NIH/NHGRI/ELSI grant # 1RO1HG01582, for which we are deeply grateful.

7. McGill epidemiologist Abby Lippman labeled the process *geneticization* (Lippman 1991). In our fieldwork we have found that both this terrain and Lippman's concept itself are contested.

8. Veterans associations (Young 1995) and Alcoholics Anonymous (Powell 1987) provide examples of such sociality forged earlier in the twentieth century.

9. One encounter at an LPA session for parents also illustrates parental interest in limb lengthening. At a session billed as a "Teen Panel," at which short-statured teens answered questions from an audience of average-statured parents, one parent asked if any of the teens had considered, or would consider, limb lengthening. All four of the young women on the panel vigorously shook their heads no. One of them spoke quite emphatically, saying, "No, no way. I have too many things I want to do with my life. I don't have time."

10. We are indebted here to the sociologist Troy Duster (1990), who suggests that eugenics is already embedded in contemporary genetic practices through an ideology of choice: with the new genetics, eugenics will come not through state policy but through "the back door," through individual choice.

11. Biotechnological individualism and the reign of free marketplace "choice" seems apparent in, for example, Eugenia Kaw's 1993 description of Asian-American women, who may deeply identify with their cultural roots yet seek to transcend racial identity and exercise choice by choosing cosmetic surgery that anglicizes their eyes. In her work on changing attitudes toward the body, the historian Joan Brumberg (1988, 1997) describes a shift away from moral self-control to control of the unruly body. Especially for women, control over diet, exercise, and, for those who can afford it, plastic surgery enables individuals to choose the bodies they will accept as their own.

12. Biomedical and biotechnical interventions may well have other meanings in different national and local contexts. For example, Taussig's work concerning Dutch genetic medicine shows that normalcy, rather than perfectibility, is strongly marked and desired (1997). Lynn Morgan's 1997 analysis of sonography in Ecuador also points toward the context-specific interpretations attached to biotechnological interventions.

13. In this essay, we use only first names when we use pseudonyms.

14. On the hidden costs of standardization, see Starr 1991.

15. *LPA Today* 35, no. 3 (May-June 1998): 7.

16 *LPA Today* 35, no. 3 (May-June 1998): 7.

17. The molecular biotechnology lab where Deborah Heath carried out fieldwork in 1992 and 1994 was working on a prototype for the biochip at that time. Among rival groups working on the same technology was the biotechnology company Affymetrix, which is now in the forefront of microarray technology (http://www.affymetrix.com/technology/synthesis.html; accessed in June 1999).

18. Our debt to Latour and Woolgar (1979) should be evident here.

19. The ideology underlying contemporary genetic counseling, offered in a mode known as nondirective counseling, is one based on the idea that knowledge enables individuals to make informed choices. Taussig (1997) has argued that this knowledge is not always perceived as enabling choice and in some cases is experienced as constraining choice.

20. We and our informants have no evidence that there is any truth to such stories. In fact, the research position held by Clair Francomano, the physician whose story opens this essay, makes it very likely that she would know if any such cases had occurred.

REFERENCES

Adoption and births. 1998. *LPA Today* 35, no. 3:7.

Althusser, L. 1971. Ideology and ideological state apparatuses. In *Lenin and philosophy and other essays,* 121–73. New York: Monthly Review Press.

Barash, D. 1998. DNA and destiny. *New York Times,* 16 November, A25.

Bellah, R., R. Madsen, W. Sullivan, A. Swidler, and S. Tipton. 1985. *Habits of the heart: Individualism and commitment in American life.* New York: Perennial.

Berman, M. 1982. *All that is solid melts into air: The experience of modernity.* New York: Simon and Schuster.
Bevir, M. 1999. Foucault and critique: Deploying agency against autonomy. *Political Theory* 27, no. 1:65–84.
Brumberg, J. 1988. *Fasting girls: The emergence of anorexia nervosa as a modern disease.* Cambridge: Harvard University Press.
———. 1997. *The body project: An intimate history of American girls.* New York: Random House.
Canguilhem, G. 1989 [1966]. *The normal and the pathological.* New York: Zone.
Dagit, D. 1998. From Russia with love: An adoption adventure. *LPA Today* 35, no. 3:8.
Duster, T. 1990. *Backdoor to eugenics.* New York: Routledge.
Edwards, J., S. Franklin, E. Hirsch, F. Price, and M. Strathern. 1993. *Technologies of procreation: Kinship in the age of assisted reproduction.* Manchester: Manchester University Press.
Foucault, M. 1988. *Care of the self: The history of sexuality.* New York: Random House.
Franklin, S. 1997. *Embodied progress: A cultural account of assisted conception.* New York: Routledge.
Friedman, L. 1990. *The republic of choice: Law, authority, and culture.* Cambridge: Harvard University Press.
Giddens, A. 1984. *The constitution of society.* Berkeley: University of California Press.
Grass, G. 1959. *The tin drum.* Trans. R. Manheim. London: Secker and Warburg.
Heath, D. 1998a. Locating genetic knowledge: Picturing Marfan syndrome and its traveling constituencies. *Science, Technology, and Human Values* 23:1.
———. 1998b. Bodies, antibodies, and modest interventions: Works of art in the age of cyborgian reproduction. In *Cyborgs and citadels: Anthropological interventions in the borderlands of technoscience,* ed. G. Downey and J. Dumit. Santa Fe, N.M.: School of American Research.
Heigi, V. 1994. *Stones from the river.* New York: Poseidon Press.
Hubbard, R. 1990. *The politics of women's biology.* New Brunswick, N.J.: Rutgers University Press.
Hubbard, R., and E. Wald. 1993. *Exploding the gene myth.* Boston: Beacon.
Kaw, E. 1993. Medicalization of racial features: Asian American women and cosmetic surgery. *Medical Anthropology Quarterly* 7, no. 1:74–89.
Kicher, P. 1992. Gene. In *Keywords in evolutionary biology,* ed. E. F. Keller and E. A. Lloyd, 125–28. Cambridge: Harvard University Press.
Lagerkvist, P. 1967. *The dwarf.* London: Chatto.
Latour, B., and S. Woolgar. 1979. *Laboratory life: The social construction of scientific facts.* Beverly Hills: Sage.
Lippman, A. 1991. Prenatal genetic testing and screening: Constructing needs and reinforcing inequities. *American Journal of Law and Medicine* 17, nos. 1–2:15–50.
Macpherson, C. B. 1962. *The political theory of possessive individualism: Hobbes to Locke.* Oxford: Oxford University Press.
Martin, E. 1994. *Flexible bodies: Tracking immunity in American culture from the days of polio to the age of AIDS.* Boston: Beacon.
Mawer, S. 1999. *Mendel's dwarf.* New York: Penguin.
Morgan, L. 1997. Imagining the unborn in the Ecuadoran Andes. *Feminist Studies* 23, no. 2:323–51.

Olby, R. 1990. The emergence of genetics. In *Companion to the history of modern science,* ed. R. C. Olby, G. N. Canton, J. R. R. Christie, and M. J. S. Hodge, 521–36. London: Routledge.

Paul, D. 1995. *Controlling human heredity, 1865 to the present.* Atlantic Highlands, N.J.: Humanities Press.

Portin, P. 1993. The concept of the gene: Short history and present status. *Review of Biology* 68:172–222.

Powell, T. 1987. *Self-help organizations and professional practice.* Silver Springs, Md.: National Association of Social Workers Press.

Rabinow, P. 1996. *Essays on the anthropology of reason.* Princeton: Princeton University Press.

Ragoné, H. 1994. *Surrogate motherhood: Conception in the heart.* Boulder, Colo.: Westview.

Rifkin, J. 1987. *Time wars.* New York: Henry Holt.

Rothman, B. K. 1990. *Recreating motherhood: Ideology and technology in patriarchal society.* New York: Norton.

———. 1998. *Genetic maps and human imaginations: The limits of science in understanding who we are.* New York: Norton.

Shiang, R., L. M. Thompson, Y. Z. Zhu, D. M. Church, T. J. Fielder, M. Bocian, S. T. Wonokur, and J. J. Wasmuth. 1994. Mutations in the transmembrane domain of FGFR3 cause the most common genetic form of dwarfism, achondroplasia. *Cell* 78, no. 2:335–42.

Starr, S. L. 1991. Power, technologies, and the phenomenology of standards: On being allergic to onions. In *A sociology of monsters: Power, technology, and the modern world,* ed. J. Law. *Sociological Review Monograph* no. 38. London: Routledge.

Strathern, M. 1992. *Reproducing the future.* New York: Routledge.

Taussig, K. S. 1997. Normal and ordinary: Human genetics and the production of Dutch identities. Ph.D. diss., Johns Hopkins University.

Taussig, K. S., R. Rapp, and D. Heath. 1999. Translating genetics: Crafting medical literacies in the age of the new genetics. Paper presented at the annual meetings of the American Anthropological Association, 21 November, Chicago, Illinois.

Tocqueville, A. de. 1945 [1835]. *Democracy in America.* New York: Vintage.

Virilio, P. 1986. *Speed and politics.* Trans. M. Polizzotti. New York: Columbia University Press.

Weber, M. 1958 [1904–05]. *The Protestant ethic and the spirit of capitalism.* New York: Charles Scribner's Sons.

Young, A. 1995. *The harmony of illusions: Inventing post-traumatic stress disorder.* Princeton, N.J.: Princeton University Press.

Chapter 4

The Commodification of Virtual Reality

The Icelandic Health Sector Database

Hilary Rose

When newspapers around the world reported, "Iceland sells its people's genome," it read to many, not least Icelanders themselves, as if Brave New World had finally arrived. It is now clear that the remarkable events on this small Nordic island must be understood as part of a much wider shift. As the big pharmaceutical companies, venture capital, and the state gravitate toward predictive medicine and pharmacogenomics, Iceland may be the first example of pharmacogenomics in action, but unquestionably it is not going to be the last.

There is a distinct irony to recent developments in pharmacogenomics: This potentially immense innovation, actively pursued by global pharmaceutical companies and venture capital, requires as its precondition a universal health care system.[1] Only the old welfare states have universal health care records. Not for the first time does the relationship between the organizational structures of health care provision and the development of genetics come into visibility and importance.[2] For pharmacogenomics, only the old welfare states offer what they speak of in their depoliticized language as a "good" population.[3]

Although the conflict over the Icelandic database broke in 1998, its origins go back to the summer of 1994. Then two Harvard-based clinical neurologists, the Icelander Kari Stefansson and his U.S. colleague Jeff Gulcher, were visiting Iceland to collaborate in a study of multiple sclerosis (MS) with an Icelandic neurologist, John Benedikz. The research project was to look for a possible genetic predisposition to the disease. In "helicopter science" mode, the researchers flew in during the summer, secured as many samples as possible from patients and their families, and then returned to the Medical School at Harvard to do the lab work.[4]

Stefansson's ambitions and vision, however, were much wider than searching for the genetics of one disease entity. Although not a geneticist by training, he was the first biomedical researcher to see the potential significance of Iceland and its genome to investigate the genetics of common or complex diseases. Furthermore, he saw how to exploit the joint interest of both the state and venture capital in the new genetics. Having spent two decades in the entrepreneurial culture first of Chicago, then of Harvard medicine (there is one marvelous quote in a *New Yorker* article where he explains how he did not work at Milton Friedman's university for nothing), Stefansson was uniquely well placed to understand the research and commercial possibilities offered by this small, rich, and relatively isolated North European population (see Specter 1999).

It was out of this vision that deCode Genetics was born, as a biotechnology company physically located in Iceland. From its inception, Stefansson had two very different, though interconnected, objectives. The first was to establish a commercial laboratory to carry out biomedical research in Iceland. Like any other commercial biotechnology company working on human genetics, this laboratory would seek to collaborate with clinicians interested in specific diseases and work to develop new DNA diagnostic tests and drugs. What made deCode different from the routine biotech company was its second and more ambitious objective of the Health Sector Database (HSD), which would link its clinical and research data to both the Icelandic health care system records and the genealogies. An Icelandic cultural passion since the time of the sagas, these genealogies constitute a narrative of both personal and national identity. Iceland was interpreted as the perfect location for this massive, linked network because of the nation's small size, high quality universal health care, medical records dating back to 1915, purported genetic homogeneity, and large and well-documented tissue bank serving as a potential repository for much of the nation's genetic record. These data, combined with the existence of the distinctive genealogies, were seen as offering uniquely favorable conditions for turning the hot ideas of preventive genetic medicine into commercially viable products. The deCode database, called the Health Sector Database, is a project that seeks to construct and commodify bioinformatics as much as it is one that commodifies the human body through the tissue bank and that brings into existence new biotechnological products.

This effort also raises the possibility of designing radically new ways of managing and delivering health care. The possibility of tailoring drugs to patients with particular genotypical profiles offers economic efficiency coupled with less discomfort and danger to patients.[5] Given the pressure on health care resources, this looks like a win-win promise for patients and the entire drug production and prescribing system. There are countervailing arguments from within both biology and the social sciences that this approach to health care overvalues the significance of the genetic profile in the production of health and disease. However, such criticism does not diminish the attractiveness of the genetic fix within an increasingly geneticized culture. DeCode shrewdly positioned its proposal to be attractive to several powerful players: venture capital, the welfare state, the health management organizations, and the insurance industry. All are potential buyers of this new commodity—bioinformatics—although their purposes will be very different.

The continuing controversy over the Icelandic program operates at two very different levels, the ethical and the technological. The first includes all doubts raised as to whether this immense biotechnological innovation is socially responsible. The technological argument asks whether the program, on its own terms, will work. Within Europe, this is the first time that the private sector has taken a central role in a major biotechnological innovation that affects an entire system of medical care and research. The judgment on its workability will now be made primarily by the market itself.[6] I do not address the technological questions here, or the impersonal judgments of the market, but rather focus on elucidating the ethical issues, as Icelanders themselves express them.

In my discussions with Icelandic women, I found that many interpreted the questions raised by the Health Sector Database in intimate terms, focusing on what it would mean for their families. They expressed concerns about privacy, viewing the idea of regional data inputting centers as possible local nodes for the dangerous leaking of confidential information. Health status and age were important variables in such responses: generally, those who were vulnerable in some way were more likely to be skeptical about participating in the project. As my chapter suggests, potential research subjects negotiate genomics within frames that may be very different from those present in the political debates. But, like the scientists, they are important if invisible stakeholders in the currently unfolding saga.

PATIENT GROUPS AS A SITE OF CONFLICT

Stefansson has continued as an active clinical researcher in the field of MS. He has been active within the Icelandic MS society, keeping in touch with the 250 patients and families who have continued to provide many samples over the years. Thus the initial clinical interest in locating a genetic predisposition to MS has been both a springboard and a continuing presence within the deCode story. Currently, the MS society shares both Stefansson's view that the emphasis should be on gene hunting and his belief that this will yield a successful therapeutic intervention for an intractable disease. An alternative objective—that the society should put its main effort into fighting for better health and welfare services for MS patients now—has been discounted.[7]

It would be a mistake to underestimate the cultural legitimacy that such patient groups and their families can offer research clinicians. Their attitudes and activities can, for example, increase or decrease the cultural geneticization of the disease. Funding patient groups by pharmaceutical companies is not an entirely philanthropic activity but is also a pragmatic means of extending and shaping markets. At the same time, patient groups frequently need external resources, so they are relieved when companies take an interest and indeed may go out of their way to invite companies in. Their problem is how to manage the relationship so that their interests rather than the company's interests take priority.[8] Patient groups are only beginning to be problematized and studied (see chapter 3, this volume); all too often the genetic literature treats such groups as if they were not an important site of conflict.

Stefansson's first move was to raise money to establish a research company. He was highly successful, raising $12 million in single-dollar shares over a few months. The new firm deCode was registered in Delaware in August 1996. These initial venture capitalists were predominantly from the United States: Alta Partners, Atlas Venture Partners, Polaris Venture Capital, Arch Venture Partners, Falcon Technologies, and Medical Science Partners, together with Advent International, which has a U.K. partner called Vanguard Medica.[9] The larger investors joined the board, but, other than Sir John Vane from Vanguard Medica and Stefansson, who serves as deCode's chief executive officer, there are no researchers on the board.[10] Stefansson also recruited Vigdis Finnbogadottir, the former president of Iceland. The ambiguous message about deCode's national identity was thus constructed with some ingenuity. Physically located in Iceland, it was nonetheless a U.S. registered company with a board dominated by U.S. and British venture capitalists.

The successful branding of deCode as Icelandic and as Stefansson's personal project is key to its popular acceptability. Social theorists have long regarded the social democratic welfare states as the historic settlement between the two great classes and the state. Forgotten in that account was that all three parties to the settlement took for granted that they had a shared nationalism from which the social democratic project drew its strength. For reasons of geography and history, a progressive civic nationalism is still vibrant within Icelandic culture, and Stefansson has managed brilliantly to locate deCode and the Health Sector Database inside a narrative of both scientific and national progress. The general public sees his charismatic nationalism and his enthusiasm for scientific innovation as exactly what Iceland needs.[11]

THE "GOOD POPULATION" FOR PHARMACOGENOMICS

Iceland is unique as a European nation-state in that its small size (275,000 inhabitants mostly living in and around the capital, Reykjavik) and watery isolation have insulated it from territorial conflicts that have ravaged the rest of Europe. Today, geothermal energy secured by harnessing the volcanoes on the island has helped generate a high per capita income, low unemployment, a strong welfare state, good health care, and high quality education. Iceland's life expectancy and low infant mortality have been bettered only by Japan at the height of that country's economic boom. DeCode portrayed the Icelandic population as the exemplary "good population" for pharmacogenomics.

The deCode documents speak of the Icelandic population as not only highly educated but also cooperative—by implication, with scientific and technological research. Judging by the evident enthusiasm for new consumer technology, from four-by-fours to cell phones and the Internet, Icelanders are now not only a wealthy population but distinctly technophile. Unquestionably, the majority of the population supports the database project, though the opinion polls hint at rather less enthusiasm among the young. This cultural enthusiasm for science and technology and its fruits is not shared by most other Europeans. Hazards, from Chernobyl to mad cow disease, have given them a sharp sense of the risks as well as the benefits of science and technology, a perception dramatically expressed in their hostility to genetically manipulated food, considered a possible risk to human health and the environment. Indeed, social theorists from Ulrich Beck to Anthony Giddens in recent years have developed the concept of "the risk society" to grasp this changed social understanding of risk.[12]

But for those, like Icelanders, living in earthquake and volcanic zones and avalanche country, it is different: the biggest threats still come from an ever-present potent nature, not culture.[13] Here, science and technology are seen as harnessing the powers of nature, and technical and social progress are still happily married. Endlessly, Stefansson has drawn on the notion that "science is progress; it cannot and must not be stopped." Iceland's openness to technical innovation has provided the project with a peculiarly friendly niche.

Icelanders were also seen as a "good" population because of their claimed homogeneity. In the mass media, the stereotypical representation of an island people descended solely from blue-eyed, blond Vikings has frequently spun out of control. In remarkably racist language, U.S. journalists describe the Icelanders as a "nation of clones," proposing "everyone in Iceland is related to everyone else" and "all of them are descended from the same few Vikings." The "blond and blue-eyed" DNA of the Icelanders is apparently part of its corporate value (see Mawer 1999; and Marshall 1998).

DECODE STARTS WORK

Stefansson returned to Iceland in 1997, and by November of that same year he had established deCode as a commercial research lab in the industrial district of Reykjavik. DeCode had soon spent more on research than the Icelandic government's annual research budget—some $65 million. In Stefansson's strategy, speed was of the essence: speed first in getting the political and venture capital support in place, then speed in getting the biotechnology company established, and lastly, speed in getting the Health Sector Database up and running under exclusive control.

In a small society it was relatively easy for Stefansson, as a member of the well-connected cultural elite, to cultivate the government politicians concerned with the economy, not least the prime minister David Oddsson. As center-right politicians, they were attracted by Stefansson's highly market-oriented vision that seemed to offer so many economic, as well as health, benefits to Iceland. Oddsson in particular publicly declared himself willing to sweep away ethical constraints that might impede the advance of the new technology. By the autumn of 1997, Stefansson was ready to approach the Ministry of Health with the proposal for the HSD.

The first time that anyone other than deCode staff, key health ministry personnel, and senior members of the government heard about the proposed HSD bill was at a meeting on March 23, 1998, six months after deCode had faxed a draft of the bill to the government. On March 31, 1998, only eight days after the oral presentation to the expert group, Ingibjorg Palmadottir, Minister of Health, introduced the HSD bill to the Althing, the Iceland legislative body. At this point the Icelandic public learned about it for the first time. As framework legislation, this first draft (and indeed the subsequent bill and final law) left much of the detail to be filled in after negotiations led by the Ministry of Health.

Initially, the bill offered little protection to patients: it would be relatively easy to identify individuals from the data collected. The database was to be funded privately.[14] In return, the company that won the license was to have monopoly access and was to receive the right to market the data for up to twelve years for use in genetic discovery, diagnostics, drug development, and health-care management. Both deCode and Stefansson were referred to directly in the gloss to the bill, even though the tender for the license was to be put out to competition.

Given that there was no prior consultation concerning the core ideas of the HSD, the negative reception from much of the clinical and research community was predictable. Thus, while the general public was by and large untroubled by the HSD and saw it in the same favorable light as did the government, many of the relevant professional groupings—clinicians, nurses, scientists, and patients' rights lawyers—found the bill ethically unacceptable in principle, and poorly drafted. Icelandic lawyers specializing in patient advocacy saw the HSD legislation as reversing the gains achieved by legislation passed in 1997 setting out patients' rights. And for many clinicians, their sense that they held professional responsibility for patients' records meant that they necessarily had to oppose entering them wholesale into the database.

The HSD also mobilized international criticism, particularly from the genetic research community and from medical associations from the national to the Icelandic World Medical Association. Some Icelandic molecular biologists, such as Bogi Anderson, working in the United States were incensed by the project and intervened in the debates frequently. Another United States–based molecular biologist, angered by the prospect of the deCode monopoly, established a rival biotech company in Iceland, in collaboration with an Icelandic entrepreneur. The U.S. molecular biologist Leroy Hood, himself no stranger to commercial funding, was quoted as expressing similar doubts about a monopoly licensee. In the run up to the December legislation, the breast-cancer geneticist Mary-Claire King and the lawyer and ethicist Henry Greely wrote a joint letter to the prime minister of Iceland urging him to reconsider. Meanwhile, the population geneticist Richard Lewontin urged that, subject to the views of Icelandic geneticists, there should be a boycott of Icelandic genetics. While few individual scientists from the United Kingdom expressed their concerns so publicly, an editorial in *Nature Biotechnology* proposed that British Biotech and deCode were two clear examples of how not to do biotechnology.[15]

THE DECEMBER 1998 HEALTH SECTOR DATABASE LAW

Despite these international concerns, the HSD became law on December 17, 1998. Like its precursors, it was framework legislation and, in consequence, much remained to be worked out between the ministry and the successful firm. The legislation offered citizens the right to opt out by mid-June 1999, well before the database was up and running, and protected their right to opt out later; however, the law does not specify whether those opting out later can withdraw their data already incorporated in the database.

When the Icelandic government was asked by the European Bioethics Committee whether data could be withdrawn, the official reply was that this was "subject to negotiation."[16] However, this claim has to be set against the record of the earlier debate in the Althing, during which the minister and government supporters had explicitly stated that data once entered could not be withdrawn.[17] Further, data on dead people was to be entered automatically; their families, who might have opted out themselves, were to have no say. Nor had the situation of vulnerable groups such as children, those with learning disabilities, severe mental illness, or elderly people with dementia been specifically considered. It was, for example, merely assumed that families would consult their children before deciding. It was also assumed that parents would agree about what was best for their children.

In this situation, a new association of Icelanders concerned about the ethical questions raised by science and technology, Mannvernd, was formed to oppose the HSD proposal. The Mannvernd critics attacked the opt-out concession as inadequate, pointing out that it would be easy for well-informed citizens—such as Mannvernd members themselves—to opt out, but more complicated and difficult for those who were less privileged or less determined. There was no provision, the critics argued, for an informed discussion of the benefits and risks of staying in or opting out, benefits and risks that would be different for different individuals. The Director General of Public Health was given general responsibility for informing the citizenry. However, as critics pointed out, the Director General also draws on the data for health care management purposes and, thus, is not entirely disinterested. That he also sat on the scientific advisory board of deCode further blurred the boundaries between public and commercial interests.

Once the legislation had been passed, the Director General of Public Health informed the population by sending a pamphlet to every household explaining the legislation and the citizen's right to opt out. The legislation recognized that it would take a little time to receive the tenders and award the contract, therefore giving citizens six months to opt out. The deadline was mid-June, and anyone who opted out before that time would have no data entered into the HSD. After that, those opting out could block fresh data from being entered but could not withdraw data already entered.

This leaflet was a minimal substitute for the kind of public education the critics of the legislation demanded. Of my informants not directly involved in the conflict, most said they had never seen the modest-sized green pamphlet and expressed surprise on being shown it. Some of those who had not seen it speculated whether some other member of the household had thrown the pamphlet away, along with junk mail. Most had picked up what they knew of the issues from the debates that filled the widely read newspaper *Morgunblad* or from television or radio coverage.

Meanwhile, the National Bioethics Committee, set up as part of the patients' rights legislation, and which had systematically opposed the HSD, was suddenly abolished in August 1999. It was replaced by a new committee composed of civil servant professionals together with a nurse appointed by the Director General, who, it was suggested, could be seen as a patients' representative. The Director General argued that the change was necessitated by the inability of the previous committee to make up its mind. However, it did not escape public comment that the new committee would be more malleable than the old one, which had been composed of individuals selected by the minister from a list nominated by the professions and by university departments. In light of these changes, many clinicians and human rights' lawyers became concerned that the previous year's legislation on patients' rights was now at risk.

The position of two categories of Icelanders in the legislation, children and the dead, aroused particular concern, because the civil right of children to opt out, and the right of anyone to choose not to be included after death, was effectively erased. The dead are automatically included by the legislation, regardless of their own views or the views of their surviving children or other family members who, in the case of genetic risk, may be directly affected. As the legislation and the politicians' statements stand, the only Icelanders who have the unqualified right to opt out (that is, who can prevent data about themselves from ever entering the database during their lifetime) are those who reached age eighteen by mid-June 1999. This unqualified right could be exercised only during a six-month period following the December legislation. Although a citizen may cease to have data entered, no one is allowed to withdraw data already entered. Thus, children under eighteen in June 1999, and even children as yet to be born, lost their unqualified right to opt out. Those children left in by their parents but who, on maturity, want to opt out, cannot withdraw their data. Even to a nonlawyer, this seems to be an arbitrary destruction of children's rights. I raised this point in my talk in a feminist studies department, where it produced considerable discussion and led to a newspaper article inviting the attention of the Children's Ombudsman.[18]

COMMODIFICATION AND GENDER

One of the most distinctive features of the Althing debate on patients' rights was that it temporarily raised the question of who owned the patient's records: the patient or the government.[19] Eventually the debate about ownership was sidestepped, and the classical noncommodified concept was restored: records were the confidential documentation of the clinical transaction and were in the custodial care of clinicians. The first step in a process of commodification is to admit the concept of ownership into the thinking about some entity or process, whether it is access to information or fishing, which had hitherto been seen in Iceland as outside the commodity relation. This anticipatory ownership debate had culturally if not legislatively let the commodity genie out of the bottle.

The commodification of nature, whether green or human nature, is scarcely new. Yet the capitalist modernity and continuous technological innovation have intensified it. Within economics, commodities are taken for granted as objects subject to supply and demand, but the social birth of a commodity is typically surrounded by intense moral debate, and debates over the commodification of nature are entrenched in Icelandic culture. The narrative of the constitutional breach of fishing quotas (so important in Icelandic history) has been mobilized by the opposition to draw parallels with the HSD and its commodification of human bodies and information.

Protagonists of the HSD also draw on the commodification narrative; they compare the exploitation of Iceland's genes as the country's equivalent of Norway's successful exploitation of its oil.

Those who drew on the commodification of nature, invoking either fish or oil, tended to be men. Commodification lay in the public arena to be fought over publicly. Women were less concerned by the commodification issue than by what the deCode project meant for preventing and treating currently intractable diseases and, most important, what it might mean privately for their families.

In discussion groups where they had space and time to explore these issues without being drowned in technical talk, women raised again and again their concerns for the psychic as well as the physical well-being of their family members. One with a history of breast cancer described first how she had decided to opt herself and her son out and then her struggle to persuade her father to join them. He was in favor of the HSD and erroneously saw breast cancer as a woman's disease. She explained to him that men could have breast cancer too, and reminded him that he had had a bypass. Supposing, she asked, this reflected a genetic predisposition to heart disease: did he really want to give his grandson the burden of the genetic knowledge of that risk as well as the cancer risk from her? Under almost any interpretation of the planned HSD process, this feedback to an individual patient was an almost impossible outcome. But that was not the point, because for her it was a matter of being cautious, of protecting her son, in a context in which her trust in institutions was being eroded. The commercialism of the HSD erased her sense of being cared for by clinicians committed to her and her family's well-being.

She was not alone: women complained that their concerns, those concerns of love and responsibility for their children, were not reflected in the media debates; instead, they said, fish quotas, deCode, and sports had occupied most of *Morgunblad*'s pages all year. How could they begin to decide, when there was no discussion of the issues that troubled them?

There was also a difference in terms of people's experience of disease, a difference shaped too by age and gender. Young people with a history of good health found it very hard to connect to the debate; most retreated to benign good intent for less fortunate others. They felt they had nothing to hide, no cause of health concern in the future, so for them there was no problem in participating. Others were less sure: they saw genetic information as a whole as potentially damaging to employment possibilities, not least if the data were sold to employers or insurers.

Where there was a family history with little experience of serious illness, both women and men were by and large enthusiastic. For them the HSD was part of the story of biomedical progress. They saw it as a source of hope, and they were also philanthropic. Their support was an expression of their civic nationalism.

However, where people had experience of chronic disease requiring long-term medical care, or had genetic disorders in their families, they were much more cautious. They saw the cumulative power of the project in a rather different way. Partly they knew much more about diseases and databases because their records were already in collections such as the heart database or the cancer database, but they saw these databases as associated with their clinical care and they felt directly and safely supported by the research and the researchers.[20] Like the woman with cancer above, they distrusted the commercialism of deCode and suspected its potential role as a means of cutting back on existing levels of health care.

Although ostensibly the database is encrypted and information can flow only upward, several of the women were unconvinced that the information would remain as purely statistical data and felt that it would, because of Iceland's small size, be identifiable. It was not the Cambridge computer security expert advising the Icelandic Medical Association who had convinced them of this, but their own knowledge of a different technology: the telephone. They recalled how Iceland's telephone system had worked in the past, with all calls routed through the local switchboard operator, who would listen in avidly. And they noted Stefansson's promise during the general election campaign (where he appeared alongside government candidates) that inputting would be done at the regional level. What Stefansson saw as a smart move to promise employment to the regions, the women saw as local node points for dangerous leaking of confidential material. Others felt confident in the guarantees given by the legislation that leaks would be severely punished.

Nonetheless, women endlessly returned to the question of whether the existence of HSD information could harm their children. One woman with two severe diseases, and with a history of family predisposition to one of them, was tremendously concerned for the welfare of her children. Using language very similar to that of mothers whose babies had been diagnosed as carrying the familial hypercholesterolaemia gene defect studied by Theresa Marteau and her colleagues, she saw and resisted the determinism on behalf of her children.[21] She and other women like her saw genetic knowledge as threatening and likely to make their children fatalistic.

The women who had experienced domestic violence and sexual abuse raised very different issues relating to the state surveillance of health. One recalled that in the past getting married had meant showing your medical record to someone in authority.[22] It was a minute before I recognized that this was an indirect way of raising the specter of the eugenic past. State eugenics, controlling women's reproductive capacities on the grounds of their fitness to mother, had not entirely disappeared from the cultural memory of women.[23] Few raised the concerns, strongly articulated by feminists and the disability movement elsewhere, about a new kind of consumer eugenics, but they had a personal, close sense of the social control of the body through which they interpreted the HSD project.

Many women who had experience of being categorized as suffering from mental disease, or had experienced sexual abuse and violence, said they wanted to keep themselves and their children out of the database. They wanted as few people as possible to know their painful secrets: confidentiality was immensely precious to them. One woman, however, took a totally different position. She could not come to the meeting I attended but sent her views through a friend. The friend explained that she wanted her own and her children's records to be included, so that the abuses they suffered would be recorded. She wanted to end the social process whereby the victims of sexual violence hide themselves or are hidden away in the name of privacy. If men fathered babies by sexually abusing their own daughters, then it would be an act of justice for the DNA record to show it. There was sympathy and admiration but no takers for her position. Some said that they had already decided to opt out even before they were invited to come to the discussion. But all said that this was the first time they had had the chance to discuss frankly and confidentially what the database meant for them. Such discussion is the ethical heart of informed consent.

CONCLUSION

Believing in the optimistic future foretold by deCode and their government, many Icelanders invested heavily in shares of deCode at U.S.$58. By October 2002, share prices had dropped to under U.S.$2. Because the Icelandic economy is massively fish-dominated (89 percent of the GNP is fish related), few other countries would share its vulnerability to the global collapse of the biotechnology sector through the fortunes of just one company. Nonetheless, the economic lesson is disturbing. Commodification within capitalist modernity is accelerating, powered both by the technosciences set to dominate the twenty-first century and by the relentless energy of venture capital. These technosciences are biotechnology and informatics. Thus, while the Icelandic controversy has been conceptualized as an extension of the commodification of nature through biotechnology by both the Icelandic anthropologist Gisli Palsson and the biologist Richard Lewontin, my own reading is that this is only partially the case (see Palsson and Rabinow 1999; Lewontin 1999). With the Health Sector Database, the most intense focus of the commodification process is on information, albeit information about the human body. Biotechnology (using informatics) is bringing into existence an entirely new class of information—genetic information—but it is informatics itself that enables old forms of information, the medical records and the genealogies, to be brought into relation with the new, creating a historically new and marketable commodity.

The Icelandic project has been steamrollered through. Despite any potential commercial advantage this gives or appears to give, the strategy carries too many problems for both society and science. The process has been so accelerated that it has been impossible to explore fully and calmly the fundamental question of what such a centralized database might or might not contribute to understanding the issues of health and disease. This is a serious deficit for the healthy development of genetics, public health, and democracy.

Those of us who observe rather than live with the HSD conflict must be aware that legislation was necessary only because a purely market approach to genomics was adopted. In a more consensual, hybridized model of genomics, the innovation might simply have been added to the research and health policy agenda by experts, with little or no public consultation. There would be no moment when the idea of a genomics database could be accepted or rejected by democratic process. Such expert-driven technological innovation is a conspicuous feature of the old welfare states, particularly Britain, with its highly secretive political culture, which it is so painfully trying to move beyond.

This old tradition is in need of serious challenge and overhaul. The debate over the Icelandic HSD has exposed contemporary genomics to vigorous public debate. Ethical issues relating to the commodification of information and nature are made starkly manifest in these exchanges. The Icelandic HSD has put these issues on the international cultural and political agenda. Iceland's highly visible conflict over commodification has the merit of helping other countries increase both the transparency and the democratic accountability of their biomedical innovation

NOTES

I am immensely grateful to all those Icelanders who gave me their time, shared their thoughts, and provided a range of published and unpublished material. The web sites of Mannvernd, deCode, and the Icelandic government have been an invaluable and continuing resource. Those who know the difficulty of maintaining privacy in a small society will understand my profound gratitude to those who trusted me with their personal narratives of pain, sickness, and disability. Without the unstinting help of three Icelandic colleagues in science studies and feminist studies, I could never have achieved so much during my short visits in 1999. I had met Dr. Eirikur Baldursson in Gothenberg, where I had been on his Ph.D. examining committee while a guest professor in science theory and feminist studies. I met Dr. Skuli Siggurdson (history of science and technology) in Berlin, where I was giving a seminar at the Max Planck Institute, and Dr. Gudrun Jonsdottir (feminist studies and social work) at Bradford University, where she was giving a seminar.

This project could be completed in the time allowed for it only because of the technology of e-mail, which made this European networking fast and inexpensive. My thanks to those colleagues from both the United Kingdom and Iceland, together with the U.S. editors of this volume who have read and painstakingly commented on the text. As usual, any errors that remain are my own.

Last, I am grateful to the Wellcome Trust for supporting this research and for finding a fast track to enable this study to happen while events were still unfolding. A full report of this study is available in a PDF file on the web site of the Wellcome Trust: http://www.wellcome.ac.uk/en/l/awtupd01q2.html, *Professor Hilary Rose, The Icelandic Health Sector Database,* April 20, 2001 (accessed on 16 December 2002).

1. Some firms such as U.S. Myriad Genetics are making similar studies within marketized medicine, but they are examining the very distinctive Mormon community, about which a good deal of genetic data has already been gathered and in which genealogy is an intense cultural commitment. (Kari Stefansson is on the Myriad board.) Other similar private sector studies include Gemini Genetics' study of twins (Cambridge, U.K.), Signal Gene's examination of genes of the descendants of French settlers in Quebec, and Newfound Genomics' analysis of the DNA of Newfoundlanders. See *SmartMoney.com,* 12 July 2000.

2. This was seen in microcosm earlier in London, when the reorganization of the NHS into trusts did immense damage to clinical genetics. Leading London-based geneticists such as Professors Bobrow, Davies, and Williamson left for Oxford, Cambridge, and Australia.

3. However, what constitutes "good" in a population is fluid. Thus, whereas deCode emphasizes the advantages of the small size (275,000) and homogeneity of Iceland's population, the proposal from Smith Kline Beecham to draw on U.K. records and populations construes the 59 million people in the socially and genetically diverse United Kingdom as even better good, and the Medical Research Council and the Wellcome Trust have decided that 500,000 is the ideal population. The "good" population for pharmacogenomics seems remarkably flexible, requiring only universal medical care and well-cared-for tissue banks.

The deCode *Non-Confidential Corporate Summary* puts this clearly: "The problem having the greatest impact, however, is finding and securing a good population" (1998, para. 3.2).

4. Given the interest in the Icelandic database, helicopter geneticists are being replaced by helicopter social scientists and an array of the world's journalists. As the key players are few, the situation in Iceland resembles the old anthropological joke about how every Hopi family has their own resident anthropologist. Fortunately, the less prominent actors have not been interviewed as intensely.

5. That pharmacological drugs are the fourth leading cause of death in U.S. hospitals realizes Ivan Illych's prophetic diagnosis of iatrogenic disease.

6. In October 2002, deCode fired one-third of its labor force and its shares dropped to below U.S.$2. Currently the market's view is negative.

7. In May 1999 the welfarists challenged the geneticizers at the committee elections. The latter won out, as Stefansson is very much a guru within the Society.

8. My own work on people with familial hypercholesterolaemia studied a voluntary group that had been transformed by substantial pharmaceutical company injec-

tions—from a shoestring operation literally meeting around the vicar's wife's dining table to a professionally led organization with its own offices, cars, etc.

9. These companies are interlinked and include some of the industry's most powerful venture capital firms. For example, Advent International, Vanguard Medica's partner, is partly owned by Hoffman La Roche, and Atlas Venture Partners partly owns deCode Genetics, deCode, Exelixis Pharmaceuticals, and Exelixis. The drug companies themselves continue to pursue mergers; thus in January 2000 the merger of Hoffman La Roche and Hoechst—now Aventis—was reported in the *Economist.* Though my account speaks of deCode as one entity, it should be more precisely understood as a group of closely linked companies; deCode Genetics is how it is listed on NASDAQ.

10. Sir John Vane was awarded a Nobel Prize for his work carried out at the Wellcome Laboratories.

11. Nordic press commentary has been quicker to spot and denounce "genetic nationalism."

12. See Beck 1992 and Giddens 1991. The weakness of the theory of the risk society is that it turns on the assumption that the old forms of risk, primarily poverty, have been overcome by the welfare state. While this may well have been true for Germany in the 1980s, when Beck wrote his book, German unification and the accelerated rolling back of the welfare state under Margaret Thatcher in the United Kingdom has meant that the risk of poverty is still starkly present. Thus the new risks from science and technology are added to the old risk of poverty. By contrast, Iceland still has a strong welfare state and a thriving economy, hence its citizens have little currently to fear from poverty. At the start of the new millennium, Iceland offers the perfect antithesis to Beck's 1980s risk society.

13. I make this distinction even while I accept that the boundary between "nature" and "culture" is under continual negotiation and, in consequence, is not fixed and static.

14. The licensee was to pay for the license itself plus all costs of setting up and running the database, including the cost of informing the public. And, when the HSD was functioning, the licensee was to make additional payments, to be agreed upon, to the government, these to be ring fenced for health services and health research and development. Icelandic Act on the HSD, 1998.

15. *Nature Biotechnology* 16 (1998): 1017–21.

16. Reply from Iceland, Rvi 49, after the hearing to Qvi 48, *Steering committee on bioethics report of the hearing of Icelandic experts concerning the law on a health sector database,* Strasbourg: Council of Europe, 1 March 1999.

17. At least one legal case has been undertaken on behalf of an individual patient to withdraw medical and genetic data held by deCode.

18. This issue of opting out is under intense and unresolved debate. The Icelandic Director General of Public Health claims that not being able to withdraw unidentifiable data is normal in epidemiological research (*Eurogapp,* 6 April, p. 29). Tom Meade, Professor of Epidemiology and chair of the Wellcome–Medical Research Council Committee developing the British study, takes a contrary view. Asked about the committee's view on the right to withdraw by a sociologist who had attended a seminar in which this Iceland HSD study was discussed, Meade hesitated but said he

thought it was ethically necessary. HUGO Public Lectures on the Impact of the Human Genome, Oxford, May 2000.

19. There was a similar conflict in the United Kingdom over the ownership of anonymized data. Eventually the Court of Appeal sanctioned the use of anonymized patient data (Robbins 2000).

20. There are two strong nonprofit databases in Iceland for heart disease and for cancer. The latter enabled two competing teams to carry out the DNA family studies that enabled both the Sanger Center and U.S. Myriad Genetics to identify BRCA 2, which was the second identified breast cancer genetic sequence. One shared its results with the nonprofit consortium led by the Sanger Center; the other shared sufficient results with Myriad to enable Myriad to make a precise identification. This positional hint enabled Myriad to patent BRCA 2 (on November 23, 1995); it already had patented BRCA 1. The Sanger Center was at that point against patenting genes as discovery not invention. In the light of Myriad's aggressive approach to patenting genes, the Sanger Center has changed its policy (Professor Michael Stratton, HUGO Public Lectures on the Impact of the Human Genome, Oxford, 1 June 2000). Eminent U.S. women who have had breast cancer have become the center of a vocal international lobby against patenting genes, but events have overtaken this hostility.

21. V. Senior, T. Marteau, and J. Weinman, 1998, Will genetic testing for predisposition for disease result in fatalism? A qualitative study of parents' responses to neonatal screening for familial hypercholesterolaemia (mimeographed).

22. The women's group conversations were mostly in Icelandic with whispered translations to me.

23. During the years of compulsory sterilization that Iceland had shared with the other Nordic countries, which ended only in the mid-1970s, some seven hundred people, mostly women, were sterilized. This collusion between state and clinicians, which denies patients' individual rights, is precisely what, after Nuremberg and Helsinki, informed consent tries to constrain.

REFERENCES

Beck, U. 1992. *The risk society: Towards a new modernity.* London: Sage.

deCode Genetics, Inc. 1998. *A Non-Confidential Corporate Summary.* Reykjavik: deCode Genetics, Inc., 7 June.

Giddens, A. 1991. *Modernity and self-identity.* Cambridge: Polity.

Lewontin, R. C. 1999. People are not commodities. *New York Times,* 23 January.

Marshall, E. 1998. Iceland's blond ambition. *Mother Jones* (May-June): 53–57.

Mawer, S. 1999. Nation of clones. *New York Times,* 23 January.

Palsson, Gisli G., and P. Rabinow. 1999. Iceland: The case of a national Human Genome Project. *Anthropology Today* 15, no. 5:14–18.

Robbins, T. 2000. Court sanctions use of anonymized patient data. *British Medical Journal* 320 (8 January): 77.

Specter, M. 1999. Decoding Iceland. *New Yorker* (18 January): 43–51.

SECTION B

Animal Species/Genetic Resources

Chapter 5

Kinship, Genes, and Cloning

Life after Dolly

Sarah Franklin

The birth of Dolly, the famous cloned Scottish sheep, was first reported on February 23, 1997, in the British Sunday *Observer* by its science editor, Robin McKie. Later that week, the means of her creation were officially documented in the British science journal *Nature,* in an article by Ian Wilmut and his colleagues titled "Viable Offspring Derived from Fetal and Adult Mammalian Cells."[1] As with another famous British birth—of the world's first test-tube baby, Louise Brown, in June 1978—Dolly's viability instantly became the subject of worldwide media attention and public debate. Her birth was seen to alter the landscape of future reproductive possibility and to raise, once again, questions about the ethics of artificially created life.

In the first full-length account of the making of Dolly, *Clone: The Road to Dolly and the Path Ahead,* the *New York Times* science journalist Gina Kolata describes the cloning of Dolly from an adult cell as comparable in scientific importance to the splitting of the atom, the discovery of the double helix, and the elimination of smallpox (Kolata 1997). According to the scientists who created her, Dolly inaugurates a new era, "the age of biological control" (Campbell, Wilmut, and Tudge 2001). Prominent ethicists, philosophers, and scientists have spoken out about cloning, testified before Congress, and published their views in editorials and anthologies. Advisory and legislative bodies worldwide have issued reports and recommendations.[2] Controversy continues to surround the question of whether humans should be cloned and the debate has now been extended to the ethics of cloning human tissue by means of stem cells and the emergent science of tissue engineering.

Anthropology and feminist theory raise a different set of questions about the cloning of Dolly, questions of kinship, gender, and biology. In this chapter I explore the notion of viable offspring with respect to the relationships between kinship, genealogy, and property. Using Dolly as a kind of shepherd, or guide, I examine her creation to see how scientific knowledge comes to be embodied, how biology is seen to be authored, and how in turn such acts of creation are protected as forms of property. Dolly's coming into being disrupts the traditional template of genealogy: she was born from a new kind of cellular assemblage, in which donor cytoplasm effectively "reprogrammed" her nuclear DNA, in a sense, to go back in time and become newly embryonic. Dolly's biology is as cultural as her ontology is historical, and she is one of a number of new animal kinds, or breeds, that instantiate larger changes in what Foucault denominated "the order of things" connecting life, labor, and language. If Dolly were a sentence, we would need a new syntax to parse her, because her counterfactual existence troubles existing grammars of species, breed, property, and sex.

These troubles are not new; indeed, many of them are quite ancient. Like other livestock, Dolly embodies a commercial purpose. In Dolly, however, genealogy is reconstituted as a unique and unprecedented conduit for the production of biowealth, and she thus requires some altered templates of theoretical explanation to address the significance of her making, her marking, and her marketing as a successful product.[3] Like older breeds, Dolly was created to explore new possibilities of making animal reproduction more efficient. In the process, she has transformed the landscape of animal reproduction.

Viable describes Dolly in several senses. She is viable in the biological sense of being capable of life outside the womb. She is also viable in the wider sense of being capable of success or continuing effectiveness: she is viable in the corporate sense of a successful plan or strategy. Her existence confirms the viability of a particular scientific technique, the technique of cloning by nuclear transfer using fully differentiated adult cells, which, until she was born, was widely believed to be impossible. Dolly's ability to survive, to function normally, and to reproduce naturally guarantees other kinds of viability: the viability of artificially created life, for example, and the viability of the stock options of her parent company, PPL Therapeutics, who financed her creation. Dolly is livestock in a overdetermined sense: she is viable not only as a single animal but also as a *kind* of animal, a new species of what might be described as breedwealth.[4] Above all, she is a newly viable form of genetic capital, in sheep's clothing.[5]

Dolly's birth offers further confirmation that biological reproduction can become an engine of wealth generation and capital accumulation. Cloning and cell fusion have become increasingly significant means of reproduction in the era of the polymerase chain reaction, immortal cell line banking, and genomic libraries. Dolly is owned as an individual animal, much as any farmer owns livestock. But she is much more valuable as an animal model for a technique owned as intellectual property, by means of a patent that covers the technique of nuclear transfer.[6] In addition, ownership of Dolly involves the production of what might be thought of as new forms of biological enclosure—that is, by the refinement of specific biotechnological pathways that reliably deliver certain kinds of functionality. For example, the means of reactivating the recombined cells out of which Dolly was made involved identifying the significance of particular stages in the cell cycle and learning how to manipulate these stages using electricity. The ability to enclose distinct components of the emergent biotechnological tool kit as private property thus involves a combination of skill, ingenuity, secrecy, and legal instruments such as patents in order to create new forms of biowealth. Anthropologically, such alterations in the fungibility of animal genealogy pose questions not only about the production of new forms of genetic capital but also about the basis for distinguishing among animal kinds—a question that in turn leads back into familiar questions about the connections between so-called biological differences, the formal categorizations based on sex, gender, kinship, and descent.

GENETIC CAPITAL

The profitable reproduction of animals as livestock has depended on specific technological innovations and market refinements. Writing of the eighteenth-century livestock breeder Robert Bakewell, the historian Harriet Ritvo describes an important shift through which this "master breeder" altered the ways in which prized animals came to be valued as individual repositories of genetic capital. Bakewell's development of careful pedigree recording enabled him to transform the livestock market, so that he could effectively rent out his animals for stud duty. To bring about this shift in the buying and selling of animal reproductive capacity, Ritvo argues, Bakewell needed to transform the conceptual basis of livestock breeding. She claims that he accomplished this transformation by means of a shift in the definition of the genetic capital from the breed as a whole to the reproductive power of a single animal:

> Bakewell claimed that when he sold one of his carefully bred animals, or, as in the case of stud fees, when he sold the procreative powers of these animals, he was selling something more specific, more predictable, and more efficacious than mere reproduction. In effect, he was selling a template for the continued production of animals of a special type: that is, the distinction of his rams consisted not only in their constellation of personal virtues, but in their ability to pass this constellation down their family tree. (1995: 416)

The shift here involves enabling a part to stand for a larger whole. It could be described as metonymic in the sense that the individual comes to be so closely associated with the breed as a whole that it can stand in its stead. More specifically, the shift is synecdochic, in the sense that *the substance from which it is made* can stand for an object itself, as in steel for sword. The accomplishment of this change in kind described by Ritvo, whereby a single animal could become a template for an entire type or breed, was accomplished through careful written records—that is, through the establishment of the studbook as a marketing device. The maintenance of such records enabled a differentiation to be drawn between male animals that were "good sires" and those that were not. In turn this differentiation enabled a reduction of the male animal to a template of his kind. It also depended on the redefinition of the breed, or breeding group, as a lineage. And these *conceptual* changes enabled an exchange—of the stud fee for generations of careful breed selection.[7]

The point of all of this was its profitability. Instilling new property values in animals, and establishing a market in which to sell them, enabled Bakewell to increase the value of his breeding stock by four-hundredfold within thirty years. Ritvo asserts, "So complete was the conceptual transformation wrought by this redefinition of an animal's worth, that at a remove of two centuries it may be difficult to recover its novelty" (1995: 417). Moreover, these eighteenth-century breeding innovations established Britain as "the stud stock farm of the world," a legacy still manifest in animals such as Dolly.[8]

As Ritvo observes, it is entirely taken for granted today that breeds are the result of careful selection, in-and-in breeding (breeding of parents and offspring) to improve the line, and the application of breeding principles to the improvement of stock by their owners. It is equally readily accepted that some animals are qualitatively and quantitatively better breeders than others, and that this quality affects their monetary value. What her analysis reveals most compellingly is the large amount of conceptual apparatus that must exist in relation to the animal for its biology to emerge as obvious in this way, or indeed for the biology of a prized ram to emerge at all. A breed is thus a biotechnological assemblage, its very constitution a discursive formation, its genome a manifestation of the breeder's art.

Dolly extends the uses of breeding in some important new directions. The definitive technology she embodies is the technique of nuclear transfer—the form of cell fusion through which Dolly was cloned.[9] Dolly's physical viability has now authenticated this technique and its profitability, much as the performance of Bakewell's Dishley rams secured the viability of an earlier form of breedwealth in livestock husbandry, and Louise Brown's viability confirmed the success of in vitro fertilization.

Like the studbook, nuclear transfer also effects a reduction of the animal to its heritable traits. But with Dolly there are several important differences. First, it is the female animal, and not the male, whose DNA serves as a template. And second, it is not the animal herself but a laboratory technique that provides the means of reproduction. These shifts, like those described by Ritvo, are both technological and conceptual. In the industrial version of breedwealth established by Bakewell, the individual animal provided both the template and the means of reproduction: the package being sold included both its genes and its generative power. In the case of Dolly, neither her own genes nor her own generative capacity are valuable. She embodies value only as an animal model for a patent application, providing living (and extensively DNA-tested) proof that Ian Wilmut's technique can be successful. The viability of the means of reproduction used to make her, nuclear transfer technology, is the source of new genetic capital—which is why intellectual property rights were sought not for Dolly herself but for nuclear transfer technology. In this sense, cloning by nuclear transfer enables genetic capital to be made alienable from the animal herself—and doubly so. This has significant consequences for how both reproduction and genealogy can be owned, marketed, and sold, and for what they mean and how they are (dis)embodied.

These shifts have implications for both genealogy and gender. Unlike Bakewell's Dishley rams, Dolly is distinct from the source of her reproductive value, which has, in a sense, been seconded to establish the viability of a technique of reproductive biology. Her own ability to reproduce is merely a subordinated sign of her individual viability as a product of corporate bioscience. Dolly is a successful trial run.

In sum, she is the cookie, not the cutter. PPL Therapeutics is the world leader in transposing human genes into animals in order to harvest peptides from their milk and make new drugs from them. The aim of producing Dolly was to demonstrate the viability of a technique that *bypasses* her own reproductive capacity, which is too inexact. Cloning by nuclear transfer is useful to such endeavors because, unlike conventional breeding, it enables exact reproduction of an animal's complete nuclear genetic blueprint. In a sense, nuclear transfer decontaminates mammalian reproduction: we might say it eliminates nuclear waste. This innovation is valuable because it enables a new form of pure reproduction in higher mammals, removed from the genetic "noise" of the rut. The problem with conventional breeding, of course, is that it is unreliable, inefficient, and thus costly. Every time a breeder mates a prized animal, the recombination of genes that is an unavoidable component of sexual reproduction is the equivalent of a genetic lottery: you never know what kind of match, or mismatch, will result.

Nuclear transfer eliminates the genetic risk of sex, producing an exact replica of the desired genetic traits.[10] Through this means, argue the team who produced Dolly, the precise genetic composition of prized individual animals will be both preserved in perpetuity and more efficiently reproduced in other animals. Indeed, nuclear transfer makes it possible for any animal, male or female, wild or domesticated or even extinct, to become a perpetual germ-line repository, a pure gene bank, because the gametes—the eggs and sperm—are no longer necessary to reproduction. A single animal can be cloned to produce an entire herd of identical animals that would otherwise take years to establish. These animals can also be improved with the addition of precise genetic traits, including those from other species. In sum, the value of nuclear transfer is so obvious it had to be invented. While compressing genealogical time, it also offers total nuclear genetic purity, in perpetuity and under patent.[11]

Nuclear transfer technology thus offers a specific redefinition of breedwealth, or livestock, by introducing new recombinant models of genealogy, species, and reproduction. The principle of nuclear transfer is the exact reverse of Bakewell's contribution and inverts what we might describe as the modern industrial model of breedwealth into its fragmented, postmodern successor project. If the studbook was a way to transform an animal's genealogy into a source of individual value, nuclear transfer is a way to depart from conventional genealogical spatiality and temporality altogether. Dolly's pedigree is removed from natural time, or the time of genealogical descent. Her mother is genetically her sister, as are her offspring.[12] She was produced from the nucleus of a mammary cell, amplified from a frozen tissue sample taken from a pregnant Finn-Dorset ewe, who had died six years earlier. This nucleus was inserted into an enucleated donor egg cell from a Scottish Blackface sheep. The resulting embryo was carried by two more sheep, the second of which gave birth to Dolly. Dolly instantiates a new form of commodifying genealogy because she establishes a new form of genealogy altogether.

So what are the implications of this enterprised-up genealogy for other naturalized categories, such as gender, sex, and species—all of which have depended on the orderly brachiations of the unilinear, bilateral, and unified genealogical descent system Darwin envisaged as the real tree of life? If Dolly is the product of a fertile union among several females—if she is the offspring of a kind of same-sex tissue merger—does this mean biological sex difference in reproduction has become obsolete? Have we seen the transcendence not only of sexual difference but of reproductive difference as well? One reading of the Dolly episode might lead to the suggestion that maternity has triumphed over paternity in a kind of recapitulation of the ancient matriarchy theories so influential in early feminism.[13] And how appropriate that sheep are a matrilineal species, each flock with its wise and woolly head ewe—just like in the film *Babe.* But the triumph-of-the-genetrix reading of cloning, which might be celebrated as the ultimate female-defined reproduction, is in tension with another possibility: that paternity has not so much been displaced as dispersed into acts of scientific creation and principles of legal ownership. It may be that the stud has vanished, but there are other father figures.

Dolly's conception raises paradoxical implications for the meanings of maternity, gender, and sex. For although the nuclear transfer technique is designed to produce female sheep from other female sheep, this occurs under the sign of familiar forms of paternity. The best transgenic ewes can be used to create the equivalent of stud lines for entire flocks. Because all, or many, of their adult cells can be used for reproduction, they surpass even the much-celebrated heights of male sperm production, with nearly every cell in their body potentially a new ewe. But these ewes are not analogous to superstuds, because their embodiment of a unique genetic template has been separated from their ability to pass it on. The whole point of a stud line derives from the idea of the unique genetic capital of a prized individual combined with that animal's capacity to pass these traits on down the family tree.[14] This was Bakewell's contribution, as outlined by Ritvo, whereby the reproductive power of a specific animal could be sold as a template. Nuclear transfer technology anachronizes this connection in the same stroke with which it eliminates conventional genealogical time, order, and verticality altogether.[15]

Such observations lead to questions about paternity and property, to Dolly's "parent" company, and to her "scientific" father. Nuclear transfer is a device for seeding a corporate plan for the production of biowealth in the form of what PPL Therapeutics describes as "bioreactors": in this case, the sheep that will function as living pharmaceutical producers by excreting valuable proteins in their milk. Dolly's own reproductive capacity, now proven in the form of her own viable offspring, becomes a publicity stunt for the more important offspring known as nuclear transfer. Dolly's lambs provide further "proof" that cloning is a perfectly natural, sound, and healthy means of reproduction (and what an attractive advertisement they are, timed perfectly to arrive each year at Easter). Ironically, Dolly's lambs do service for the scientific paternity of her own creation, which lies with Wilmut and his colleagues at the Roslin Institute in Scotland, who designed the blueprint of the technique that made her a viable offspring to begin with. Dolly's own maternity is as inconsequential in itself as are her healthy eating habits: just one more sign that she is a perfectly sound animal. It might be said her maternity is a paradoxical stamp of approval for her thoroughly man-made viability.[16]

The significance of paternity in the context of Dolly's creation is also evident in the patent application for specific uses of nuclear transfer technology. As Mark Rose (1993) has suggestively chronicled, the establishment of copyright was explicitly argued by analogy to paternity. An author's original works were an inviolable possession of their creator, just as his children belonged to him because he was their procreator. Offspring of the brain and of the loin, argued prominent literary figures such as Daniel Defoe, derive from individual acts of creation and must be protected as such. *Plagiarism* derives from the Latin word for kidnapping.

The invisibility of the maternal in such an argument directly anticipates the situation with Dolly. Defoe's argument that authors are essentially the fathers of their texts restages a perennial fantasy of male-birthing from which the maternal is excluded. It is an exclusion that recalls a phrase in Zora Neale Hurston's ethnography, *Tell My Horse.* Hurston describes the use of the expression "the rooster's egg" to describe children of white fathers and black mothers who were defined as white by virtue of their paternity.[17] The subordination of maternity in the attempt to secure racial privilege is mocked by the figure of the rooster's egg, marking this denial of maternity as an absurdity, a fantasy, and a lie. The invisible, or subordinated, maternal in the context of copyright was directly paralleled on Bakewell's farm, where the female animal was irrelevant, and only the male line "counted" for stud fees. Dolly's subordinated maternity thus repeats this long-standing pattern of maternal erasure, one that in her case is compounded by the explicit display of her recuperated maternity to confirm the skill of her creator.

It is the skill of this original creator, as an innovator, which is protected under the patent for nuclear transfer, that Dolly authenticates as the viable offspring of pater Wilmut. To be patentable, an invention must be original, of utility, and nonobvious—and nuclear transfer is all of these, although, like much contemporary patented biowealth, it relies closely on designs that are "found in nature," notably the cell cycle. This form of ownership does not explicitly accrue to Dolly herself, who is but its means of realization or its proof. What the patent protection secures in Dolly's case is the capacity for her maternity to be distributed. Her reproduction becomes partible: she is newly profitable because she is multiply divisible, and it is her divisibility that makes her newly fungible. Hortense Spillers famously described in the same way the distributed maternity of female slaves, whose reproductive capacities their nineteenth-century masters could either sell or use themselves. The production of Dolly similarly conjoins commercial and biological enclosure by isolating particular reproductive pathways and creating a market in access to them. In both cases reproduction must be separated from genealogy—a feat particularly evident in cloned transgenic animals.

The popular association of cloning with slavery shares this recognition of the shame and disempowerment that occasions the loss of reproductive power.[18] It might be argued that animals have long been owned in this way, their reproductive power part and parcel of their value. But, as Ritvo shows, this is not self-evidently the case. The capacity to own, market, and sell the reproductive powers of animals has changed dramatically over time and has done so in close association with redefinitions of other forms of property, such as intellectual property. Moreover, the reconceptualization of property is itself technologically assisted, through inventions such as studbooks, pedigrees, and patents. Today, frozen cell lines, molecular biology, and nuclear transfer are part of a wider set of conceptual and technological transformations in the capacity to own, manipulate, and profit from the reproductive power of animals, plants, and microorganisms. This phenomenon can only be described as an intensification of the politics of reproduction and an enterprising-up of genealogy. And just as capital is changing, so the new biology does not guarantee the same syntax it used to guarantee for other domains: what does it mean when genealogy can be remade through technique? What happens when the means of reproduction themselves can be owned under a patent? What is Dolly's proper gender, or sex, if instead of being born she was made?

Using the patented transgenic oncomouse as one of her guides, or figures, in *Modest_Witness@Second_Millennium,* Donna Haraway describes what she calls a "shift from kind to brand" (1997: 65–66). Borrowing from, and mutating, Marilyn Strathern's work on kinship in *After Nature* (1992), Haraway describes kinship as "a technology for producing the material and semiotic effect of natural relationship, of shared kind" (53). She describes kinship "in short" as "the question of taxonomy, category and the natural status of artificial entities," adding that "establishing identities is kinship work in action" (67). In the context of such denaturalized animate entities as oncomouse, Haraway argues that "type has become brand," and that the brand has become a kind of gender. The brand becomes for Haraway a kind of hypermark establishing kind and type in a *semantics of propriety* that is explicitly postnatural.

Haraway's shift from kind to brand thus describes the way in which the production of a certain type of animal, such as the oncomouse, occurs out from under the sign of natural history and instead beneath its brand name. This interpretation thus literalizes the brand slogan of Dupont, "Where better things for better living come to life," which Haraway first brought to her reader's attention in 1992, in the article "When Man™ Is on the Menu," in which she claimed that the new cyborg animals of corporate biotechnology "will be literate in quite a different grammar of gender" (1992: 42).

Haraway's 1992 essay appeared in the same Zone anthology, titled *Incorporations,* that carried Paul Rabinow's essay arguing that the new genetics represent the apotheosis of modern rationality, in that "the object to be known—the human genome—will be known in such a way that it can be changed." In this article Rabinow made the often-requoted prediction that,

> in the future, the new genetics will cease to be a biological metaphor for modern society and will become instead a circulation network of identity terms and restriction loci, around which and through which a truly new type of autoproduction will emerge, which I call "biosociality." . . . In biosociality, nature will be remodeled on culture understood as practice. Nature will be known and remade through technique and will finally become artificial just as culture becomes natural. (1992: 241–42)

For Rabinow, the nature/culture split will disappear in a penultimate collapse of the very distinction out of which modernity emerged as a discursive condition in the first place.[19] For Haraway, nature is not so much displaced as reanimated, acquiring a new capacity to mark a different set of relations in the context of corporate technoscience, in which unnatural relations such as transgenics reappear as naturalized kinds through brands.

There is no doubt that Dolly is the founder animal for a new species of product in which family resemblance is at a premium. She is not branded as such, but she secures a patent application through what might as well be her brand slogans: Made in Scotland, Designed by Roslin, and Brought to You by PPL Therapeutics. As the technology for making cloned transgenics improves, there will emerge successor generations of products in a commodity lineage of designer sheep. Global marketing strategies, such as those used by Intel, Nokia, and BMW, borrow from familiar kinship idioms to provide analogies for the ways in which products are "related," but what is more revealing is how these analogies can also operate in reverse. In other words, the brands and trademarks connecting products to their parent company stand in for shared substance, forming the basis of kin-relatedness as a familiar form of propriety-by-descent. These commodity descent lines are therefore instantiations of a different kind of substantial connection, which is established through trademark or brand as its mark. What is interesting is that, as Strathern argues, such analogies can be reversed: the traffic can make a U-turn. Hence, whereas genitorship has historically been the model for the naturalized propriety of copyright, we might argue that commercial propriety can now engender and naturalize paternity. Possession itself can figure technoscientific fatherhood.

Thus not only nature but also paternity is "known and remade through technique," to redirect Rabinow's apt phrasing. Haraway's "shift from kind to brand" also points to this collapse of the commercial and the paternal. But now, as distinct from earlier episodes, it is the means of reproduction itself, and not merely its offspring, that paternity defines as its own. This made-in-the-lab paternity may in fact perfectly instantiate what Rabinow describes as "the truly new form of autoproduction" that is "the apotheosis of modern rationality."

Like maternity, nature does not so much disappear as become a kind of trope in the context of contemporary biotechnology (see further in Franklin, Lury, and Stacey 2000; Franklin in press). The same can be said for kinship and gender, which become much more like brand in their capacity to signify difference—through relations of enterprise and propriety rather than through relations such as genealogical descent. Now that animals such as Dolly are both born and made, they not only embody nature "remade through technique" but also the "shift from kind to brand" in their corporately owned and redesigned corporeality.

In sum, the gender of the new genetic capital is very familiarly paternal, but this repeat of an ancient tradition has taken a few new turns. For one, the means of reproduction have been removed from the animal and placed under the sign of patent. For another, Dolly's own maternity does service to the value of nuclear transfer as a means of both producing and protecting genetic capital. And all of this is possible because reproduction has been removed from genealogical time and space, becoming, through new technologies, no longer either vertical or bilateral. Life after Dolly is both differently viable and newly profitable.

Dolly also shows us some important dimensions of what happens to gender when it is made, not born. She helps us to ask what happens to what Monique Wittig (1992) calls "the mark of gender" when that marking occurs through branding as a proprietary relation. In examining how brands are naturalized as what Haraway calls genders, there are important questions to be asked about how nature comes to signify in a postnatural culture. Does this model of gender simply give us more of them? If gender becomes a commercial equation, is it easier to opt out altogether? Is cloning a form of gender trouble?

In genealogical terms, nuclear transfer effects a ninety-degree turn, as a result of which the notion of descent is no longer the equivalent of gravity. Instead, enterprised-up genealogy is newly flexible, so that it is more subject to redesign and freed from the narrow trammels of species-specific reproductive isolation to become newly promiscuous: a mix 'n' match *recombinatoria,* not unlike alchemy.

CONCLUSION

Examining Dolly as I have done suggests she belongs to what Foucault might have described as a new order of things, in which life, labor, and language have been transformed in their constitutive relations. Never concerned with nature and culture per se, Foucault took from his predecessor Georges Canguilhem a historical and philosophical question about the relation of knowledge production to life-forms and, indeed, of epistemology to life itself. Always attentive to the constitutive power of knowledge in its many forms (disciplinarity, governmentality, classification, surveillance), and its myriad corresponding objects (prisons, clinics, museums, bodies, sexualities), Foucault stressed in his writing the importance of the transformation of consanguinity into population, and sovereignty into regimes of public health. Dolly perfectly instantiates this same constellation, and simultaneously inaugurates its transformation: she is, after all, part of a corporate plan to put human genes into animals in order to be able to derive profitable pharmaceutical products from their milk. Her coming into being is as a new life-form belonging to the future of medical treatment, wired to the human genome on the Internet, in which the genetic specificity of the individual will replace the formerly generic model of the human used to develop new drugs in the past.

Known and remade through technique, Dolly embodies changes in both knowledge production and governmentality. She is the viable offspring of the epistemological recalibration of biology by technology, as a result of which it is less important to know what she *is* than what she *does*. Though it is now proven feasible, cloning by nuclear transfer remains poorly understood scientifically. An ongoing discrepancy separates the Lego-like logic of millennial biotechnology—with its daunting technical language full of noun-verb hybrids for components that allow pieces to be put together and pulled apart—from the self-evident complexity of the relationalities out of which genetic expression emerges. The very term *genetic information* is a fiction, like *numeric value:* it makes sense only if you take for granted everything needed to explain it.

What holds Dolly together is, consequently, not Foucault's order of things connected to the "life itself" he claims is the foundational concept of modern biology, but Lifeitself™, as in the Dupont slogan "Where better things for better living come to life."[20] The new order of things instantiated through biotechnology has been vastly enabled by a loosening of patent law, which, from the early 1980s onward has with increasing liberality allowed life-forms to be patented not only when they are nonobvious inventions but also, ever more frequently in the age of genomics, simply when they are useful techniques. This mechanism of the nation-state to promote industry through the patent, and its officers, and to conjoin labor and life into a productive force, is precisely aimed to fuel market speculation and encourage venture capital in a market dominated by multinational pharmaceutical giants, to create a situation one journalist has compared to the sixteenth-century competition between France, England, and Spain to claim the New World.[21]

The density and power of the capital resource Lifeitself™ asks that it be understood as part of a historical transformation of a distinctive kind. The splicing together of human genes with those of other species into a new *ars recombinatoria* of life-forms that no longer belong to natural history or genealogy as we have known them means that none of the naturalized categories hold still in relation to what used to be seen as their given attributes. Is cloning by nuclear transfer sexual reproduction or not? How many parents does Dolly have? Kinship and gender, those serviceable anthropological digging tools, offer one way of thinking about what happens to these categories as kinds of kinds, or as the grammatical categories of a sociality understood to be glued together in some way by relationships established through reproduction and sex. In seeking to understand the recalibration of life itself in the context of biotechnology, the question has to be asked, What happens when we understand genes as themselves the vehicle for cultural expression?

NOTES

1. Dolly was more than six months old at the time of her birth announcement: she had come into the world in a shed in a small Scottish village on June 5 the year before.

2. I have provided a list of several of these reports and anthologies about cloning in the reference list for this essay, which is part of a larger project on kinship and cloning supported by a fellowship from the Leverhulme Trust.

3. Although Dolly continues a long tradition of animal breeding for human purposes and thus is hardly unique in her embodiment of human technical and discursive markers, this chapter focuses less on such continuities and more on the ways in which cloning constitutes a distinctive moment in animal manufacture. A different chapter might emphasize a reverse set of claims about Dolly's links to historical traditions of animal breeding. I use the terms *biowealth* and *breedwealth* to emphasize these connections.

4. The ability to control animal breeding is one of the main definitions of domestication. Human control over animals, often expressed as dominion, has been linked to wealth generation since the emergence of what are now called breeds or breed lines. *Breedwealth* is a term that emphasizes both the commercial motivations of "the breeder's hand" and the intensification of commercial interest in the cellular and molecular biology of animal reproduction.

5. Part of Dolly's parent company was purchased in 1999 by the Geron Corporation, which specializes in medical applications of cloning and has developed techniques for stem-cell amplification to generate replacement organ tissue. This application of cloning by nuclear transfer, and its potential use as a form of assisted conception, is the most likely means by which human cloning will be inaugurated.

6. Dolly's creation is covered by two patent applications filed by the Roslin Institute: PCT/GB96/02099, titled "Quiescent Cell Populations for Nuclear Transfer," and PCT/GB96/02098, titled "Unactivated Oocytes as Cytoplast Recipients for Nuclear Transfer." These applications are filed in most countries in the world and cover all animal species, including humans. The Roslin Institute's policy is to license its patents by field of use.

7. This is necessarily a very brief summary of Ritvo's argument. Her work is greatly important in understanding not only the emergence of animal pedigrees but the importance of many domesticated species to Darwin's models of evolution.

8. As Cooper claims in his midcentury evaluation of Bakewell, "There are in fact only two breeds today not of British origin, namely Friesian cattle and Merino sheep, which have a truly international status" (1957: 90). The Roslin Institute in Scotland is itself heir to this same lineage, as a direct descendent of the Imperial Bureau of Animal Breeding and Genetics, created in 1929.

9. Dolly is not properly described as a clone, and the term *clone* does not appear anywhere in the *Nature* article by Wilmut and colleagues announcing her birth. She is the result of a merger between the cells of two animals, whereas a clone is, in the strict botanical sense, an entity grown from a single cell of its pro-

genitor (*cloning* comes from the Greek for "twig" and is perhaps most accurately used to describe the way a gardener grows a new hydrangea from a single twig of a parent plant).

10. The exact genetic traits sought by PPL Therapeutics are transgenic. The first cloned transgenic sheep, named Polly, was announced in July 1997. Polly was created by a version of the technique used to create Dolly, namely the technique used to produce Megan and Morag, the sheep born at Roslin in 1996, using fetal rather than adult cells. Polly carries not only the targeted human gene but also the marker for it. The Roslin web pages explain that "earlier techniques have been hit-or-miss for mixing animal DNA[,] but cloning should make that process more precise."

11. I exaggerate deliberately to make the point that the promise of nuclear transfer corresponds with a commercial logic that is, by definition, hyperbolic. It is important to qualify many of the claims made about cloning and stem cells not only in terms of their low success rates and worryingly high levels of pathology but also because many decades will likely pass before any widely available therapeutic benefits are derived from this highly publicized area of scientific research.

12. Although it is tempting to use traditional kinship categories to play with Dolly's family tree, it is misleading to do so, insofar as these terms assume certain kinds of genetic relationship even as they often depart from them entirely (such is the admirable flexibility of kinship categories in general). Dolly has her "own" DNA and is a genetically distinct individual, at the same time that the blueprint for her nuclear genome was inherited from only one parent.

13. Philip Kitcher (1998), for example, supports cloning for families on behalf of stable lesbian couples who would like to have a child, and who could, if one partner donates the egg and the other the nucleus, more closely emulate the heterosexual ideal of conjugal and procreative unity (arguably not the most widely shared aspiration among lesbian couples). This example is only one of many in which a technique often described as bringing about the end of sex is readily resituated within normative family values.

14. As Ritvo explains, Bakewell used progeny tests to chart the performance of his studs to discover their "hidden" qualities. In addition to seeking purity of descent (preserved through in-and-in breeding of parents and offspring), he sought what is technically known as prepotency, which Ritvo defines as "a heritage sufficiently concentrated and powerful to dominate the heritage of potential mates" (1995: 419). This is only one example of some of the many ideas about inheritance that continue to influence the breeder's art. For example, even though Bakewell's celebrated Dishley sheep did not prove to have much staying power as a breed, their best-known descendants, the Blue-Faced Leicesters, are still used primarily to produce "tups," young rams which are sold to be used for crossbreeding with other sheep.

15. It is tempting to note that the transgenic possibilities opened up through sheep-human combinations create a new kind of ewe-man genome initiative, but to suggest such a merger is to overlook the technical complexities that continue to beset this field of endeavor.

16. Dolly is herself better known for her habit of stamping her foot to indicate disapproval, the standard threat gesture of the ewe. From the beginning treated with special care, Dolly is reported to be well aware of her stature, and to respond with an irritated stamp of the hoof to transgressions such as an inadequate dinner.

17. The expression is also the title of a collection of essays by Patricia Williams (1995).

18. Interestingly, the use of the term *clone* to denote loss of reproductive propriety is also evident in the marketplace, where it denotes an illegitimately copied product, as in a "Gucci clone," or the risk of illegitimate product-use to markets, as in mobile phone fraud. Genetic markers are used by companies such as Monsanto to prevent "cloning" of their agricultural products, in both the scientific and commercial sense. Older associations of clones with drones or slaves are based on the stigma attaching to illegitimate, unnatural, or diminished origins.

19. In contrast, Bruno Latour argues this division was only an enabling fiction for modernity to begin with, hence his title claim *We Have Never Been Modern.*

20. I am borrowing back and remutating the term *life itself* from Haraway's description of it as "a thing-in-itself where no trope can be admitted" or as "a congeries of entities that are themselves self-referential and autotelic," like Dawkins's selfish gene—in sum, a kind of fetish (1997: 134–35). I argue that not only is the fetishism of life itself as a commodity in evidence but specifically its removal from genealogy, which has consequences for what propriety, enterprise, or commerce can connect.

21. Writing in *Wired,* the journalist Michael Gruber suggests that "the 21st century will be more like the 16th than the 20th, with biology standing in for the New World. The pharmas and the big chemical companies are the great expeditionaries—Cortés, Pizarro, de Soto, Raleigh, and so on. Government regulatory agencies are—what else?—the European imperial powers. The pharmas are after treasure, of course. The regulators want to keep control, which they express as an overarching social good—back then it was Defense of the Realm and Propagation of the Faith: today it's Public Health" (1997: 198).

REFERENCES

Campbell, K., I. Wilmut, and C. Tudge. 2001. *The second creation: The age of biological control by the scientists who created Dolly.* London: Heineman.

Cooper, M. McG. 1957. Present-day evaluation. In *Robert Bakewell: Pioneer livestock breeder,* ed. H. C. Pawson, 89–95. London: Crosby Lockwood and Son.

Croke, V. 1999. Tufts-Genzyme team cites cloning advance. *Boston Globe,* 27 April, A1, A13.

Foucault, M. 1970. *The order of things: An archaeology of the human sciences.* New York: Pantheon.

Franklin, S. In press. Rethinking nature/culture: Anthropology and the new genetics. *Anthropological Theory.*

Franklin, S., and M. Lock, eds. In press. *Remaking life and death: Toward an anthropology of bioscience.* Santa Fe: School of American Research Press.

Franklin, S., C. Lury, and J. Stacey. 2000. *Global nature, global culture.* London: Sage.

Gruber, M. 1997. Map the genome, hack the genome. *Wired* (October): 153–56, 193–98.

Haraway, D. 1992. When Man™ is on the menu. In *Incorporations,* ed. J. Crary and S. Kwinter, 38–43. New York: Zone.

———. 1997. *Modest_witnesssecond_millennium: FemaleMan©_meets_OncoMouse™*. New York: Routledge.

Kitcher, P. 1998. Life after Dolly. In *The human cloning debate,* ed. G. McGee, 107–124. Berkeley: Berkeley Hills Books.

Kolata, G. 1997. *Clone: The road to Dolly and the path ahead.* London: Allen Lane.

Latour, B. 1993. *We have never been modern.* Trans. C. Porter. Cambridge: Harvard University Press.

Rabinow, P. 1992. Artificiality and enlightenment: From sociobiology to biosociality. In *Incorporations,* ed. J. Crary and S. Kwinter, 234–52. New York: Zone.

Ritvo, H. 1995. Possessing mother nature: Genetic capital in eighteenth-century Britain. In *Early modern conceptions of property,* ed. J. Brewer and S. Staves, 413–26. London: Routledge.

Rose, M. 1993. *Authors and owners: The invention of copyright.* Cambridge: Harvard University Press.

Spillers, H. 1987. Mama's baby, papa's maybe: An American grammar book. *Diacritics* 17, no. 2:65–81.

Strathern, M. 1992. *After nature: English kinship in the late twentieth century.* Cambridge: Cambridge University Press.

Williams, P. 1995. *The rooster's egg: On the prevalence of prejudice.* Cambridge: Harvard University Press.

Wilmut, I., A. E. Schnieke, J. McWhir, A. J. Kind, and K. H. S. Campbell. 1997. Viable offspring derived from fetal and adult mammalian cells. *Nature* 385:810–13.

Wittig, M. 1992. *The straight mind and other essays.* New York: Harvester Wheatsheaf.

Chapter 6

For the Love of a Good Dog

Webs of Action in the World of Dog Genetics

Donna Haraway

To put this in more enthusiastic terms, the dog is a veritable genetic gold mine.
J. P. SCOTT AND J. H. FULLER, *Genetics and the Social Behavior of the Dog*

PREHISTORIES

Story. A Political Awakening

Born in 1944, I grew up in Denver in the 1950s. McCarthyism passed me by, but the new leash law really got my attention. While my adult peers were once red-diaper babies radicalized by blacklists, my earliest political passions were of a lower order on the great chain of chromatic consciousness. When I had to fence my "intact" male Dalmatian-cross mutt—despite getting every adult I knew to promise to vote against the leash law—my political soul came of age. The adults lied, the law passed, the dog was restricted, and my notions of nature and culture got their first rude reworking. That lesson in cross-species democracy, mendacity, freedom, and authority is my key for this essay.

What happens when the mongrel fields of biological and cultural anthropology of genetics are approached through the genome of "man's best friend" instead of "man"? My goals are modest. I am at the beginning of a project that promises to require all the cartographical resources I can learn how to deploy. But here, I will draw a low-resolution linkage map. I want to suggest how large and rich the world of the dog genome is, how many kinds of investment—emotional, intellectual, ethical, communal, institutional, narratival, financial, and political—are made in canine genetics, how full of fascinating actors these companion-animal genetic worlds are, and how some vexing questions in science studies and anthropology might be approached with a canine eye. In particular, this essay will develop the notion of an "apparatus of naturalcultural production," through which the subject-making object called the dog genome collects up the passions and skills of a mangy crowd of human and nonhuman actors. I am interested in the ways that trading zones and boundary objects are constructed to facilitate traffic among scientific professionals and lay people, commercial and academic sites, conservation biologists and dog breed club members, magazine writers and population biologists, and so on. I must pay special attention to how the traffic between popular stories and scientific theories ties together historically specific dogs and humans over the species-life of both sorts of genetically diverse social mammals.

Institution of the Kennel Clubs: The Galapagos Islands of Canine Evolution

What is a dog breed? And what does a nineteenth-century object like a breed have to do with a postmodern marvel like the genome? Accompanied by their equally weedy companion species, *Homo sapiens,* dogs are globally distributed. Both species appear at first glance to be highly diverse, or polytypic, as the comparative anatomists and anthropologists say. Interestingly, however, for dogs as for humans this phenotypic diversity seems to rest on modest genetic diversity compared to other widespread large-bodied mammals (see Alan M. Templeton, chapter 12, this volume).

From the remarkably consistent genetic evidence (including comparisons of mitochondrial DNA, Y-chromosome DNA, and nuclear gene DNA), dogs and humans both appear to have breeding habits that keep the genes flowing back and forth through evolutionary time (say fifty thousand to one hundred thousand years)—yielding what geneticists call a trellis model (Templeton, chapter 12, this volume) instead of a divergent tree model for their patterns of genetic relatedness. There is a lot of traffic in genes in a trellis, yielding populations that regularly show more intrapopulation heterozygosity for genes of interest than interpopulation divergence in the repertoire of available alleles. Populations do differ genetically, with difference being a function of geographical distance. Natural and cultural selection as well as genetic drift operate in dogs, as in people, producing some important genetic difference among populations; but enough back-and-forth gene mixing has characterized the history of both species to render them strikingly genetically unified—swimming promiscuously in the same global gene pools. As a consequence, the measured genetic distances fall far below values that, to a biologist, would allow one to talk about races in people or subspecies in dogs.[1] Distinct kinds of dogs, linked by various rates of partly human-controlled gene exchange, have existed for a very long time all over the world; but the institutionalized breed has a recent pedigree. The studbook, the written breed standard, the breed club, and the dog show constitute a historically specific genetic technology for the production of dogs in urban industrial society. It is a technology that has reshaped dogs across all landscapes, urban and rural, and across all canine jobs in these societies. If dogs are perhaps the chief agents of their own original domestication from their wolf forebears, as I show in the second half of this essay, modern breeds are most certainly a social invention for which particular sorts of humans call the shots. And ironically, as breed clubs have controlled dog reproduction in the last one hundred years or so, dogs have become textbook lessons in the loss of interbreed genetic diversity; the breeds consequently have come to diverge genetically and differ from each other.

Purebred dogs belong to bounded populations whose members are registered in a closed studbook and whose breeding stars are selected by interpretations of a closely guarded written standard. Even if the numbers of a breed total in the tens of thousands, the effective population size in the evolutionary biologist's sense might be fewer than a dozen animals. If the original group of registered dogs was already more or less related, only a few of those founders were used much for breeding. Some male dogs were wildly popular as studs, and various forms of inbreeding were regarded as the best ways to fix desired types and stabilize an identity for kennels and strains. All living dogs in a numerous breed might relate back to a tiny ancestral population. A breed numbering in the thousands *could* be as inbred as the modern cheetah; they could almost be clones.[2]

Of course, breeds differ from each other in genes that are deliberately selected for or against in breeding programs, but the troubling story for genetic diversity comes from another aspect of those breeding practices—namely, the unintentional and random loss of alleles at unselected loci. The very *random* loss of alleles that necessarily characterizes fixed small populations over time means that different breeds lose different alleles. That is, genetic diversity is increased between breeds (populations) in a perverse—but mathematically correct—sense of no longer sharing the same range of alleles at more loci. At the same time, genetic diversity within breeds is decreased, producing homozygosity at more loci, to the point of causing an international research and regulatory emergency, not to mention canine suffering and human grief. The old trellis is morphing into a young tree bearing genetically dangerous fruit.

If I have learned nothing else from my participation in the Internet discussion lists of dog breeders, owners, and biological professionals; my prowling through the canine newsletter, magazine, training manual, and scholarly and popular book literatures; my researching the breed and applying for a purebred Australian Shepherd puppy produced from working lines; my living with an adult Australian Shepherd–Chow mix from a ranching family; and my talking to breeders, trainers, hunters, service-dog handlers, conformation show competitors, dog-sport participants, rescue activists, and other contemporary dog people, I have learned to eschew ideological—or academic—reductionism about these cross-species communities. Very few of these canines and humans, either the heroes or the villains, breathe the rarefied air provided by a purely critical analysis of the institution of the modern breed.

My best information so far comes from communities committed to Australian Shepherds, Border Collies, Golden Retrievers, Basenjis, and Great Pyrenees. These dogs do different jobs and must respond with often specialized skill and judgment to all sorts of people, other dogs, equipment, and machinery and to varied species and landscapes. For a quick glimpse of practices, including interpretations of written breed standards that are neither arbitrary nor reducible to class symbolic action, consider the many self-critical, ethical breeders of sport or working dogs. They tend to raise one or two litters a year for sale, and they work assiduously to produce dogs who can perform physically and mentally with good health for many years. They care about temperament, physical qualities (including conformation), both context-specific trainability and independent judgment by the dogs, and working drive and skill. The best among these breeders evaluate dogs carefully and place them in pet, show, and working homes only after a process of evaluating the people as well. They keep lifelong tabs on their dogs, insist on sale contracts for dogs that should not be bred (or people that should not be breeders), and specific spay-neuter arrangements in an effort to keep the dogs out of ignorant breeding practices in homes or worse in commercial puppy mills. Breeders sometimes promise lifelong willingness to take back a dog in order to keep their dogs out of shelters. These people and their friends often spend money and time—lots of it—rescuing abandoned or abused members of their breed from puppy mills and shelters.

These breeders are caught up nonetheless in the dilemmas of the claims about increasing incidence of genetic disease and the increased surveillance of canine health. Activists among them set up health and genetics committees in their breed clubs and work to reform kennel clubs, their own breed club, and their own breeding practices. These activists, mostly self-educated in science, ask hard questions about the adequacy of data that strike at the heart of knowledge production in technoscientific worlds. My larger project must ask multilayered questions about the material-semiotic practice of "love of the breed" that permeates dog worlds. My destination is the genome and its associated discourses of health and diversity.

BIRTH OF THE KENNEL

Story. Overhearing Prozac

Toward the end of her sixteen-year life, my half–Labrador Retriever, Sojourner, and I frequented her vet's office. I had read Michel Foucault, and I knew all about biopower and the proliferative powers of biological discourses. I knew modern power was productive above all else. I knew how important it was to have a body pumped up, petted, and managed by the apparatuses of medicine, psychology, and pedagogy. I knew modern subjects had such bodies, and that the rich got them before the laboring classes. I was prepared for a modest extension of my clinical privileges to any sentient being and some insentient ones. I had read *Birth of the Clinic* and *The History of Sexuality,* and I had written about the technobiopolitics of cyborgs. I felt I could not be surprised by anything. But I was wrong. I had been fooled by Foucault's own species chauvinism into forgetting that dogs too might live in the domains of technobiopower. *Birth of the Kennel* might be the book I need to write.

While Sojourner and I waited to be seen, a lovely Afghan hound pranced around at the checkout desk as his human discussed recommended treatments. The dog had a difficult problem—obsessive self-wounding when his human was off making a living, or engaging in less justifiable nondog activities, for several hours a day. The afflicted dog had a nasty open sore on his hind leg. The vet recommended that the dog take Prozac. I had read *Listening to Prozac,* so I knew this drug promised, or threatened, to give its recipient a new self in place of the drab, depressive, obsessive one that had proved so lucrative for the nonpharmaceutical branches of the psychological professions. For years, I had insisted that dogs and people were much alike, and that other animals had complex minds and social lives, as well as physiologies and genomes largely shared with humans. Why did hearing that a pooch should take Prozac warp my sense of reality in the way that makes one see what was hidden before? Surely, Saul on the way to Damascus had more to his turnaround than a Prozac prescription for his neighbor's ass!

I was hooked into the mechanisms of proliferating discourse that Foucault should have prepared me for. I was on the road to the fully embodied modern dog-human relationship. Drugs, restraints, exercise, retraining, altered schedules, searching for improper puppy socialization, scrutinizing the genetic background of the dog for evidence of canine familial obsessions, wondering about psychological or physical abuse, finding an unethical breeder who turns out inbred dogs without regard to temperament, getting a good toy that would occupy the dog's attention when the human was gone, accusations about the workaholic and stress-filled human lives that are out of tune with the more natural dog rhythms of ceaseless demands for human attention—all these discursive moves and more filled my enlightened mind. There could be no end to the search for ways to relieve the suffering of dogs and to help them achieve their full canine potential. Furthermore, I am convinced that that is the ethical obligation of the human who lives with a member of a companion species. I can no longer make myself feel surprise that a dog might need Prozac and should get it.

Neither can the author of *The Dog Who Loved Too Much,* Dr. Nicholas Dodman (1996), who explained the psychopharmacological treatment of canine behavior disorders in his popular advice book. That thread led to a wealth of canine advice literature; and browsing any good Internet book site or perusing a hard-copy pet catalog—or stopping off at a PetsMart superstore—lets me survey a vast array of materials that could improve canine lives.

A committed dog companion can get fun-and-games workbooks (including a lesson on multispecies dunking for apples—points get deducted for tooth marks or half-eaten apples) to teach dogs and humans to play together in neighborhood communities or in the privacy of their homes. I gave such a book to my husband for his birthday, and we began training all the members of our household to have a good time together. It all starts with a lot of mouth massage. The more high-culture types among us can subscribe to an excellent literary newsprint magazine, *The Bark,* published in Berkeley. Resemblances to the underground newspaper the *Berkeley Barb* are deliberate. *The Bark* ran the article "Dogs in the Visual Arts."[3] Some of my humanist-scholar colleagues are not ready for this.

Health manuals and self-help literatures abound, and those wishing to avoid the supposedly toxic foods and overvaccination doctrines of the evil official profit-making dog world will not lack for a text named something like *Our Dogs, Ourselves.* Others can find guidance for evaluating a scientific amino acid balance from the Ralston Purina web site, which also gives information for puppy raising and a link to the latest in genetic research. Those responsible about dental health will do more than get their pooch's teeth cleaned annually—tooth-healthy chew toys are on the market. Sports enthusiasts can find clubs, manufacturers, Internet build-it-yourself sites, and county fairground practice days. We belong to the NBA (Nothin' But Agility).[4] Conspiracy theories, government cover-ups, scientific progress, elaborate commodity culture, the war between animal rights and animal welfare discourses, team sports, grief groups, adoption bureaucracies, fetal and uterine-contraction monitors belted to a bitch and equipped with a remote modem tied into twenty-four-hour computer data analysis for problem pregnancies, a dog-cloning project promoted and closely followed on the Internet, abuse recovery therapies: nothing is lacking in contemporary dog natureculture.

To drive the point home, consider my colleague, Professor Angela Davis, whose impeccable red-diaper credentials extend to an honorary degree awarded by Lenin University, Tashkent, Uzbekistan, in 1972. She certainly noticed McCarthyism, not to mention race, gender, class, and sexuality oppression, in cold-war America. As adults, Angela and I share more than an antiracist feminist theory commitment to intersectional analyses of inequalities. She gives me names of dog acupuncture practitioners and training consultants to help along faltering child-dog friendships. She purchased a special wheeled cart for back leg support to help her aged dog walk in her last months. And, healing in a leg cast from injuries sustained while running with her young dogs, she admitted wistfully from behind piles of dissertation chapters that, in her alter ego, she imagines herself a dog breeder.

The Internet culture of canine genomics is rich and important. Cindy Tittle Moore compiled dog-related e-mail sites from 1995 to 1999.[5] My twelve-point, Times, single-spaced printout of her copyrighted list required forty-three pages. The wired dog world mediates international and local exchanges among actors that either could not occur without the Internet or would occur more slowly and less publicly. Play among metastatic commercialism, thick research cultures, and animated professional-lay exchanges characterizes dog gene links. Freelance writers for *DogWorld* (especially Susan Thorpe-Vargas, a Samoyed breeder with a Ph.D. in immunology; John Cargill, an Akita person and a statistician; and D. Caroline Coile, a Saluki breeder with a doctorate in neuroscience and behavior) seed the popular canine knowledge terrain with publications on genetic health and diversity. *DogWorld* has published at least eighteen articles on genetics in relation to health and breeding since 1996 (e.g., Cargill and Thorpe-Vargas 1996, 1998, 2000; Coile 1997; Padgett 1996–97). Older registries for compiling databases on inherited conditions face extensive changes in their practice as a result of the revolution in molecular biology and dog politics.

Consumer culture permeates genetic culture and vice versa. The scramble for dog genes is a scramble to survive for competing biotechnology companies. Meanwhile, a giant in the commercial revolution that defined middle-class dog culture after World War II—the dog food company Ralston Purina—is a mover and shaker in the genetics revolution. New kinds of surveillance—epitomized in mandatory DNA-testing for litters to verify pedigrees and in proliferating gene tests for inherited conditions—discipline the lives of dogs and people. Biosociality is the fluid in which dog and human subjects gestate, as they meet at national specialty shows by the gene-testing apparatus for progressive retinal atrophy.

Contesting for the meanings of genetics has become an obsession in dog worlds. The sense of a state of emergency pervades much of the discourse. In the face of ongoing inbreeding practices that would curl an evolutionary biologist's hair—and in the face of levels of denial that ensure good incomes to therapists into the future—dog gene discourse is volatile. Millenarian thematics borrow from the rhetorics of endangered species, planetary biodiversity loss, and postcolonial criticisms of typological racism. Like much in technoscientific culture, the discourse is simultaneously practical and apocalyptic—and compelling in both registers. On the e-mail lists, some breeders feel attacked when population genetics is the topic. Others (or the same people in another mood) energetically try to learn what for them is a new language written in the ciphers of statistics, along with the language of molecular genetics written in the technical and commercial codes of DNA. Both languages carry major implications for the breeders' practice of "love of the breed." On the Canine Genetics Discussion Group list (CANGEN), "lay people" and "professionals" vigorously work to educate each other about their realities and, perhaps, to shape a better shared reality in the process. "Lay people" welcome "expert" discussions, but not without interrupting and demanding translations of jargon and verbal explanations of equations, and not without contesting the data, models, and theories. Geneticists seek "lay people" as their collaborators in research projects and vice versa. And "lay people" can be impressively literate in the languages and practices of genetics, while genetics professionals can be amateurs and seekers in dog worlds. Dog natureculture has been a cross-generic symbiosis from the start and, during this genetic turn, is so perhaps more than ever.

Agents in Their Own Story

Dogs are agents in cross-species worlds. They motivate their humans, even as their humans learn to draw from new bait bags to move their dogs to perform desired actions. This fundamental point can be illustrated in many ways, but I will content myself here with a return to scientific origin stories. The origin of dogs might be a humbling chapter in the story of *Homo sapiens,* one that allows for a deeper sense of coevolution and cohabitation and a reduced exercise of hominid hubris in shaping canine natureculture. Even as Man the Hunter was retired from the ecological theater and the evolutionary play a couple of decades ago (try not to notice his distressing radioactive half-life into the new millennium in the form of evolutionary psychology), the noble dog-wolf as hunting companion to this mythic hominid personage has a shit-eating grin for more reasons than one.

Accounts of the relations of dogs and wolves proliferate, and molecular biologists tell some of the most convincing versions. Robert Wayne and his colleagues at UCLA studied mitochondrial DNA (mtDNA) from 162 North American, European, Asian, and Arabian wolves and from 140 dogs representing 67 breeds, plus a few jackals and coyotes (Vilá et al. 1997). Their analysis of mtDNA control regions concluded that dogs emerged uniquely from wolves—and did so much earlier than scenarios based on archaeological data suggest. The amount of sequence divergence and the organization of the data into clades support the idea that dogs emerged more than one hundred thousand years ago, with few separate domestication events. Three-quarters of modern dogs belong to one clade; that is, they belong to a single maternal lineage. The early dates give *Canis familiaris* and *Homo sapiens sapiens* roughly the same calendar, so folks walking out of Africa soon met a wolf bitch who would give birth to man's best friends. Building a genetic trellis as they went, dogs and people walked back into Africa too. These have been species more given to multidirectional traveling and consorting than to conquering and replacing, never to return to their old haunts again. No wonder dogs and people share the distinction of being the most well-mixed and widely distributed large-bodied mammals. They shaped each over a long time. Wayne argues that to domesticate dogs took a lot of skill or it would have happened more often. His story bears the scent of the anatomically wolfish hunting dog, and this dog is a human-made hunting tool–weapon. In this version, morphologically differentiated dogs did not show up in the fossil or archaeology record until twelve to fourteen thousand years ago because their jobs in settled post–hunter-gatherer, paleoagricultural communities did not develop until then; so, they got physically reshaped late in the relationship.

People call the shots in a story that makes domestication a one-sided human social invention. But archaeozoological expert Susan Crockford disagrees. She argues that human settlements provided a species-making resource for would-be dogs in the form of garbage middens and—we might add—human bodily waste. If wolves could calm their well-justified fear of *Homo sapiens,* they could feast in ways familiar to modern dog people. Scott Weidensaul states, "Crockford theorizes that, in a sense, wild canids domesticated themselves" (Weidensaul 1999: 57). Crockford's argument turns on genes that control rates in early development and on consequent paedomorphogenesis. Both the anatomical and psychological changes in domesticated animals compared to their wild relatives can be tied to a single potent molecule with stunning effects in early development and in adult life—thyroxine. Those wolves with lower rates of thyroxine production, and so lower titers of the fright-flight adrenaline cocktail regulated by thyroid secretions, could get a good meal near human habitations. If they were really calm, they might even den nearby. The pups who were the most tolerant of their two-legged neighbors might make use of the caloric bonanza and have their own puppies nearby as well. A few generations of this could produce beings remarkably like current dogs, complete with curled tails, a range of jaw types, considerable size variation, dogish coat patterns, floppy ears, and—above all—the capacity to stick around people and forgive almost anything. People would surely figure out how to relate to these handy sanitary engineers and encourage them to join in tasks, like herding, hunting, watching kids, and comforting people. In a few decades, wolves-becoming-dogs would have changed, and that interval is too short for archaeologists to find intermediate forms.

Crockford made use of the forty-year continuing studies of Russian fur foxes, beginning in the 1950s, which have been in the recent popular science news (Weidensaul 1999; Trut 1999; Browne 1999; Belyaev 1969). Unlike domesticated animals, wild farmed foxes object to their captivity, including their slaughter. In what were originally experiments designed to select tamer foxes for the convenience of the Soviet fur industry, geneticists at the Siberian Institute of Cytology and Genetics found that by breeding the tamest kits from each fox generation—and selecting for nothing else—they quickly got doglike animals, complete with nonfox attitudes like preferential affectional bonding with human beings and phenotypes like those of Border Collies.[6] By analogy, wolves on their way to becoming dogs might have selected themselves for tameness.

With a wink and a nod at problems with my argument, I think it is possible to hybridize Wayne's and Crockford's evolutionary accounts and so shamelessly save my favorite parts of each—an early coevolution, human-canine accommodation at more than one point in the story, and lots of dog agency in the drama of genetics and cohabitation. First, I imagine that many domestication sequences left no progeny, or that offspring blended back into wolf populations outside the range of current scientific sensors. Marginally fearless wolfish dogs could have accompanied hunter-gatherers on their rounds and gotten more than one good meal for their troubles. Denning near seasonally moving humans who follow regular food-getting migration routes seems no odder than denning near year-round settlements. People might have gotten their own fear-aggression endocrine systems to quell murderous impulses toward the nearby canine predators who did garbage detail and refrained from threatening them. Paleolithic people stayed in one place longer than wolf litters need to mature, and both humans and wolves reuse their seasonal sites. People might have learned to take things further than the canines bargained for and bring wolf-dog reproduction under considerable human sway. This radical switch in the biopolitics of reproduction might have been in the interests of raising some lineages to accompany humans on group hunts or perform useful tasks for hunter-gatherers besides eating the shit. Paleoagricultural settlement could have been the occasion for much more radical accommodation between the canids and hominids on the questions of tameness, mutual trust, and trainability.

Above all, this origins story must engage with the question of reproduction. It is on this matter that the distinction between dogs and wolves really hinges; molecular genetics may never show enough species-defining DNA differences. Rather, the subtle genetic and developmental biobehavioral changes through which dogs got people to provision their pups might be the heart of the drama of cohabitation. Human baby-sitters, not Man the Hunter, are the heroes from dogish points of view. Wolves can reproduce independently of humans; dogs cannot. Even Italian feral dogs still need at least a garbage dump (Boitani et al. 1995).[7] Ray Coppinger and Richard Schneider summarize the case: "In canids with a long maturation period, growth and development are limited by the provisioning capacity of the mother. . . . Wolves and African hunting dogs solved the pup-feeding problem with packing behavior, in coyotes the male helps, and jackal pairs are assisted by the 'maiden aunt.' The tremendous success of the domestic dog is based on its ability to get people to raise its pups" (1995: 36). People are part of dogs' extended phenotype in their Darwinian, behavioral ecological, and reproductive strategies. It might prove to be a bad bargain; for the cost of puppy sitting is high with the American Kennel Club (AKC), reproduction technology like Whelp Wise, and Perkin-Elmer and other pharmaceutical and biotechnology giants in the loop, but surely not as high as that paid by the remaining wolves in relation to the depredations of the two-legged planetary social mammal.

ACCOUNTING FOR GENES: C. A. SHARP AND THE *DOUBLE HELIX NETWORK NEWS*

With narrative agency secured to the dogs, and maybe more than that, it is time to conclude with the story of a remarkable dog person, C. A. Sharp, whose practice is a microcosm of the themes of my project.[8] Sharp began breeding Australian Shepherds in the late 1970s and served on the genetics committee of the Australian Shepherd Club of America (ASCA) from the late 1970s until 1986, when the board eliminated the committee in a controversial and poorly explained move. In the winter of 1993, she began writing and distributing the *Double Helix Network News (DHNN)*. The first issue of the *DHNN* described itself as a "kitchen-table" enterprise. By 1999, about 150 people—mostly breeders, a few dog research professionals, and one or two ringers like me—subscribed. As she learned desktop publishing, Sharp emphasized networking, sharing information, educating each other, dealing with what she called the ostrich syndrome among breeders about genetic disease, and practicing love of the breed through responsible genetics.

With a B.A. in radio, television, and cinema studies from Fresno State University and a job as an accountant, Sharp has never claimed scientific insider status. She properly claims expert status of a rich kind, however; and she is regarded as an expert in both the breeder and professional scientific communities. She coauthored a paper in the early 1990s with the veterinary ophthalmologist L. F. Rubin on the mode of inheritance in Aussies of the eye defect Collie eye anomaly (CEA). She also engaged in collaborative research with Dr. John Armstrong of the University of Ottawa on the relation of longevity with coefficients of inbreeding in Aussies, until his sudden death in the summer of 2001. She has functioned as a clearinghouse for genetic data on her breed, performed pedigree analyses for specific conditions, and taught breeders the rudiments of Mendelian, molecular, and population genetics and the practical steps that both show- and working-dog breeders can and should take to detect and reduce genetic disease in their lines. She mediates among communities of practice from her location as a self-educated, practically experienced, savvy activist who is willing and able to express controversial opinions within linked social worlds.

Two chapters from Sharp's history suffice to suggest ways of seeing the stakes in the contemporary naturalcultural worlds of genetics. Her involvement in determining the mode of inheritance of an eye disorder in her breed shows how lay agency can work in canine genetics research and publishing. And her participation in the Canine Genetics Discussion Group e-mail list, CANGEN, maps a mutation in her intellectual and moral field, with a changing emphasis from disease-linked genes to genetic diversity in the context of widespread turn-of-the-millennium attention to evolution, ecology, biodiversity, and conservation.

Sharp's interest in the genetic basis of eye disorders dated to 1975, when her first bitch was a puppy. She went to an All-Breed Fun Match near Paso Robles, which turned out to have an eye clinic. Sharp asked what it was about and had her dog checked. "I just got interested and started educating myself," she says. She made it a point afterward to get her dogs' eyes checked annually, which meant going to clinics at the local Cocker Spaniel club or hauling dogs a few hours away to Stanford to a veterinary ophthalmologist. She started reading in genetics, guided by an Aussie person named Phil Wildhagen—"who is quite literally a rocket scientist, by the way," Sharp laughs gleefully. About 1983, the Genetics Committee of ASCA put out a call for people to assist it in gathering data. "One thing led to another," she says, "and I was on the committee."

This was the period when the Genetics Committee was shifting its attention from coat color, which had been of particular interest during the 1970s when what counted as an Aussie was codified in the written breed standard, to the more controversial topic of genetic disease. A breeder gave the Genetics Committee two puppies affected with collie eye anomaly, a condition Aussies were not supposed to have. This breeder also went public with the fact of CEA in her dogs and was vilified for her disclosure by Aussie people terrified of this kind of bad news in the breed. Sharp began writing a regular column in the *Aussie Times* for the Genetics Committee (Sharp 1998).

Starting with the original donated pair, the committee conducted a series of test matings to determine the mode of inheritance. Involving a couple of dozen dogs and their pups, these crosses were conducted in the kennels of two committee members, including Sharp, at their expense, which amounted to several thousand dollars. Most of the affected test puppies were placed in pet homes, with advice to spay or neuter. Some were placed in a university for further research work. The committee collected pedigree data and Canine Eye Registry Foundation (CERF) exam sheets on their test matings and on dogs brought to their attention by a growing number of interested Aussie breeders touched by the *Times* column and word-of-mouth. The pattern of inheritance indicated an autosomal recessive gene. It was now *technically* possible to take action to reduce the incidence of the condition.[9] But *real* possibility is another matter.

First, it was not only Aussie breeders who denied the existence of CEA in these dogs. Simply put, according to Sharp, "Collie eye anomaly in Aussies wasn't 'real' when we started working with it." For example, Sharp brought a couple of puppies from test matings to an eye clinic at a show in Fresno, only to be told by the ophthalmologist that Aussies did not have the condition. Sharp got the exam by mobilizing her technical vocabulary—a familiar move for lay activists in AIDS advocacy, breast cancer politics, and technoscience in general: "'Their mother has an optic discoloboma; [another relative] has choroidal hyperplasia; please check these dogs. . . . Grumble, grumble,' then he checked the puppies." Sharp recalled breeders around the country telling her about attempting to get genetic advice from vets who told them to relax—Aussies do not have CEA; it is not in the literature. Finally, armed with "nearly 40 pedigrees with varying degrees of relationships, plus the test-mating data, I went in search of an ACVO [American College of Veterinary Ophthalmology] vet who might be interested in what I had" (Sharp 1998). Sharp emphasizes that she could not make CEA "real" on her own—"certainly not with a B.A. in Radio, Television, and Cinema." The data had to be published in the right place by the right person. "It's not recessive until someone out there says it is; then it's recessive," she says. "Out there" meant inside institutionalized science. No science studies scholar is surprised by this social history of truth, or by the recognition of it by a savvy lay knowledge producer working within a clerical culture.

The popular but controversial ASCA Genetics Committee had ceased to be; so Sharp began looking for a collaborator to legitimate the data and analysis she had already collected. She talked to several likely scientists, but they had other priorities. Frustrated, Sharp recalls insisting, "Look, until one of you people writes it up, it isn't real." Effective corrective action depended on the reality of the fact. The chain finally led to Dr. Lionel Rubin at the University of Pennsylvania, who was publishing a book on inherited eye disease in dogs (Rubin 1989). The book was already in galleys, so the Aussie story did not make that publication. Sharp assembled the data and did the genealogy charts from the test matings arranged by the committee and turned this over to Rubin, who hired a professional pedigree analyst for the final charts. From the time Rubin began working with Sharp, publication took two years (Rubin, Nelson, and Sharp 1991). With a proper pedigree at last, CEA in Aussies as an autosomal recessive condition was on its way to becoming a fact.[10]

But the reality of the fact remained tenuous. Sharp notes that the demand for independently replicated experiments seems to have kept the fact out of the Aussie section of the ACVO handbook that came out after 1991. Sharp emphasizes that such expensive, ethically fraught research on a large companion animal is unlikely to be replicated: "It wouldn't have happened the first time if those of us out here in the trenches had not been interested enough to gather the data." But she argues, "Why can't the ACVO say it's *prob-*

ably recessive?" She adds, "At least when someone out there asks me now, I can send them a copy of the paper." The newest bible of inherited dog problems does include the fact Sharp's network made real (Padgett 1998: 194, 239). Not surprisingly, Sharp had consulted George Padgett of Michigan State University, an important institution in the apparatus of dog genetics natureculture, when she designed her pedigree analysis service and data system for Aussie breeders once the first phase of the research had indicated the mode of inheritance. Padgett confirmed that her approach was scientifically sound, and Sharp put the service in place a year or so before she started the *Double Helix Network News*.

Sharp relates with pride that the veterinary ophthalmologist Greg Acland at Cornell told her that the Aussie CEA study provided one of the most impressive data sets on the mode of inheritance of a single-gene trait anywhere in the dog literature. The CEA recessive gene "fact" is stronger in a robust network that includes Rubin, Padgett, Acland, and Sharp's expert lay practices. This is no surprise to a reader of *Science in Action* (Latour 1987). This is the stuff of objectivity as a precious, situated achievement (Haraway 1988). This is also the stuff of "science for the people"—and for the dogs. Mendelian genetics is hardly a new science at the beginning of the twenty-first century, but sustaining and extending its knowledge-production apparatus still takes work.

But making the fact hold inside official science was not enough. Inside the Aussie breed communities is an equally crucial location for this fact to become real, and so potentially effective. Denial here takes a form different from that in the scientific communities, and so the material-semiotic rhetorics for persuading the fact into hard reality have to be different. Sharp's practices in the *DHNN* are part of the picture. While she set up her pedigree analysis service, a group of committed breeders in Northern California took an extraordinary step. They developed a test breeding program and forms to document the breedings. Most important, they went public with the results. According to Sharp, "As a group, they purchased a full-page ad in the breed magazine admitting they had produced CEA and listing the names of their carrier dogs. In a subsequent ad they told about the test-breeding they had done to clear their related stock" (Sharp 1998). Their group action forestalled the kind of attack that had been made on the donor of the first pair of affected puppies given to the genetics committee. This time, the grumblers were relegated to the underground, and the test breeders reshaped the explicit community standard of practice. The standard might not always be followed, but the reversal of what is secret and what is public in principle was achieved.

One final bit helped stabilize CEA as a fact in the Aussie world: emotional support for people who find the disease in their lines. Dog people tend to see any "defect" in their dogs as a "defect" in themselves. This kind of matter is fundamental to the apparatus of situated medical knowledges in both inter- and intraspecies contexts, if usually skirted in orthodox accounts of the care and feeding of biomedical facts. Genetic disease is stigmatizing to the flesh and the soul. Dogs and people are companions in that drama. Sharp could not be the emotional support person in the Aussie genetic disease world. "When people call me about genetic problems in their Aussies, I'm the 'expert,' not a kindred spirit," she says (Sharp 1998). Thus, Sharp asked the Northern Californians who went public with their dogs' and their own names to function as a support group, which she referred to quite literally as grieving breeders.[11] Biosociality is everywhere.

In 1999 Sharp received far fewer reports of CEA in Aussies than she had seven or eight years before. Getting puppies checked through CERF is now standard ethical practice, and serious breeders do not breed affected dogs. Puppy buyers from such breeders get a copy of the CERF report along with their new dog, as well as strict instructions about checking the eyes of breeding stock annually if the new pup does not come with a spay-or-neuter contract. Facts matter.[12]

The world of disease-linked genes is, however, only one component of the story of dog genetics, especially in this era of biodiversity discourse. No matter how extensive the DNA-testing apparatus becomes, or how full the computerized and internationally available Ralston Purina genetic family registry gets, or how successful the canine genetic mapping projects are, or how effective action is to keep crucial genetic markers in the public domain, or how earnest breeders get about open inherited-disease registries and carefully chosen matings, disease-related genes are not the right port of entry to a universe of consequential facts for dog people practicing love of the breed. Enhancing and preserving genetic diversity is not the same thing as avoiding and reducing genetically linked illness. The discourses touch in many places, but their divergences are reshaping the intellectual and moral worlds of many dog people. Sharp's story is again instructive.

Sharp was a subscriber to an Internet discussion group called K9GENES. On that list, the population geneticist and rare-dog-breed activist Dr. Robert Jay Russell, president of the Coton de Tulear Club of America, criticized breeding practices that reduce genetic diversity in dog breeds and the AKC structure that keeps such practices in place, whether or not the kennel club funds genetic disease research and mandates DNA-based parentage testing. Russell's controversial postings were blocked from the list several times, prompting him to log on under a different e-mail account and reveal the censorship.

These events led to the founding in 1997 of the canine diversity genetics discussion group CANGEN, moderated by Dr. John Armstrong at the University of Ottawa, to allow free genetics discussion among breeders and scientists. Armstrong also maintains the Canine Diversity Project web site, where one can get an elementary education in population genetics, read about conservation projects for endangered canids, consider activist positions on dog breeding operating outside the kennel clubs, and follow links to related matters. Concepts like effective population size, genetic drift, and loss of genetic diversity structure the moral, emotional, and intellectual terrain. CANGEN is an impressive site, one where it is possible both to observe and interact with other dog people learning how to alter their thinking—and possibly their actions—in response to each other. The list started with thirty members, and Armstrong expected it to reach one hundred. Taxing its computer resources at the University of Ottawa, in spring 2000 CANGEN had three hundred subscribers. Acrimonious controversies have surfaced in the discussion group, and some participants complain that threads get ignored. Breeders periodically express a sense that they are treated disrespectfully by some scientists (not to mention vice versa), though of course breeders and scientists are neither exhaustive nor mutually exclusive categories on CANGEN. Subscribers, scientists or not, occasionally leave the list in a huff or in frustration. A few dogmatists dedicated to the Truth as revealed to themselves cut a wide swath from time to time. All that said, in my opinion CANGEN remains a rich site of discussion among diverse actors.

Sharp welcomed the higher level of scientific discourse and the emphasis on evolutionary population genetics on CANGEN. She felt challenged by the statistical arguments and wanted to explore the practical consequences for the kind of breeding advice she gives in the *DHNN*. After the summer 1998 issue, the newsletter shifted direction. Sharp began with an article explaining the doleful effects of the "popular sire syndrome" on genetic diversity and made clear that line breeding is a form of inbreeding. In fall 1998, she explored how severe selection against disease-linked genes can worsen the problem of the loss of genetic diversity in a closed population. She cited with approval the success of the Basenji club in getting AKC approval for importing African-born dogs outside the studbook, a daunting endeavor given AKC resistance.

Sharp's winter 1999 *DHNN* feature article was introduced by a quotation from a fellow CANGEN member who has been especially outspoken, Dr. Hellmuth Wachtel, Free Collaborator of the Australian Kennel Club and member of the Scientific Council of the Vienna Schenbrunn Zoo. Sharp explained genetic load, lethal equivalents, population bottlenecks, genetic drift, coefficients of inbreeding, and fragmented gene pools. In the spring 1999 *DHNN,* Sharp published "Speaking Heresy: A Dispassionate Consideration of Outcrossing"—an article she expected to make "the excretory material hit the circulatory apparatus." Love of the breed is messy.

The new genetics is not an abstraction in dog worlds, whether one considers the politics of owning microsatellite genetic markers, the details of a commercial gene test, the problem of funding research, the competing narratives of origin and behavior, the pain of watching a dog suffer genetic illness, the personally felt controversies in dog clubs over breeding practices, or the crosscutting social worlds that tie different kinds of expertise together. When I asked Sharp what she thought breeders, geneticists, dog magazine writers, and others might be learning from each other on CANGEN or other places, she zeroed in on the rapid and deep transformations in genetics over the last decades. Her growth in genetic knowledge, she suggested, including her ability to handle the whole apparatus of molecular genetics, was natural and continuous—until she got on CANGEN. "The only epiphany sort of thing I've been through was when I got on CANGEN and started reading all the posts from the professionals. . . . I knew there were problems with inbreeding, but I didn't have a grasp about what the whole problem was until I started learning about population genetics." At that point, the analogies with wildlife conservation and biodiversity loss hit home, and she made the connection between her dog work and her volunteering as a docent at her local zoo. Citizenship across species ties many knots.

. . .

The epiphany for me in my shaggy dog story about webs of action is that anthropology in the age of genetics is about an old symbiosis—among knowledge, love, and responsibility. Like the story of human genetics analyzed by M. Susan Lindee (see chapter 2, this volume), dog genetics is a social network as much as a biotechnical one. Neither microsatellite markers nor thirty-generation pedigrees fall from the sky: they are the fruit of historically located, naturalcultural work. Breed standards, dog genomes, and canine populations are material-semiotic objects that shape lives across species in historically specific ways. This essay has asked how heterogeneous sorts of expertise and caring are required to craft and sustain scientific knowledge. The story of C. A. Sharp navigates the linkages of lay and professional work. Genetic flows in dogs and humans have implications for meanings of species and race; origin stories remain potent in scientific culture; and molecular high technology can be mobilized to sustain ideas of diversity and conservation, while mutations in cold war politics make tame Russian foxes speak to Anglo breed club dogs. Internet sociality shapes alliances and controversies in dog worlds, and popular and commercial practices infuse technical and professional worlds and vice versa. Dogs appear more than once in this essay as lively actors.

None of this is breaking news in science studies, but all of it holds my attention as a scholar and a dog person. Interested in the symbioses of companion species of both organic and inorganic kinds, I end with fusions. The passage of the leash law enclosed the commons of my childhood dog-human world. The proprietary regimes and DNA-testing surveillance mechanisms at the turn of the millennium map and enclose the commons of the genome and mandate new kinds of relations among breeders, researchers, and dogs. Local and global crises of the depletion of cultural and biological diversity lead to novel kinds of enclosure of lands and bodies in zoos, museums, parks, and nations. No wonder that I am looking in the story of dogs and people for another sense of a common life and future. And so this essay ends where the sticky threads of DNA wind into the frayed planetary fibers of human and nonhuman naturalcultural diversity so crucial to cross-species cohabitation.

NOTES

Special thanks to Susan Caudill, Angela Davis, Sarah Franklin, Val Hartouni, Nancy Hartsock, Rusten Hogness, Gary Lease, Karen McNally, C. A. Sharp, and Linda Weisser—plus General Spots, Alexander, Sojourner, Roland, Cayenne, Hieronymous, Willem, and Bubbles.

1. There are several ways to measure genetic distance, and the precise measure chosen can affect the result. However, for humans, all the measures agree. Dogs are less studied: I have not found comparative data on genetic distances separating dog populations around the world. In chapter 12 of this volume, the human geneticist Alan Templeton uses *genetic distance* to mean "the extent of genetic differentiation between two populations in terms of the alleles that are unique to each population and the extent to which shared alleles have different frequencies" (237). Referring to people, Templeton stresses, "Indeed it is hard to find any widespread species that shows so little genetic differentiation among its populations as humans" (238–39). I suspect he would find humanity's peer in dogs. Coppinger and Schneider summarize data from dog molecular genetics to conclude that "there is less mtDNA difference between dogs, wolves, and coyotes than there is between the various ethnic groups of human beings" (1995: 33). Vilá and colleagues (1997, 1999) find no way to separate dog breeds from each other with mtDNA data. These authors provide a context of species biology for debates about genetic diversity and the genetic basis of breed behavioral and structural specialization. Dogs have long been subject to systematic selection by people for specialized behavioral and morphological features; there is no parallel to this kind of selection in *Homo sapiens*' evolutionary history. Thus, we should expect genetic specialization in dogs for behavior, even in the context of a trellis shape to their population genetic history. Still, dogs, not to mention people, likely remain potentially generalists to a high degree. This point is worth remembering in the face of the differences in appearance (height, shape, etc.) that dogs and people show as individuals and as populations.

2. Jeffrey Bragg (1996) argues that the registered Siberian Huskies of Canada and the United States are in this dangerous genetic condition. The solution of importing unregistered dogs into breeding programs from the "landraces" of Russian Siberian Huskies has met fierce resistance from the Canadian Kennel Club. A resident of the Yukon Territory, and chair of the Working Canine Association of Canada, the controversial Bragg writes about Siberian Huskies and breeds Seppala Siberian Sled Dogs. See the Canine Diversity Project (http://www.magma.ca/~kaitlin/diverse.html; accessed on 10 December 2002). Not all breeds are in such straits. Breeds can have significant numbers of founder animals and effective population

sizes, as well as breeders who emphasize moderation and versatility in the interpretation of the standard and in judging both conformation and performance. Even so, the argument that all registered breeds continue to suffer genetic damage built into their apparatus of naturalcultural production is gaining a hearing among dog people. But few dog people are fluent in the discourses of population genetics and biodiversity. Asking how and if these people become influential in their breed clubs, and how and if they change their own and others' breeding and purchasing practices—and make demands on veterinarians, biotechnical food and pharmaceutical companies, kennel clubs, university research apparatuses, funding organizations, and geneticists—is central to my research.

3. *The Bark,* no. 7 (1999). http://www.thebark.com (accessed on 10 December 2002).

4. http://www.users.aol.com/nbagility (accessed on 30 December 1999).

5. http://www.k9web.com/dog-faqs/lists/email-list.html (accessed on 10 December 2002).

6. Like much in the former U.S.S.R., this trickster drama of worker safety, industrial efficiency, and evolutionary theory and genetics in the far north devolved in the post–cold war economic order. Since the salaries of the scientists at the Genetics Institute have not been paid, much of the breeding stock of tame foxes has been destroyed. The scientists scramble to save the rest—and fund their research—by marketing them in the West as pets with characteristics between dogs and cats. A sad irony is that if the geneticists and their foxes succeed in surviving in this enterprise culture, the population of remaining animals bred for the international pet trade will have been genetically depleted by the slaughter necessitated by the rigors of post-Soviet capitalism and commercializing the animals not for fur coats but as pets.

7. Australian Dingoes and New Guinea Singing Dogs are another matter, not discussed here.

8. Thanks to Sharp for an interview, Fresno, California, 14 March 1999, and for permission to quote. Unless otherwise indicated, quotes in this section are from this interview.

9. For principles of test breeding and CEA pedigree analysis, see *DHNN,* summer and spring 1993.

10. CEA can have other modes of inheritance, and its mode of inheritance is unknown in several breeds in which the symptoms occur. An apparently similar condition does not necessarily relate to the same alleles or even loci in different breeds (or mixes). Sharp is attempting to get Aussie people to cooperate with Greg Acland at Cornell and OptiGen in his effort to develop breed-specific CEA gene tests. Collies and Border Collies will soon have their test, but Aussies will be left without if inaction persists. Similarly, VetGen is attempting to develop breed-specific DNA tests for certain kinds of epilepsy, and Sharp's efforts to get Aussie people to open up about their epileptic dogs and provide pedigrees and samples for research have not been successful. Her files on epilepsy in Aussies grow thick, but this breed's culture does not seem ready for an open registry and activist research on the problem. VetGen will discontinue its work on epilepsy in Aussies unless data are forthcoming. Lay cooperation here is a fundamental part of scientific knowledge production. As of October 2000, such cooperation in relation to VetGen's epilepsy study had improved but was still anemic compared to action taken on behalf of other breeds.

11. "The CEA 'support group,' always informal, does not really exist anymore. Over the years, folks have wandered out of the breed or on to other things, but it was helpful at the time." C. A. Sharp, pers. comm., 13 April 1999.

12. With about 1 percent of Aussies affected with CEA, CERF reports indicate that the gene frequency is fairly steady; 10–15 percent of Aussies may be carriers. Sharp, pers. comm., 13 April 1999. Detecting carriers requires a gene test, and research is under way at Cornell, with Greg Acland as principal investigator (*DHNN* 10, no. 4 [fall 2002]).

REFERENCES

Belyaev, D. K. 1969. Domestication of animals. *Science Journal* (U.K.) 5:47–52.

Boitani, L., F. Francisci, P. Ciucci, and G. Andreoli. 1995. Population biology and ecology of feral dogs in central Italy. In *The domestic dog: Its evolution, behaviour, and interactions with people,* ed. J. Serpell, 217–44. Cambridge: Cambridge University Press.

Bragg, J. 1996. Purebred dogs into the twenty-first century: Achieving genetic health for our dogs. Canine Genetic Diversity Web Site, http://www.magma.ca/~kaitlin/diverse.html.

Browne, M. W. 1999. New breed of fox as tame as a pussycat. *New York Times,* 30 March, D3.

Cargill, J., and S. Thorpe-Vargas. 1996. A genetic primer for breeders. *DogWorld* 81, no. 5.

———. 1998. Devising a genetics game plan. *DogWorld* 83, no. 9:20–24.

———. 2000. Seeing double: The future of canine cloning. *DogWorld* 85, no. 3:20–26.

Coile, D. C. 1997. Tipping the genetic scales. *DogWorld* 82, no. 10:40–45.

Coppinger, R., and R. Schneider. 1995. Evolution of working dogs. In *The domestic dog: Its evolution, behaviour, and interactions with people,* ed. J. Serpell, 21–47. Cambridge: Cambridge University Press.

Dodman, N. 1996. *The dog who loved too much.* New York: Bantam Books.

Haraway, D. 1988. Situated knowledges. *Feminist Studies* 14, no. 3:575–99.

Latour, B. 1987. *Science in action.* Cambridge: Harvard University Press.

Padgett, G. A. 1996–97. Canine genetic disease: Is the situation changing? Pts. 1–4. *DogWorld,* 81, no. 12:44–47; 82, no. 1:26–29; 82, no. 3:24–26; 82, no. 4:36–39.

———. 1998. *Control of canine genetic diseases.* New York: Howell.

Rubin, L. F. 1989. *Inherited eye diseases in purebred dogs.* Baltimore: Williams and Wilkins.

Rubin, L. F., B. Nelson, and C. A. Sharp. 1991. Collie eye anomaly in Australian shepherd dogs. *Progress in Veterinary and Comparative Ophthalmology* 1, no. 2:105–8.

Scott, J. P., and J. H. Fuller. 1965. *Genetics and the social behavior of the dog.* Chicago: University of Chicago Press.

Serpell, J., ed. 1995. *The domestic dog: Its evolution, behaviour, and interactions with people.* Cambridge: Cambridge University Press.

Sharp, C. A. 1993 ff. *Double Helix Network News.* E-mail: helixqnis.net.

———. 1998. CEA and I. Canine Diversity Project, http://www.magma.ca/~kaitlin/diverse.html.

Trut, L. N. 1999. Early canid domestication: The fox-farm experiment. *American Scientist* 87:160–69.

Vilá, C., J. E. Maldonado, and R. K. Wayne. 1999. Phylogenetic relationships, evolution, and genetic diversity of the domestic dog. *Journal of Heredity* 90, no. 1:71–78.

Vilá, C., P. Savolainen, J. E. Maldonado, I. R. Amorim, J. E. Rice, R. L. Honeycutt, K. A. Crandall, J. Lundeberg, and R. K. Wayne. 1997. Multiple and ancient origins of the domestic dog. *Science* 276:1687–89.

Wayne, R. K. 1993. Molecular evolution of the dog family. *Trends in Genetics* 9:218–24.

Weidensaul, S. 1999. Tracking America's first dogs. *Smithsonian* 30, no. 3 (March): 44–57.

Chapter 7

98% Chimpanzee and 35% Daffodil

The Human Genome in Evolutionary and Cultural Context

Jonathan Marks

One of the most overexposed factoids in modern science is our genetic similarity to the African apes, the chimpanzees and gorillas. It bears the precision of modern technology; it carries the air of philosophical relevance. It reinforces the cultural knowledge that genetics reveals deep truths about the human condition, that we are but a half step from the beasts in our nature.

But how do we know just how genetically similar we are to them? What is that estimate based on? What real significance does it have for our conceptions of ourselves in the modern world and for the role of genetic knowledge in shaping those conceptions? This is where genetics and anthropology converge, the gray zone of "molecular anthropology," technologically molecular and intellectually anthropological, in principle at least.

I attempt in this essay to do something that is classically anthropology. I take a well-known natural fact and show it to be a construction of the social and cultural order and, in that capacity, in need of deconstruction.

HISTORY

Our biological similarity to the apes was known long before there were geneticists. To eighteenth-century scholars, apes had roughly the same status as Bigfoot does today: they lived in remote areas and were seen only by untrained observers. Consequently, reports about them differed widely in quality and reliability.

These creatures were situated on the boundary between personhood and animalhood and, as a result, were immensely interesting. That boundary is of course the domain of powerful mythological motifs in all cultures, for the distinction between person and animal allows us to situate ourselves in the natural order, to make some sense of our place in it. And the mythology is just as powerful in the scientific culture, as the scientific literature will easily attest: these creatures are both "us" and "not-us," and *we need to know what they really are* (Corbey and Theunissen 1995; Haraway 1989).

The maturation of biological systematics—how we formally organize and partition nature—came through the work of a Swedish botanist and physician, Carl Linnaeus. Biologists as far back as Aristotle had classified animals; but Linnaeus succeeded in imposing regularity and rigor on the process. An inspiring teacher and prodigious writer, Linnaeus supervised more than 180 doctoral theses during his academic career and took an active role in writing them as well. His students and colleagues sent him plant specimens from all over the world, and Linnaeus fit them all into the system of nature. It was said that God created, but Linnaeus arranged.

His most famous work, *System of Nature,* went through twelve editions as the authoritative guide to the arrangement of plants, animals, and minerals ordained by God and discerned by the author. This work constitutes the bulk of his legacy to the history of biology, and modern biological classification officially dates itself from the tenth edition (1758).

What sense did Linnaeus make of the relationship between people and apes? The father of zoological classification was so confused that he simply divided reported apes into two sets, the more anthropomorphic and less anthropomorphic. He designated the former as a second species of humans (*Homo troglodytes,* which he also called *Homo nocturnus*—nocturnal, cave-dwelling man) and placed the latter in another primate genus *(Simia satyrus).* Paradoxically, not only were the apes simultaneously very much like us and very much not like us, but they were now formally both human and nonhuman at the same time (Bendyshe 1865).

Biologists since the mid–eighteenth century have sought to highlight one or the other side of this paradox. Thus, we follow Linnaeus today in classifying humans as "merely" another species of primate, specifically merely another species of apelike creature. The official version follows paleontologist George Gaylord Simpson's 1945 monograph on classifying mammals, placing us in the superfamily Hominoidea, along with the lesser apes (gibbons) and great apes (chimpanzees, gorillas, and orangutans).

Alternatively, we could choose to emphasize the Otherness of humans by dividing the primate group fundamentally into Quadrumana and Bimana, or two-handed and four-handed, as was popular among Linnaeus's close intellectual descendants. Primates, being generally arboreal creatures, are different from other mammals in the anatomy of their hands and feet, which are grasping structures and lack the claws that other arboreal mammals use. Humans, of course, retain this heritage in their hands, but not in their feet: we are not gracefully four-handed, but pitiably two-handed. This classification would acknowledge the unique aspects of the human feet, which are specialized unlike those of any other primate and have lost the ability to grasp, except in a very rudimentary manner. It would not deny the common ancestry of humans and apes, but would merely highlight the divergence of humans from their ape ancestry.

We could even distinguish humans from other multicellular life altogether, as the subkingdom Psychozoa (mental life), which the zoologist Julian Huxley proposed in the 1950s. A species that lives completely by its wits—that is to say, one that relies entirely on the technological products of its societies for individual survival—is quite different from other life on earth. Perhaps that is worth acknowledging zoologically (Huxley 1957).

Humans are marked by a large number of physical, ecological, mental, and social distinctions from other life. These are not necessarily improvements, of course—we have no more objective way to evaluate "improvement" in the natural world than we do in the cultural world—but merely differences. Other primate species spend time on two legs, and other vertebrate species are bipedal (birds and kangaroos come readily to mind), but not in the same manner as humans. Other species communicate, but not via the absurdly arbitrary and symbolic media we call language. Other species modify natural objects and use them to aid in feeding (Jane Goodall's observations of chimpanzees stripping twigs and using them to fish for termites are classic), but none relies on its technology to survive as humans do. And in no other species does technology take on an evolutionary trajectory of its own, a result of the social cycle of invention, adoption, spread, and modification. Other species appear to grieve, but none weeps as humans do—and certainly not over imaginary events, like *Les Misérables* or *Love Story.*

What does genetics have to say about all this?

Nothing.

MOLECULAR GENETICS

Sameness/Otherness is a philosophical paradox resolved by argument, not by data. Genetic data tell us precisely what we already know, that humans are both very similar to, and diagnosably different from, the great apes.

But genetics is able to put a number to that similarity. It is not uncommon to encounter the statement that we are something like 98 percent genetically identical to chimpanzees. You can count the number of base differences among the same region of DNA in humans and chimpanzees and gorillas, and add them up. Or you can do the same thing to the products of the DNA, protein structures. All such comparisons invariably yield the result that humans and chimpanzees (and gorillas) are extraordinarily similar. But that was known to Linnaeus without the aid of molecular genetics.

What is new? Just the number.

If you compare a human and a chimpanzee, it is easy to see that they are remarkably similar in body structure. Every bone of the chimpanzee body corresponds almost perfectly to a bone in the human body—but differs ever so slightly and diagnostically, in ways that are generally related to the human habit of walking upright. And if not related to our bipedal habit, any detectable difference is very likely related to either of two other human physical specializations, our reduced front teeth and our enlarged brain.

The problem is simply that it is difficult to say just how similar a particular chimpanzee body part and its human counterpart are, percentagewise. A percentage, after all, is a scalar, one-dimensional measure, while a body part is a three-dimensional entity. Indeed, four-dimensional, if you consider the developmental aspects of growth—as people mature, they do not merely expand, they also change in form.

If you were to compare human and chimpanzee structure to that of, say, a snail, it would be patently obvious that humans and chimps are over 99 percent identical in practically every way. The human and chimp have bones, they lack a shell, they have limbs, they do not leave a trail of slime as they move; every nerve, every sinew, every organ is almost the same in a chimp and human and very different, if present at all, in the snail.

Exactly how similar they are, of course, is elusive. It would be rather old-fashioned and premodern to say, "Humans and chimpanzees are really, really, really similar," even though it is a true statement.

What genetics offers is the opportunity to place a hard number on two-way comparisons, by virtue of the fact that the genetic instructions (and their primary products) are composed of long chains of subunits, differences that can be tabulated and numerically manipulated. Thus, one can look at a short region around one of the genes for hemoglobin, as my colleagues and I did in 1986, and find a DNA sequence that reads GCTGGAGCCTCGGTGGCCAT in a baboon and GCTGGAGACTCGGTGGCCAT in an orangutan (Marks et al. 1986). One difference in twenty possibilities; the very linearity of DNA sequences makes them easy to compare. And with a much longer DNA sequence, you would expect to get a more precise estimate. In this case, for example, we found a difference between a baboon and an orangutan of about 5 percent in this small region, and a more general level of difference of about 8 percent.

But the comparison can also be misleading in two important ways.

To begin with, such comparisons of DNA sequence ignore qualitative differences, those of kind rather than amount. To take the smallest case, consider a different sequence of twenty DNA bases from the same region: CCTTGGGCCTCCCGCCAGGC in the baboon and CCTTGGGCTCCCGCCAGGCC in the orangutan. If you look at them in parallel rows you find them to be different:

CCTTGGGCCTCCCGCCAGGC
CCTTGGGCTCCCGCCAGGCC

But if you look more carefully, you might observe that the general gestalt of the sequences is roughly the same. If there is one *C* too many in the middle of the top sequence, or one too few in the bottom, the match becomes far more complete. If we look at it again, inserting a small gap for one base too many or too few, we see the similarity that was previously hidden from view:

```
CCTTGGGCCTCCCGCCAGGC
CCTTGGGC TCCCGCCAGGCC
```

So we may infer an insertion in one lineage or a deletion in the other, in order to make this sequence look maximally similar. But how can we know for sure? This is a very different kind of inference from the DNA base substitutions we were tabulating a few paragraphs ago. This involved ignoring the actual census of differences in favor of a gestalt similarity and then retabulating the differences—a highly subjective procedure, although probably right.

So we have overridden the *observation* of seven differences in twenty with the *inference* of one difference, of a different sort, in twenty. Though again, this is not to say that it is illegitimate. The question is, what does it do to the number, our precise estimate of the degree of genetic difference? First, it inserts an element of subjectivity masked by the number itself; and second, it sums together DNA base substitutions and DNA base deletions, as if they were biochemically identical and quantitatively equivalent. In fact they are neither; this is a classic case of apples and oranges.

And that was an easy one.

Actually, the molecular apparatus has complex ways of generating insertions and deletions in DNA, which we are only beginning to understand. For example, a stretch of DNA from a ribosomal RNA gene is forty bases long in humans and fifty-four bases long in orangutans. The sequences on either side match up perfectly. How do we know what bases correspond between the two species, how do we decide how many substitutions have occurred, when obviously some have been inserted and deleted as well? The authors of the original study inferred five gaps and six base substitutions (Gonzalez et al. 1990), but it could just as easily be two gaps and nine substitutions or five gaps and three substitutions (Marks 2002). While we might, by Occam's razor, choose the alignment with the smallest numbers of mutational events, we still have to decide whether a gap "equals" a substitution, or whether a gap should be considered rarer and, therefore, worth, say, five substitutions.

```
Human       CCTCCGCCGCGCCG    CTCCGC GCCGCCGGGCA          CGGCC          CCGC
Orangutan   CC            GTCGCCTCCGCCACGCCGCGCCACCGGGCCGGGCCGGCCCGGCCCGCCCCGC

Human            CCTCCGCCGCGCCGCT        CCGCGCCGCCGGGCACGGCCCCGC
Orangutan   CCGTCGCCTCCGCCACGCCGCGCCACCGGGCCGGGCCGGCCCGGCCCGCCCCGC
```

Human CCTCCGCCGCGCCG CTCCGCGCCGCCGGG CAC GGCC CCGC

Orangutan CCGTCGCCTCCGCCACGCCGCGCCACCGGGCCGGGCCGGCCCGGCCCGCCCCGC

The problem is that we cannot tell which DNA sequence alignment is right, and the one we choose will contain implicit information about what evolutionary events have occurred, which will in turn affect the amount of similarity we tally (Mindell 1991). How similar is this stretch of DNA between human and orangutan? There may be eight differences or eleven differences, depending on how we decide the bases correspond to each other across the species—and that is, of course, assuming that a one-base gap is also equivalent to a five-base gap and to a base substitution. This is the fundamental problem of homology in biology: What is the precisely corresponding entity in the other species?

In a more general sense, however, the problem of taking quantitative estimates of difference between entities that differ in quality is prevalent throughout the genetic comparison of human and ape. The comparison of DNA sequences presupposes that there are homologous sequences in both species, which of course there must be if such a comparison is actually being undertaken. But other measurements have shown that a chimpanzee cell has 10 percent more DNA than a human cell (Pellicciari et al. 1982). (This does not mean anything functionally, since most DNA is functionless.) How does one work that information into the comparison or into the 98 percent similarity?

In the example above, the problem began with the assumption that forty DNA bases of human sequences were homologous to fifty-four DNA bases of orangutan sequence. A simple estimate of similarity and difference necessarily must be confounded by variation in size of the entities being compared.

If we compare the genes for α-globin on chromosome 16 between human and chimpanzee (half of hemoglobin, the molecule that transports oxygen and carbon dioxide in our blood), we find a near identity of the base sequences. But we also find that, with rare exceptions, humans have two copies of the α-hemoglobin gene, aligned in tandem; chimpanzees, however, have three (Zimmer et al. 1980). Or, if we look at the genes that code for the Rh blood group, again the nucleotides of the genes match up almost perfectly between the two species, but humans have two such genes and chimpanzees have six (Westhoff and Wylie 1998). How can we make simple numerical sense of that?

An odd recognition of the fact that mutational modes are complex has recently occurred. Tabulating both nucleotide substitutions and insertions/deletions, researchers have found the chimpanzee and human genomes *not* to be over 98 percent identical, but closer to 95 percent identical (Britten 2002). The problem, however, is not that the two genomes are "only" 95 percent identical, but that any tabulation of the precise amount of identity is forced to shoehorn the results of several different mutational processes into its grand tally. Neither number has the force of accuracy, because the precise number obtained depends on what one recognizes as a meaningful difference (only nucleotide substitutions, or the genomic fruit salad of changes?), how one counts it (is a three-hundred-base insertion three hundred differences or only one?), and whether there is any scientific value at all in trying to derive an official amount of genetic difference between the two species' genomes in the first place when the official amount necessarily combines differences of quantity and quality.

The second misleading area of DNA sequence comparisons entails a consideration of the other end of the scale. The structure of DNA, the famous double helix, is built up of four simple subunits. Each of our reproductive cells has a length of DNA encompassing approximately 3.2 billion of these subunits, but there are still only four of them: adenine, guanine, cytosine, and thymine, or A, G, C, and T. This creates a statistical oddity. Since genetic information is composed of DNA sequences, and there are only four elements to each DNA sequence, it follows that two DNA sequences can differ, on the average, by no more than 25 percent. Certainly a very small stretch of DNA might be zero percent similar to another very small stretch (AAAAT matches GGCCG nowhere, after all), but *on the average, two random stretches of DNA will be statistically obliged to match at one out of four places.*

In other words, two stretches of DNA generated completely at random, completely independently of one another, would not be zero percent similar, but rather, would be 25 percent similar.

But what would constitute a comparison of two DNA segments that emerged completely independently of one another? When we compare DNA sequences, of course, we are comparing corresponding DNA sequences, such as the gene that codes for the electron-transport protein cytochrome c. Such correspondence, or homology, in biology is a reflection of common descent. A human and a chimpanzee have similar genes for cytochrome c because they are descended from a common ancestor that had a gene for cytochrome c similar to both.

In fact, so do humans and fruit flies. Their DNA sequences did not emerge independently of one another but are products of the divergent histories of lineages that became separated some hundreds of millions of years ago.

In fact, so do humans and daffodils. Their DNA sequences are not independent of one another, either. The only such DNA sequences would be those that result from independent origins of DNA-based life; and then we still expect them to be 25 percent identical, by virtue of the way in which DNA is constructed.

Thus, if one compares the DNA of a human and a daffodil, the 25 percent mark is actually the zero mark, and since humans and daffodils do share a common ancestry, one would expect them to be generally more similar than 25 percent—say about 35 percent.

In the context of a 35 percent similarity to a daffodil, the 98 percent similarity of the DNA of human to chimp does not seem so remarkable. After all, humans are obviously far more similar to chimpanzees than to daffodils.

But more than that, to say humans are about one-third daffodil is more ludicrous than profound. There are hardly any similarities one can identify between a daffodil and a human being. DNA comparisons thus overestimate similarity at the low end of the scale (because 25 percent is actually the zero mark of a DNA comparison) and underestimate comparisons at the high end. At least snails, in the previous anatomical comparison, move around and eat; they cannot photosynthesize; they are much more similar to us than daffodils are. So from the standpoint of a daffodil, humans and chimpanzees are not *even* 98 percent identical: they are probably 100 percent identical. The only difference between them is that the chimpanzee would probably be the one *eating* the daffodil.

The problem is that, in being told about these data without a context in which to interpret them, we are left to our own cultural devices to impart meaning to them. Here, we generally are expected to infer that genetic comparisons reflect deep biological structure, and that 98 percent correspondence is an overwhelming similarity. Thus, "the DNA of a human is 98 percent identical to the DNA of a chimpanzee" becomes casually interpreted as: "Deep down inside, humans are overwhelmingly chimpanzees. Like 98 percent chimpanzee."

We do not really know precisely how similar the DNA of chimps and humans is, except that they are very similar, as are their bodies. Genetics appropriates that discovery as a triumph because it can place a number on it, but the number is actually rather unreliable. And whatever the number is, it should not be any more impressive than the anatomical similarity; all we need to do is put that old-fashioned comparison into a zoological context.

The paradox is not that we are so genetically similar to the chimpanzee; the paradox is that we presently find the genetic similarity to be so much more striking than the anatomical similarity.

THE NATURE OF THE COMPARISON

Where does such a number, ostensibly representing our basic similarity to another species, come from? Three sorts of data have produced these tabulations. The first is a comparison of protein structure from 1975 by Berkeley geneticists Mary-Claire King and Allan Wilson (1975). Proteins are long strands of elementary subunits that can be compared, and that are reflections of gene structure. Examining the differences between forty-four known proteins possessed by humans and chimpanzees, King and Wilson found they were 99.3 percent identical.

We know that only about 1 percent of the total DNA of a cell is actually expressed as proteins, and that protein-coding regions are the most slowly changing parts of the DNA. The reason is that mutations or changes to functional DNA are more likely to do systemic damage to the organism than are mutations to nonfunctional DNA. Consequently such mutations are most unlikely to be perpetuated in the species, as their bearers fail to thrive and reproduce as efficiently as the other creatures who lack the damaging mutations. Mutations to nonfunctional DNA occur at the same rate; but because they are not expressed, they do their bearers no harm. One would expect that DNA fraction to differ more widely across species. Consequently, the King and Wilson estimate of 99.3 percent is an overestimate of the similarity of human and chimpanzee DNA, being derived from a skewed sample of the DNA reflecting only protein-coding regions.

The second class of data was made available in the mid-1980s with the development of direct DNA sequencing technology. DNA is a long, linear molecule, again composed of simple subunits that can be compared. Bits and pieces of it have been compared between human and chimpanzee, amounting to less than 100,000 DNA bases in length. But there are 3.2 billion bases in a human genome, so obviously we have here a minuscule proportion and, again, one biased toward regions that contain functional units, genes—which tend to be where people look for DNA to sequence.

The most comprehensive comparison we have is actually infinitesimal in scope, about forty thousand bases of the region of the β-hemoglobin genes on chromosome 11. And we find human and chimpanzee, base for base, to be about 1.9 percent different (Bailey et al. 1992). The difference between the .7 percent estimate and the 1.9 percent estimate is a consequence of the fact that this DNA comparison includes much more nongenic DNA than genic DNA (which is the only class that would be translated into protein differences). Indeed, the DNA sequences of genes included in that region are virtually identical between the two species.

But again, that is a high estimate, because it focuses on a region known to contain genes and is therefore more conservative than we should expect the overall DNA to be. Genes are rare in the genome. Not only that, but we know the evolution of human hemoglobin has been sensitive to specific environmental problems—such as malaria—and therefore may have its own evolutionary idiosyncrasies.

Another piece of DNA that has been well studied is the mitochondrial DNA, or mtDNA. Whereas 3.2 billion nucleotides form the twenty-three chromosomes of a human nuclear genome (the DNA of a human gamete, half that of an ordinary cell), there is a tiny fraction of DNA located outside the nucleus. The mitochondrion, a subcellular organelle universally labeled in biology textbooks as "the powerhouse of the cell," generates metabolic energy for the physiological processes of life. It also has 16,500 nucleotides of DNA, which code for some of the molecules used by the mitochondrion.

The mitochondrial DNA, however, is not 1 percent different among humans, chimpanzees, and gorillas. It is 9 percent different (Arnason et al. 1996). Why? Because mtDNA mutates at a much higher rate than nuclear DNA. The mutations have little or no effect on the life of the organism, and so the differences simply accumulate through time. Two organisms, therefore, will be far more similar in their nuclear DNA than in their mtDNA. For example, the "mitochondrial Eve" work was based on tabulating the differences detectable among human beings (.2 percent difference), whose nuclear DNAs are so similar as to preclude that kind of study.

The point is an important one: Different bits of DNA evolve at their own rates, and therefore—as a result of the 6–7 million years of evolutionary time separating humans, chimps, and gorillas—some bits are 10 percent different, some are not comparable because they are different in kind rather than amount, and most are less than 3 percent different.

DNA hybridization was a technique popular in the 1980s. It promised a mass genome comparison, rather than a high-resolution snapshot of a single region (Sibley and Ahlquist 1984). The technique began, however, by discarding the half of the genome that was most redundant and, presumably, least full of genes. What remained was bonded to the DNA of a different species, and the temperature at which the strands of this "hybrid DNA" came apart was estimated. Since hybrid DNA is held together by fewer bonds than native DNA, the difference in thermal stability might be a crude indicator of how many DNA mutations differentiate the genes of the two species. Assuming (conveniently) that 1 degree of difference in thermal stability between human-chimp DNA hybrids and human-human DNA equals 1 percent genetic difference, they concluded that humans and chimpanzees are 1.8 percent genetically different. Yet a different study calculated the conversion ratio at 1 degree equals 1.7 percent difference (Caccone et al. 1988), and thus humans and chimps would be over 3 percent different rather than 1.8 percent. And this experiment examines only half the total DNA in the first place. So it actually boils down to a demonstration that half the DNA of humans is either 98.2 percent or 97 percent identical to that of chimpanzees.[1]

More important, DNA hybridization was based on an archaic view of the genome.[2] It underestimated the extent to which serial redundancy pervades the genetic blueprints—even "unique-sequence" genes come in clusters or families of genes structurally similar to one another, the results of ancient "rubber-stamping" duplication events. Thus, when the DNA of one species is hybridized to that of another, a gene can pair with either its perfect partner (ortholog) or a different gene from the same family in the other species (paralog). In the example here, from the genes for hemoglobin, human α-1 globin has an orthologous counterpart in the gorilla and at least six paralogous counterparts. The experiment favors pairing orthologous DNA, but a

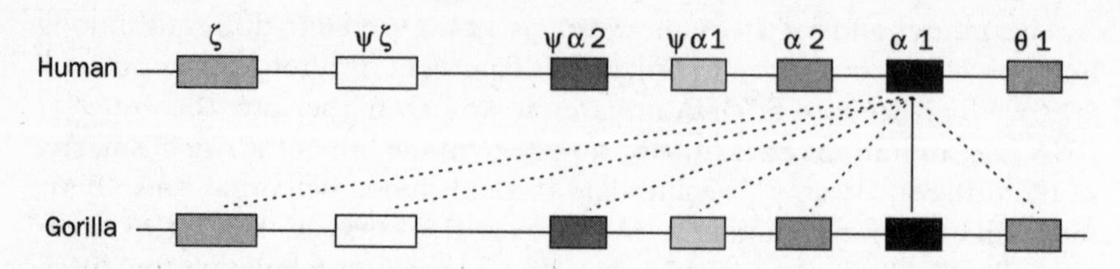

Figure 7.1 Molecular homology. Because of the prevalence of serial duplications of genetic material, a piece of human DNA (here, the α-1 globin gene) has one orthologous counterpart (solid line) and several paralogous counterparts (broken lines) in the gorilla genome.

small proportion of the hybrid DNA mixture invariably will be composed of poorly paired paralogous DNA hybrids. For close relatives, the proportion of this unwanted DNA may be greater than the small difference between the orthologous DNA segments the technique was designed to detect! (See figure 7.1.)

This estimate of genetic similarity along a single dimension consequently misses a significant component of difference—arguably the most important discovery in evolutionary biology in the last few decades—genome complexity. It is not that the human genome is more complicated than the chimpanzee's, but merely that the evolutionary processes operating to differentiate them from one another are far more diverse and extensive than mere nucleotide substitution, which leads to base-pair mismatch, would suggest.

CHROMOSOME STRUCTURE

Yet another kind of comparison involves examining the structure of the chromosomes. Since the primary function of chromosomes is to get the immensely long DNA strands through cell division smoothly by condensing them into a manageable number of structures, it does not really matter whether there are ten, twenty, or fifty of them. Any number in that range will do well. Consequently we find that nearly all mammals fall in the ten-to-fifty range, and in general, that closely related species have similar numbers of chromosomes.

Each of the twenty-three pairs of human chromosomes has a nearly identical counterpart in the apes. Perhaps a half dozen have undergone minor structural changes; but if you can recognize the standard pattern of bands on human chromosomes (known as G-bands), you can recognize chimpanzee chromosomes. It is not known precisely what the bands are, but they can be generated easily in many ways, the most common being to expose the chromosomes to a mild enzyme treatment, which breaks down some of the protein complexes in the chromosome packaging. This permits a stain to bind preferentially to certain areas, creating arrays of alternating stained and unstained regions—bands. And those arrays are nearly identical in humans, chimps, and gorillas (Marks 1983).

The exception is human chromosome 2 (the second largest chromosome, for they are numbered according to size). One can look at the chromosomes of a chimpanzee forever and never see the large pair known as chromosome 2 and found in the human. What one sees, however, is two small pairs of chromosomes in the chimpanzee, which, when joined end to end, produce a dead ringer for human chromosome 2 (IJdo et al. 1991).

Apparently a fusion occurred in the human lineage, creating chromosome 2 and reducing the count from twenty-four pairs to twenty-three pairs. Is that what makes us human? Yes and no. Yes in a narrow, diagnostic sense: given the chromosomes from a cell of any living species, if you find chromosome 2 among them, they are from a human. No in a functional sense: the fusion is not what gives us language or bipedalism or a big brain or art or sugarless bubble gum. It is simply one of those neutral changes lacking outward expression, which is neither good nor bad but merely diagnostic.

The chromosomes of humans and chimpanzees also differ in a subtle but reliable way under a slightly more complex treatment known as C-banding. C-banding seems to mark specifically a few chromosomal zones containing highly redundant "junk" DNA sequences (satellite DNA). In the human the characteristic zones are at the middle, or centromere, of each chromosome; are slightly below the centromere on chromosomes 1, 9, and 16; and make up most of the Y chromosome.

We are the only species with such a pattern. If one looks at the chimpanzee's cells using the identical procedure, one encounters the centromeric bands readily enough, but the marked regions of chromosomes 1, 9, 16, and Y do not contain satellite DNA. Not only that, but something entirely unfamiliar will be evident—bands at the tips of nearly every chromosome and even in the middle of a chromosome arm (figure 7.2). The terminal bands are present in the gorilla's cells as well; the bands have been seen in every chimpanzee and gorilla studied, and in no human or orangutan (Marks 1993).

What is the cause of the bands on the tips? Something trivial: the emergence of yet another functionless class of DNA, which somehow managed to "colonize" the ends of the chromosomes. This class of DNA packs itself more densely into the chromosome than the rest of the DNA does, and it absorbs more stain. If you had a tube of cells and did not know whether they were derived from a human or a chimpanzee, a C-band analysis would tell you immediately.

The story is recursive. We find ourselves to be always genetically very similar to chimpanzees and yet diagnosably different. This is not terribly different from the conclusions we can draw from comparing anything else

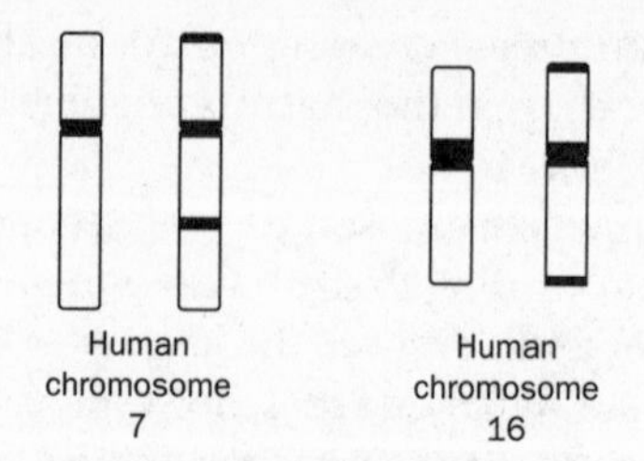

Figure 7.2 Some diagnostic differences between human chromosomes. Chromosome 7 of a human and its chimpanzee counterpart (left pair) can be readily distinguished by the terminal C-band and interstitial C-band in the chimpanzee chromosome. Chromosome 16 (right pair) has a prominent C-band below the centromere in humans, and terminal bands in chimpanzees.

between chimpanzees and humans—hair, organs, skin, muscles, bones. So the paradox of our exceeding genetic similarity seems less paradoxical. We learn that we are similar to, but invariably slightly different from, our closest relatives. It could not really be any other way, given the fact of evolution.

THE CENTRAL FALLACY OF MOLECULAR ANTHROPOLOGY

The argument I criticize here is the one that begins with our unimpeachable genetic similarity to the chimpanzee and concludes that we are therefore "nothing but" chimpanzees genetically. In fact, the data have been around for a surprisingly long time. The fact that it took decades to grasp the simple and direct implication alone might suggest that the argument is specious. By the 1920s, the similar blood reactions of human and ape sera were well-known. And shortly after the celebrated trial of John T. Scopes (convicted of teaching evolution in Tennessee in 1925), H. L. Mencken's literary magazine ran the article "The Blood of the Primates" (Hussey 1926). It was quite clear about the results: "The sanguinity of the horse and donkey, which are capable of hybridization, is less close than the kinship of Homo sapiens and the anthropoids."

The article, of course, failed to draw the conclusion that we are apes, in spite of the data. That inference would have to wait for the rise of molecular reductionism in the 1960s and expresses what we can call the Central Fallacy of Molecular Anthropology.

Ultimately this fallacy is not a genetic one but a cultural one—our reduction of life to genetics. Geneticists are among its most frequent perpetrators, and of course it is in their interest to perpetuate it. In the 1960s, when it became possible to make direct comparisons between the proteins of different species, the biochemist Emile Zuckerkandl was impressed by the fact that only one difference in the 287 amino acids in (half of) hemoglobin could be found between a human and a gorilla. So impressed, in fact, that he could proclaim that, "from the point of view of hemoglobin structure, it appears that [a] gorilla is just an abnormal human, or man an abnormal gorilla, and the two species form actually one continuous population" (Zuckerkandl 1963).

The claim was nonsense to the distinguished paleobiologist George Gaylord Simpson (1964): if you cannot tell a gorilla from a human, he suggested, you should not be a biologist. If you cannot tell them apart by their hemoglobin, *just look at something else.*

Suggesting that the relationships of our *blood* are the relations of *us,* that we are our blood, is simply a metaphoric statement, actually metonymy, the substitution of a part for the thing itself. Perhaps it is resonant with us because our brain evolved to make metaphoric connections, and because blood is such a symbolically powerful substance. Blood is, after all, a metaphor for heredity.

But it is not literally true that we are our blood; and science is supposed to be about literal truths, not literary ones.

There is one sense in which we can acknowledge that we are apes: phylogenetically. We fall within a group constituted by the great apes—those large-bodied, tailless, flexible-shouldered, slow-maturing primates: the chimpanzee, gorilla, and orangutan. And indeed, we are more closely related to the chimpanzee and gorilla than they are to the orangutan. This implies that the category of great ape is artificial, for it comprises species that are not each others' closest relatives, or rather, that it excludes a member of the group of relatives—namely, us.

We are excluded by virtue of the fact that we have diverged relatively rapidly from the classic ape form and mode of life and have evolved to fill a different niche, that of an ape which is bipedal and culture reliant. So although we fall within the great-ape category genetically and are recently descended from them biohistorically, we are nevertheless different from them by virtue of having evolved a very large number of readily observable specializations or novelties. We have left them in our wake, so to speak. The apes seem to resemble each other more than they resemble us because they did not develop the things we did.

But does this mean we are apes, as the genetic enthusiast Richard Dawkins (1993) has argued? Consider a different group of animals: say, a sparrow, a crocodile, and a turtle. Two are crawling, green-scaled reptiles, and one is a bird. As it happens (like humans and gorillas, relative to orangutans), the sparrow and crocodile are more closely related to each other than either is to the turtle. The two reptiles are more similar to one another, but they are similar simply because the birds, which originated from a group of reptiles, developed a set of specializations and flew away, leaving the green, scaly creatures behind. Since the kind of reptile from which the birds originated was related to the crocodile, it happens that the sparrow and crocodile, a few hundred million years later, are more closely related than either is to the turtle.

But does that mean that the sparrow is a reptile? No, it means it is closely related to reptiles; the reptiles subsume its group. If the word *bird* is to have any meaning, it must mean something different from the word *reptile.* Birds are birds; reptiles are a different kind of group, unified in this classification by *not* having evolved the specializations of birds.

We can go even farther back in biological history. Several hundred million years ago, a group of fish developed specializations of their limbs that enabled some of their descendants to venture out of the sea and onto land. A living representative of that group of teleost fish is the famous "living fossil," the coelacanth. Since all living land vertebrates (or tetrapods) are descended from that particular group of fish, it follows that if you compare a coelacanth, a human, and a tuna, the closest relatives are the coelacanth and the human, not the two fish.

Let us return to the apes, then. The argument is that humans are apes because we belong to the group that produced chimpanzees and orangutans; and because we are more closely related to some apes than those apes are to other apes, we fall within that category. But we also belong, in a larger sense, to the group that produced coelacanths and tuna—namely, fish—and we are more closely related to some fish (coelacanths) than those fish are to other fish (tunas). We fall (by virtue of being tetrapods) within that category as well.

In other words, *we are apes, but only in precisely the same way that we are fish* (figure 7.3).

Doesn't seem quite so profound now, does it?

This is not so much a revelation about our basic natures as a revelation about how we name zoological groups. As constituted by a group of related species that a subset has evolved away from, the species that remain unchanged will constitute a paraphyletic category. They may look similar, but they are defined on the basis of lacking the specializations of the other group—tetrapods in the case of paraphyletic fish, birds in the case of paraphyletic reptiles, and humans in the case of paraphyletic great apes.

What appeared at first to be a revelation about our animal nature is instead a revelation about how we box up nature to make sense of it. Making sense of the world through classification is a fundamentally human act, and each human group does it in a different way and thus imposes a distinctive structure and meaning on the world.

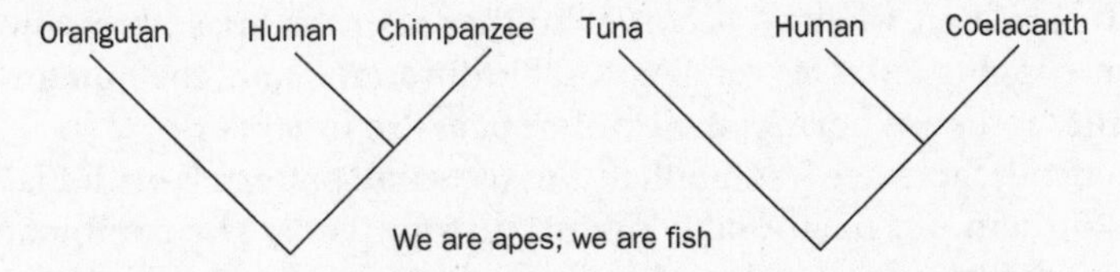

Figure 7.3 The paraphyletic nature of the category "apes" is equivalent to the paraphyletic category "fish" and the position of humans within it. Thus, the phylogenetic argument that "we are apes" because we fall into that category is precisely as valid as the argument that "we are fish."

Modern science classifies animals by two criteria, descent and divergence. This practice sometimes creates confusion, such as the occasional paraphyletic category, in which divergence takes precedence over proximity of descent. A school known as cladism prefers to classify by descent only, and thus would recognize only closest relatives—and would bury the categories great apes, reptiles, and fish. Ultimately this is a philosophical decision, is unresolvable by recourse to data, and displays the variation of ideas about classifying among scientists (Hull 1970). Scientists are a human society, and like other human societies, they make sense of their world by organizing it into groups.

But in the case of apes and humans, this has more to do with the work of Émile Durkheim and Marcel Mauss than with the work of Charles Darwin and Thomas Huxley.

CONCLUSIONS

The similarity revealed by genetics between humans and apes is interesting but not profound. The meaning imparted to that similarity is derived from (1) the nature of DNA and the scalar, decontextualized comparisons it affords, and (2) our unfamiliarity with the nature of genetic comparisons.

What seems like a paradox, the genetic similarity and the anatomical difference, is itself largely a consequence of three hundred years of anatomical study and an order of magnitude less of genetic study. In historical perspective, however, the paradox vanishes. When Edward Tyson published his *Ourang-outang, sive Homo sylvestris, or the Anatomy of a Pygmie* in 1699, he explicitly was struck by its extraordinary *similarity* to the human form. It was precisely that physical correspondence that led scholars of the eighteenth century, notably Lord Monboddo and Jean-Jacques Rousseau, to consider the chimpanzee as merely a different human variety, indeed within our own species. *When anatomical studies of the apes were a novelty, apes were considered to be astonishingly similar to humans.*

Centuries later, we have become familiar with the apes physically. They are exceedingly similar to, yet diagnosably different from, the human. Their similarities to us are boring; their differences are interesting.

Genetically, apes are less familiar, but the same pattern is evident. We are similar to them, yet diagnosably different from them. The combination of the novelty of genetics and its cultural power has permitted the observation that we are similar to the apes to acquire a new vitality, but it most often leads to recapitulating eighteenth-century philosophy. To the writers of a contemporary, popular biology book, half of our species are "demonic males," a factoid "written in the molecular chemistry of DNA" (Wrangham and Peterson 1996: 198), although without any Southern blots or sequence data to substantiate the point.

To another writer, we are a "third chimpanzee" (Diamond 1991), a conclusion easily derived by overstating the similarities between humans and chimpanzees. Indeed, when physical and behavioral comparisons were novel, Lord Monboddo came to a similar conclusion:

> That the Orang Outang is an animal of the human form, inside as well as outside: That he has the human intelligence, as much as can be expected in an animal living without civility or arts: That he has a disposition of mind, mild, docile, and humane: That he has the sentiments and affections peculiar to our species, such as the sense of modesty, of honour, and of justice; and likewise an attachment of love and friendship to one individual, so strong in some instances, that the one friend will not survive the other: That they live in society, and have some arts of life; for they build huts, and use an artificial weapon for attack and defence, viz., a stick; which no animal, merely brute, is known to do. They shew also counsel and design, by carrying off creatures of our species, for certain purposes, and keeping them for years together, without doing them any harm; which no brute creature was ever known to do. They appear likewise to have some kind of civility among them, and to practice certain rites, such as that of burying the dead. (Monboddo 1774: 275)

To a third, our newfound genetic similarity means "we are apes" (Dawkins 1993)—and in spite of the slightly ambiguous referent, this British biologist presumably means "we" *Homo sapiens.*

And to some animal rights activists, the same genetic similarity means apes should be accorded human rights (Cavalieri and Singer 1993), in spite of the facts that (1) they are not human, (2) we cannot even guarantee human rights to humans, and (3) we do not base the allocation of rights on genetic similarity (Marks 1994b).

Examining the empirically well-attested genetic similarity of humans and apes can show somewhat paradoxically that the genetic similarity is a constructed fact. Its meanings are contestable and not "read" from nature. Ultimately this may serve as an example of the reciprocal illumination shed by genetics and anthropology, which may someday result in the development of a truly molecular *anthropology*.

NOTES

The material in this paper is derived from my book *What It Means to Be 98% Chimpanzee* (Berkeley: University of California Press, 2001). My deepest appreciation goes to Alan Goodman and Deborah Heath for the invitation to participate in the symposium, their incisive comments on this paper, and their brilliant editing, as well to the other participants for their helpful input. This paper is dedicated to the memory of Ashley Montagu (1904–1999) and Sherwood Washburn (1912–2000).

1. The issue of "resolution of the trichotomy" (i.e., determination of the closest relatives among humans, chimpanzees, and gorillas) is off the present subject. While some data certainly do associate humans and chimps exclusively of gorillas (notably mtDNA [Gagneaux et al. 1999]), these data are frequently the most problematic, often being based on unarticulated assumptions or containing internal contradictions (Marks 1994a, 1995). Most molecular data actually fail to associate any pair (see Ruano et al. 1992; Johnson and Coppen 1999), and much of the newest data links chimps and gorillas, as per the traditional arrangement (Dangel et al. 1995; Deinard et al. 1998; Djian and Green 1989; Fracasso and Patarnello 1998; Livak et al. 1995; Marks 1993; Meyer et al. 1995; Retief et al. 1993). The genetic linkage of humans specifically to chimpanzees is thus not so much a constructed fact as a false one.

2. Marks et al. 1988; Sarich et al. 1989; Sibley et al. 1990; Marks 1991. Of more esoteric interest is the claim of Caccone and Powell (1989) to have replicated the findings of Sibley and Ahlquist (1984, 1987). They claimed to have obtained the same numbers as those reported by Sibley and Ahlquist and to have obtained the same tree, thus ostensibly validating both the previous study and the technique itself. In fact they matched numbers that the previous authors admitted had been falsified, which is itself quite a feat; and they matched them with a different measure (δT_m versus $\delta T_{50}H$)—as if three inches could match three centimeters. Sibley and colleagues (1990) published their true values comparable to those reported by Caccone and Powell (1989), and these neither match nor resolve the phylogeny. The technique is now in disrepute and rarely used (see Marks 1991 for a lengthy discussion).

REFERENCES

Arnason, U., X. Xu, and A. Gullberg. 1996. Comparison between the complete mitochondrial DNA sequences of Homo and the common chimpanzee based on nonchimeric sequences. *Journal of Molecular Evolution* 42:145–52.

Bailey, W. J., K. Hayasaka, C. G. Skinner, S. Kehoe, L. C. Sieu, J. L. Slightom, and M. Goodman. 1992. Reexamination of the African hominoid trichotomy with additional sequences from the primate beta-globin gene cluster. *Molecular Phylogenetics and Evolution* 1:97–135.

Bendyshe, T. 1865. The history of anthropology. *Memoirs of the Anthropological Society of London* 1:335–458.

Britten, R. J. 2002. Divergence between samples of chimpanzee and human DNA sequences is 5%, counting indels. *Proceedings of the National Academy of Sciences, USA* 99:13633–35.

Caccone, A., R. DeSalle, and J. R. Powell. 1988. Calibration of the change in thermal stability of DNA duplexes and degree of base pair mismatch. *Journal of Molecular Evolution* 27:212–16.

Caccone, A., and J. R. Powell. 1989. DNA divergence among hominoids. *Evolution* 43:925–42.

Cavalieri, P., and P. Singer, eds. 1993. *The Great Ape Project.* New York: St. Martin's Press.

Corbey, R., and B. Theunissen, eds. 1995. *Ape, man, apeman: Changing views since 1600.* Leiden: Department of Prehistory, Leiden University.

Dangel, A., B. Baker, A. Mendoza, and C. Y. Yu. 1995. Complement component C4 gene intron 9 as a phylogenetic marker for primates: Long terminal repeats of the endogenous retrovirus ERV-K(C4) are a molecular clock of evolution. *Immunogenetics* 42:41–52.

Dawkins, R. 1993. Meet my cousin, the chimpanzee. *New Scientist* (5 June): 36–38.

Deinard, A. S., G. Sirugo, and K. K. Kidd. 1998. Hominoid phylogeny: Inferences from a sub-terminal minisatellite analyzed by repeat expansion detection (RED). *Journal of Human Evolution* 35:313–17.

Diamond, J. 1991. *The third chimpanzee.* New York: Random House.

Djian, P., and H. Green. 1989. Vectorial expansion of the involucrin gene and the relatedness of the hominoids. *Proceedings of the National Academy of Sciences, USA* 86:8447–51.

Fracasso, C., and T. Patarnello. 1998. Evolution of the dystrophin muscular promoter and 5' flanking region. *Journal of Molecular Evolution* 46:168–79.

Gagneaux, P., C. Wills, U. Gerloff, D. Tautz, P. A. Morin, C. Boesch, B. Fruth, G. Hohmann, O. Ryder, and D. S. Woodruff. 1999. Mitochondrial sequences show diverse evolutionary histories of African hominoids. *Proceedings of the National Academy of Sciences, USA* 96:5077–82.

Gonzalez, I. L., J. E. Sylvester, T. F. Smith, D. Stambolian, and R. D. Schmickel. 1990. Ribosomal RNA gene sequences and hominoid phylogeny. *Molecular Biology and Evolution* 7:203–19.

Haraway, D. 1989. *Primate visions.* New York: Routledge.

Hull, D. L. 1970. Contemporary systematic philosophies. *Annual Review of Ecology and Systematics* 1:19–54.

Hussey, L. M. 1926. The blood of the primates. *American Mercury* 9:319–31.

Huxley, J. S. 1957. The three types of evolutionary process. *Nature* 180:454–55.

IJdo, J. W., A. Baldini, D. C. Ward, S. T. Reeders, and R. A. Wells. 1991. Origin of human chromosome 2: An ancestral telomere-telomere fusion. *Proceedings of the National Academy of Sciences, USA* 88:9051–55.

Johnson, W. E., and J. M. Coppen. 1999. Reconstructing primate phylogenies from ancient retrovirus sequences. *Proceedings of the National Academy of Sciences, USA* 96:10254–60.

King, M.-C., and A. C. Wilson. 1975. Evolution at two levels in humans and chimpanzees. *Science* 188:107–16.

Livak, K. J., J. Rogers, and J. B. Lichter. 1995. Variability of dopamine D4 receptor

(DRD4) gene sequence within and among nonhuman primate species. *Proceedings of the National Academy of Sciences, USA* 92:427–31.

Marks, J. 1983. Hominoid cytogenetics and evolution. *Yearbook of Physical Anthropology* 25:125–53.

———. 1991. What's old and new in molecular phylogenetics. *American Journal of Physical Anthropology* 85:207–19.

———. 1993. Hominoid heterochromatin: Terminal C-bands as a complex genetic character linking chimps and gorillas. *American Journal of Physical Anthropology* 90:237–46.

———. 1994a. Blood will tell (won't it?)? A century of molecular discourse in anthropological systematics. *American Journal of Physical Anthropology* 94:59–79.

———. 1994b. Review of *The Great Ape Project,* ed. by P. Cavalieri and P. Singer. *Human Biology* 66:1113–17.

———. 1995. Learning to live with a trichotomy. *American Journal of Physical Anthropology* 98:211–32.

———. 2002. *What it means to be 98% chimpanzee: Apes, people, and their genes.* Berkeley: University of California Press.

Marks, J., C. W. Schmid, and V. M. Sarich. 1988. DNA hybridization as a guide to phylogeny: Relations of the Hominoidea. *Journal of Human Evolution* 17:769–86.

Marks, J., J.-P. Shaw, and C.-K. J. Shen. 1986. The orangutan adult alpha-globin gene locus: Duplicated functional genes and a newly detected member of the primate alpha-globin gene family. *Proceedings of the National Academy of Sciences, USA* 83:1413–17.

Meyer, E., P. Wiegand, S. Rand, D. Kuhlmann, M. Brack, and B. Brinkmann. 1995. Microsatellite polymorphisms reveal phylogenetic relationships in primates. *Journal of Molecular Evolution* 41:10–14.

Mindell, D. 1991. Aligning DNA sequences: Homology and phylogenetic weighting. In *Phylogenetic analysis of DNA sequences,* ed. M. Miyamoto and J. Cracraft, 73–89. New York: Oxford University Press.

Monboddo, James Burnet, Lord. 1774. *The origin and progress of language.* Edinburgh: J. Balfour.

Pellicciari, C., D. Formenti, C. A. Redi, and M. G. Manfredi Romanini. 1982. DNA content variability in primates. *Journal of Human Evolution* 11:131–41.

Retief, J. D., R. Winkfein, G. Dixon, R. Adroer, R. Queralt, J. Ballabriga, and R. Oliva. 1993. Evolution of protamine P1 genes in primates. *Journal of Molecular Evolution* 37:426–34.

Ruano, G., J. A. Rogers, A. C. Ferguson-Smith, and K. K. Kidd. 1992. DNA sequence polymorphism within hominoid species exceeds the number of phylogenetically informative characters for a HOX2 locus. *Molecular Biology and Evolution* 9:575–86.

Sarich, V., C. Schmid, and J. Marks. 1989. DNA hybridization as a guide to phylogeny: A critical analysis. *Cladistics* 5:3–32.

Sibley, C. G., and J. E. Ahlquist. 1984. The phylogeny of the hominoid primates, as indicated by DNA-DNA hybridization. *Journal of Molecular Evolution* 20:2–15.

———. 1987. DNA hybridization evidence of hominoid phylogeny: Results from an expanded data set. *Journal of Molecular Evolution* 26:99–121.

Sibley, C. G., J. A. Comstock, and J. E. Ahlquist. 1990. DNA hybridization evidence of

hominoid phylogeny: A reanalysis of the data. *Journal of Molecular Evolution* 30:202–36.
Simpson, G. G. 1945. The principles of classification and a classification of mammals. *Bulletin of the American Museum of Natural History* 85.
———. 1964. Organisms and molecules in evolution. *Science* 146:1535–38.
Westhoff, C., and D. Wylie. 1998. Investigation of the RH *[sic]* locus in gorillas and chimpanzees. *Journal of Molecular Evolution* 42:658–68.
Wrangham, R., and D. Peterson. 1996. *Demonic males: Apes and the origins of human violence.* Boston: Houghton Mifflin.
Zimmer, E. A., S. L. Martin, S. M. Beverley, Y. W. Kan, and A. C. Wilson. 1980. Rapid duplication and loss of genes coding for the alpha chains of hemoglobin. *Proceedings of the National Academy of Sciences, USA* 77:2158–62.
Zuckerkandl, E. 1963. Perspectives in molecular anthropology. In *Classification and human evolution,* ed. S. Washburn, 243–72. Chicago: Aldine.

PART 2
Culture/Nature

SECTION A
Political and Cultural Identity

Chapter 8

From Pure Genes to GMOs

Transnationalized Gene Landscapes in the Biodiversity and Transgenic Food Networks

Chaia Heller and Arturo Escobar

INTRODUCTION: GLOBALIZATION AND THE DISCOURSE OF GENE ACTION

In recent years, a diverse international movement against biotechnology has emerged to contest the encroachment of global capital into agriculture and other bioscience arenas. Participants in the anti–World Trade Organization (WTO) demonstrations in Seattle during November 1999 included peasant farmers and representatives of indigenous groups from Europe and India, and consumer and ecology groups from around the world. Drawing linkages between food, land, bodily sovereignty, cultural autonomy, and identity in an age of globalization, these groups have protested the biological and cultural homogenization associated with genetically modified organisms (GMOs) and the fast-food culture of McDonald's. Participation by the French farmer José Bové in the WTO demonstrations signals a new kind of activism that is transnational and built around new biological technologies as they are deployed in systems of global capital.

In this essay, we examine two social movements linked to powerful oppositional international networks engaged in interrogating biotechnology. The first movement began in France, on January 8, 1998, when about 150 farmers from one of the largest French farmers unions, the Confédération Paysanne, attacked a Novartis conditioning and storage plant, where they found and destroyed five tons of transgenic maize. Three of the farmers were arrested. At the court hearing, with several hundred supporters outside the courthouse, expert witnesses argued for their release by building a case against genetically engineered crops.

The second social movement we consider, a movement for ethnic, cultural, and territorial rights, unfolded within indigenous and black communities of rain forest areas of Colombia. Among other actions undertaken by this movement was its mobilization in response to news of an application submitted by BioAndes, a subsidiary of Andes Pharmaceuticals, based in Washington, D.C., who requested access to biological and genetic resources in the Colombian National Park System. The application was the first in Colombia under the Access to Genetic Resources agreement reached by Andean Pact countries, and it came under intense scrutiny by national nongovernmental organizations working on the issue of genetic resources and by local residents in the affected areas. Prominent were their concerns about lack of consultation with local communities, ambiguity about the resources and areas to be covered and about the use of local knowledge, and overall disregard for local communities' collective rights to local biodiversity.

These two moments of activism and mobilization, out of the many located in various parts of the world in the late 1990s that we could cite, involve the novel intersection of genetic knowledge with forces of globalization. This intersection results in transnationalized genetic landscapes that pose unprecedented challenges to biological and anthropological investigation. We argue here that biodiversity and GMOs are not issues or objects but powerful networks for the production of nature and culture. These networks are sites of resistance to what many actors see as the extension of the commodification and technologization of nature, and the loss of local autonomy over "natural" and cultivated environments in the face of global capital and genetic rationality.

Although biodiversity and transgenic agriculture seem at first to occupy opposite ends of a spectrum that ranges from notions of the "pure" to the "unnatural" (in the genes of wild biodiversity and GMOs, respectively), the two cases are actually, and increasingly, interrelated. In both cases, as we shall see, gene technology and patents are used to consolidate power over food and nature. In both cases, too, multinational corporations and international organizations such as the WTO play a key role in brokering power in the production of new trade, regulatory, and production practices. And, while capital and governmental powers go transnational, resistance to these forces is globalizing as well. As the WTO protests in Seattle show, social movements are emerging within the international antiglobalization movement in which actors are resisting the managerial and market logic embedded in discourses of both biodiversity and GMOs.

The parallels do not end there. Both cases involve discourses of genetic essentialism and naturalism, although with contrasting emphases. Just as biodiversity is linked to the survival of biological life and the human race, GMOs are linked to the solution to world hunger, malnutrition, and environmental problems associated with chemical agriculture. Both of the movements we discuss here complicate any easy conflation of gene technology with either progress and survival or devastation. Both involve stories about life and death—global, populational, and technological. Both are also sites for the exploration of alternative practices of biological conservation and sustainable agriculture—practices developed by local communities that do not rely on genetic technologies, intellectual property rights, patents, or alternating scenarios of hope and despair. In addition, both cases illustrate the creative ways in which nonexperts recast science questions in their own cultural terms.

There are also important differences. In the case of GMOs, the managerial discourse on risk is prominent among both expert regulatory bodies and activists. While small farmers often appeal to risk discourse to bolster their claims against GMOs, they also propose alternative frames for assessing the technology that engage questions about the quality of food, life, and agriculture. In the biodiversity case, actors challenge the dominant discourses of intellectual property rights by developing alternative notions of collective and sui generis rights, so that autonomy over local territories, knowledge, and resources has become the most pressing demand. Finally, the two cases involve different expressions of a naturalistic discourse. Anti-GMO activists often appeal to a type of genetic naturalism that condemns human beings for disrupting a static, pristine nature. In contrast, biodiversity activists emphasize preserving the natural coevolution of organisms and cultures and, thus, embed human beings and culture inside natural systems.

Transnational gene landscapes are multiplicities linking organisms, ecologies, histories, and cultures in complex and unprecedented ways. Ultimately, "it is this complex assemblage of powerful tools, major social institutions, vast public and private resources, and a long-standing 'discourse of gene action' that all of us are now learning to live within, ask questions about, and, we hope, improve upon" (Fortun and Bernstein 1998: 209). The cases of biodiversity and transgenic agriculture reveal unique and original dimensions of this situation, raising key questions for anthropology in the age of genetics.

CONSERVING GENES, CONSERVING NATURE
Genetic Resources and the Politics of Biodiversity Conservation

Biodiversity is defined as the natural stock of genetic material within an ecosystem. More broadly, it encompasses diversity within species and between species and ecosystems.[1] Although defined as a scientific problem, biodiversity could also be described as the response made to a situation that bleeds out of science proper. Notions of what biodiversity is, as well as notions of what threatens it and of possible interventions, are the subjects of contentious debates. Biodiversity has both biological and cultural meanings and referents in the world.

Biodiversity as a discourse is of recent origin. This discourse has fostered a transnational network of actors that encompasses diverse practices, cultures, and stakes.[2] Each actor's identity affects, and is affected by, the network. International institutions, nongovernmental organizations, botanical gardens, pharmaceutical companies, and experts occupy the dominant sites of the network, but the truths they produce might be resisted or re-created to serve other ends—for instance, by social movements. From a dominant perspective, the aim might be to create a stable network for the movement of objects, resources, knowledge, and materials by relying on a simplified construction, most effectively summarized in Daniel Janzen's motto about biodiversity: "You've got to know it to use it, and you've got to use it to save it" (1992; Janzen and Hallwachs 1993). However, countersimplifications and alternative discourses by subaltern actors have also circulated through the network, with significant effects.

The dominant view, which could be called a globalocentric perspective, emphasizes resource management. This view is produced by dominant institutions, particularly the World Bank and the main northern environmental nongovernmental organizations (e.g., the World Conservation Union, World Resources Institute, and World Wildlife Fund), and is supported by the G-8 countries. The narrative concerning threats to biodiversity emphasizes loss of habitats, species introduction in alien habitats, and fragmentation due to habitat reduction, rather than underlying questions of power. It offers prescriptions for the conservation and sustainable use of resources at the international, national, and local levels; suggests appropriate mechanisms for biodiversity management, including in situ and ex situ conservation and national biodiversity planning; and proposes the establishment of appropriate mechanisms for compensation and economic use of biodiversity, chiefly through intellectual property rights, one of the most contentious issues (see below).[3]

This perspective is challenged by Third World national governments, which, without questioning it in a fundamental way, seek to negotiate the terms of biodiversity treaties and strategies to their perceived advantage. But the real challenge comes from progressive nongovernmental organizations and social movements. A small but increasing number of progressive nongovernmental organizations regard the dominant globalocentric perspective as a form of bioimperialism. This critique articulates a concomitant shift of responsibility from South to North, as the source of the diversity crisis.[4] It emphasizes habitat destruction by megadevelopment projects, the consumption habits of the North, and agriculture promoted by capitalism and positivistic science.

These critics are adamantly opposed both to biotechnology as a tool to maintain diversity and to the adoption of intellectual property rights as the means of protecting local knowledge and resources. Instead, they advocate forms of collective rights that recognize the intrinsic value of shared knowledge and resources. This view thus contests the most cherished constructs of modernity, such as positivist science, the market, and individual property. The nongovernmental organizations advancing this position constitute subnetworks at national and transnational levels that are still poorly understood. They are a prime example of the emergent set of transnational practices and identities that link virtual and place-based modes of activism.[5]

Social movements that explicitly construct a political strategy for the defense of territory, culture, and identity craft a second challenge to the globalocentric perspective. While sharing much with the progressive nongovernmental organization perspective, this perspective is distinct conceptually and politically and occupies a different role in the biodiversity network. Aware that biodiversity is a hegemonic construct, activists in these movements acknowledge that this discourse nevertheless opens up a space for the defense of not only genetic resources or biodiversity but also their entire life projects. The concern with biodiversity has frequently followed from broader struggles for territorial control. In Latin America, social movements have achieved significant goals in this regard, chiefly in conjunction with the demarcation of collective territories in countries such as Ecuador, Peru, Colombia, Bolivia, and Brazil. These experiences are yet to be examined ethnographically and comparatively. In what follows, we present an account of one such movement, highlighting the broader perspective on biodiversity it has developed.

From Gene Politics to Social Movements' Political Ecology

Since the end of the 1980s, the peoples of the Pacific rain forest region of Colombia have experienced an unprecedented emergence of collective ethnic identities, which have had strategic significance in culture-territory relations. This process takes place at a complex national and international conjuncture. The neoliberal effort to open the economy to world markets after 1990 and a substantial reform of the national constitution in 1991, among other things, granted black communities of the Colombian Pacific region collective rights to the territories they have traditionally occupied. Internationally, tropical rain forest areas have come to be seen as housing the majority of the biological diversity of the planet. The emergence of collective ethnic identities in the Colombian Pacific and similar regions thus reflects a double historical movement: the emergence of the biological as a global problem, and the eruption of the cultural and the ethnic as potent oppositional themes and forces.[6]

The social movement of black communities, centering on the Proceso de Comunidades Negras (PCN, Process of Black Communities, a network of more than 140 local organizations), has been prominent in these developments. The PCN emphasizes social control of the territory as a precondition for the survival and strengthening of culture and biodiversity. Activists have worked with the river communities to teach the meaning of the new constitution and the concepts of territory, development, traditional production practices, and use of natural resources. Between 1991 and 1993, this sustained effort resulted in the elaboration of a proposed law of cultural and territorial rights, as called for by the 1991 constitution (Law 70, approved in 1993), and the firming up of a series of political and organizational principles. The Third National Conference of Black Communities in September of 1993 proposed an organizing strategy "for the achievement of cultural, social, economic, political, and territorial rights and for the defense of natural resources and the environment." The conference adopted a set of politico-organizational principles that emphasized identity, territory, autonomy, and alternative development.

Because of its rich natural resources, the Pacific coast region of Colombia has been a critical site for the global negotiation of biodiversity. Activists have sought to insert themselves in discussions about biodiversity conservation, genetic resources, and the management of natural resources. River communities and PCN activists have engaged with the government-run Pacific Biodiversity Conservation Project (Proyecto Biopacífico, PBP), which accepted the black and indigenous movements as its most important interlocutors. PCN members also have participated in the transnational movements through the Convention on Biological Diversity and such movements as the People's Global Action against Free Trade. And PCN activists have run for local elections, organized locally and nationally, sought funding for territorial demarcation, and participated in the intense negotiations that took place from 1996 to 1998 over the future of the PBP. PCN activists have progressively developed what we would call a political ecology framework by interacting with community, state, nongovernmental organizational, and academic sectors.

Within this framework, the territory is seen as a fundamental and multidimensional space for creating and re-creating the ecological, economic, and cultural practices of the communities. The territory is seen as incorporating patterns of settlement, use of spaces, and practices of signification and use of resources. Recent anthropological studies documenting cultural models of nature among black river communities validate this conception (Restrepo and del Valle 1996).

In the midst of this, violence in the region—perpetrated by armed actors such as right-wing paramilitaries, guerrillas, and government forces—has escalated, some of it directed explicitly against activists and communities to discourage them from pressing for territorial demands. These tensions are related to the overall intensification of development, capitalism, and modernity in the region (Escobar and Pedrosa 1996).

One of the PBP's important contributions was to conduct explicit research on the traditional production systems of the river communities, systems geared toward local consumption rather than the market and accumulation. Low-intensity exploitation, shifting use of productive space over broad and different ecological areas, manifold and diverse agricultural and extractive activities, family and kindred-based labor practices, and horticulture have all contributed to a sustainable agriculture. Yet many of these river-dependent systems are under heavy stress, chiefly from the growing extractivist pressures (Sánchez and Leal 1995).

PCN activists have defined biodiversity as "territory plus culture" and construed the entire Pacific rain forest region as a "region-territory of ethnic groups"—that is, an ecological and cultural unit laboriously constructed through the daily practices of the communities. The region-territory is also viewed in terms of "life corridors," veritable modes of articulation between sociocultural forms of use and the natural environment. There are, for instance, life corridors linked to the mangrove ecosystems, to the middle part of the rivers extending toward the inside of the forest, and to particular activities such as traditional gold mining and women's shell collecting in the mangrove areas.

The region-territory is a management category that pertains to the construction of alternative life and society models. It is an attempt to explain biological diversity from inside the ecocultural logic of the Pacific. And it is conceived of as a political construction for the defense of the territories and their sustainability. Conversely, that which is designated simply as the *territory* is seen as the space of effective appropriation of the ecosystem—that is, a space actively used to satisfy community needs and for socioeconomic and cultural ends. For a given river community, this appropriation has longitudinal and horizontal dimensions, sometimes encompassing several landscapes and river basins. The territory thus embodies a community's life project.

The region-territory, then, is a strategy for sustainability and vice versa: sustainability is a strategy for the construction and defense of the region-territory. Said differently, sustainability cannot be conceived in terms of patches or singular activities, or only on economic grounds. It must respond to the integral and multidimensional character of the practices of effective appropriation of ecosystems. The region-territory thus can be said to articulate the life project of the communities with the political project of the social movement. Similarly, the definition of biodiversity proposed by the movement provides elements for reorienting biodiversity discourses according to local principles of autonomy, knowledge, identity, and economy. For the activists, as they theorize local practices, nature is not an entity "out there," but is produced through the collective practices of humans integrated with it (Descola and Pálsson 1996). From this perspective, the reductive view of biodiversity in terms of genetic resources to be protected through intellectual property rights is incoherent and untenable.

The struggle for territory is above all a cultural struggle for autonomy and self-determination. The strengthening and transformation of traditional production systems and local economies, the need to press on with the collective titling process, and the work of strengthening organizations and developing forms of territorial governability are all important components of an overall strategy centered on the region. The Colombian conservation establishment is primarily interested in the development of genetic resources and habitat protection—the standard issues within the global biodiversity debate. But the ecocultural demands of the activists in the PCN have played an important role in the biodiversity discussions, and their concerns have come to play a role in the strategies of formal Colombian institutions. For many on the national staff of the PBP, and for the PCN activists, the negotiations over biodiversity have been generally positive but also hard, tense, and frustrating. Through these negotiations, the PBP and the PCN have come to share the goal of "constructing a region" in ways that markedly contrast with dominant views. Together, they developed a complex view of the socioeconomic, cultural, and political forces that shape the Pacific, and their joint research amply demonstrates the compatibility of traditional systems with biodiversity.

NATURE TO DENATURE: FROM BIODIVERSITY TO GENETICALLY MODIFIED ORGANISMS

If biodiversity engages with threats to nature portrayed as real, then GMOs engage with threats posed by a nature perceived as unreal, denatured, and involving genetically modified versions of otherwise natural organisms. Both narratives reflect an advanced capitalization of nature that seeks to preserve and modify biological organisms for the production of new technologies, information, products, and markets. As in the case of biodiversity, GMOs constitute more than their physical referents, more than the novel biological organisms produced within laboratories of multinationals such as Monsanto. In addition, they represent a historically produced discourse about the improvement of nature through molecular biology, about the identification of the gene as the building block of life, and about genetics as the solution to the world's medical, agricultural, environmental, and social problems (Nelkin and Lindee 1995).

We now turn to the French anti-GMO movement and the perspectives of French family farmers. In France, the term for GMOs generally refers to genetically modified foods (rather than trees modified to grow faster or sheep modified to produce humulin or other drugs), and the debate has centered on issues of food safety and risk. Risk is in fact the dominant way to talk about GMOs in national and international circles. Establishing itself as the only accepted discourse through which actors may broker claims about particular technological practice, risk discourse produces a zone of silence within public debate, muting competing critiques of technology framed in social, economic, or cultural terms. Just as social movements struggle to counter the globalocentric view of biodiversity, activists in France counter the "riskocentric" focus on risk by promoting competing framings of GMOs. French activists have promoted competing framings of the GMO issue that decenter the dominant risk discourse by introducing a range of social, economic, and political issues linked to processes of globalization.

By the early 1990s, U.S.-based biotechnology companies, who claimed the right to market these crops in order to collect a return on nearly fifteen years of investment in product research and development (Wright 1993), began to market the first generation of transgenic crops, of which approximately half had an inserted gene for herbicide resistance.[7] Two of the most profitable first-generation crops have been Monsanto's Roundup Ready soy, which is herbicide resistant, and Novartis Maximizer corn, which is designed to resist the European corn borer beetle.[8] Having introduced transgenic varieties to U.S. markets two years earlier, Monsanto and Novartis decided to introduce their resistant varieties of soy and corn to Europe in the fall of 1996.

France has had an ambivalent relationship to agricultural biotechnology. For many French scientists, corporations, and government agents, agricultural biotechnology represents a crucial means by which France may maintain a competitive edge within an increasingly global agricultural market (Le Déaut 1998). However, with a lack of venture capital, France's technoscience infrastructure depends primarily on a public sector ill-equipped to bear the costs of developing agricultural biotechnology. For some actors within the government and public science sector, GMOs are a symbol of desired competitiveness within a global economy. For yet others, GMOs symbolize U.S. forms of aggressive capitalism and environmental irresponsibility. The influx of leftist-green ecological actors into French government and other public institutions (with the 1997 election of the socialist-Green alliance headed by Lionel Jospin) created a context, which, if not publicly critical of GMOs, is characterized by degrees of reticence and ambivalence.

French nongovernmental organizations have played a definitive role in influencing French GMO policy and public opinion. In the fall of 1996, when France imported its first transgenic products, *Libération,* a major left-leaning newspaper, published a front-page article titled "Watch Out for Mad Soy!" linking transgenic soy to the recent scandal of mad cow disease that had sensitized the population to issues of food safety. By spring of 1997, seeds had been sown for the emergence of an anti-GMO movement in France, constituted by an informal yet identifiable coalition of consumers associations, farmers unions, and ecology groups such as Ecoropa and Greenpeace France. The actor-network theory described by Bruno Latour and M. Callon can be used to map the networks of association that constitute this GMO debate (Latour 1987; Callon 1986a). In the French case, anti-GMO coalitions make up heterogeneous networks that assert and circulate their own discourses about food, nature, and globalization.

The risk discourse that dominated French discussion about GM foods during this period reflects a wider process of recasting "natural" and institution-driven dangers as a set of statistically calculable, insurable harms assumed necessary for social progress (Heller 2001). Actors come to regard categories of self, nature, and society as fields of potential liabilities and benefits to be understood through cost-benefit analysis. Risks do often have real referents in the world, of course, and technologies such as GMOs may indeed, be associated with ontologically real environmental or health dangers. But the salient process here is the *translation* of potentially real dangers into public discourses of risk-benefit analysis.

The translation of economistic cost-benefit analysis into science discourse requires a double maneuver: First, economistic notions of cost must be recast as the inevitable and objectively calculable risks portrayed as necessary for achieving the benefits of science. Second, economistic notions of profit must be recast in altruistic terms, making "benefits" now stand for universal human "progress." By framing science practice as an objective weighing of inevitable risks against the altruistic benefits of human progress, scientific risk discourse obscures profit-seeking dimensions of science practice that are becoming increasingly visible: for example, as multinational companies play greater roles in shaping genetic research and development in the biotechnology industry.

Talk of the risks of GMOs is produced and normalized by institutions, such as the WTO, which require that claims made by particular countries against food products be framed in terms of health risk. The WTO chose the Codex Alimentarius as the regulatory body for food safety (Le Déaut 1998). As a United Nations commission, the Codex establishes food safety criteria for the Food and Agriculture Organization and the World Health Organization. Heeding the Codex, the GMO report published in June 1998 by the French Parliamentary Office on Science and Technology Assessment warns that "if a country seeks to limit the importation or sale of a product, it must, to avoid being punished for protectionism, prove that the product constitutes a real risk to public health" (Le Déaut 1998). A climate of self-censorship thus emerges within the GMO debate, in which actors assume that the WTO will accept only risk-related claims, despite the fact that the WTO has not yet established criteria for GMOs (Marris 1999). Such assumptions reflect the potency of the discursive boundaries around GMOs, boundaries that mute discourses unrelated to risk within both governmental decision-making bodies and nongovernmental organizations alike. Framing social, economic, and political arguments against food products as barriers to free trade, international trade agencies are able to protect the transnational market from "biased" or "subjective" knowledge claims of individual countries.

Risk has become the euro of public debate: the single common currency accepted in European discussion of agricultural trade policy. Risk talk circulates through the anti-GMO movement in France as actors enroll scientific experts in the network to legitimize their claims against GMOs. If one strategy of activists is to appropriate this risk discourse, another is to displace the primacy of risk by presenting alternative framings of GMOs as they invoke alternative discourses of democracy, economics, food quality, and nature. For instance, consumer, ecology, and farmers organizations forge alliances by focusing on food quality and on a common call for democracy and public debate regarding GMOs. Thus discussions of quality and democracy unsettle and broaden a debate anchored in notions of risk and precaution. These strategies represent an ever differentiating repertoire of discourses, intermediaries, and identities that actors continually reshape as they position themselves in relation to an even wider network organized around the dominant discourse of GMO risk.

At the end of the 1970s, a new collective expression of peasant identity emerged in France: the Confédération Paysanne (CP). This movement signaled a politicization of agricultural policy that posits agriculture as a global problem while expressing the cultural desire to sustain what the movement defines as a "peasant way of life" (Confédération Paysanne 1997). The CP represents an increasingly powerful union of family farmers. Many of the union's founders emerged within, and ultimately rejected the industrial perspective of, France's largest farmers union, the Fédération Nationale des Syndicats et Exploitants Agricoles (FNSEA). CP founding members reclaimed the term *paysan* (peasant), endowing it with new, oppositional cultural meaning. Historically, the term had a negative connotation, referring to family farmers who were conservative, parochial, and marginalized from modern productive society. Among the last in western Europe to industrialize their agricultural practices, French *paysans* were still using farm animals to power their plows well after World War II (Hervieu 1996). The shift in French agricultural policy toward an industrial model, promoted by the FNSEA, entailed the construction of an identity for farmers who would be portrayed no longer as a distinct premodern cultural group but as an entrepreneurial sector fully integrated into modern French society (Girod 1997). The CP's rejection of the entrepreneurial and industrial farmer, and its reclamation of the term *paysan* in the 1970s, constituted a collective refusal of this project. The *paysan* invoked by the founders of the CP was not an entrepreneur capitalist but a worker who would "work the land as his tool" (Girod 1997).

In the winter of 1998, the CP launched its first successful anti-GMO direct action, putting the organization on the national and international maps. Prime Minister Lionel Jospin had authorized the cultivation of a variety of Novartis's transgenic corn on French soil without providing the extensive research trials and public debate that he had promised would occur before authorization.[9]

In protest, on January 8, 1998, José Bové and René Riesel, CP founders and leaders of the anti-GMO movement within the CP, led 150 farmers carrying bags of their own locally produced corn into a Novartis storage plant in southern France, in the town of Nérac. The farmers mixed the natural with the transgenic corn and sprayed the mixture with fire hoses to symbolically contaminate and destroy the Novartis stock.

When Bové and Riesel were arrested, they declared to the media that the trial would not be their own but would be "the trial of the GMOs." They began to publicly and effectively expand the dominant discourse surrounding agricultural biotechnology in France by introducing a distinct social and political perspective. Also being transformed in this public debate was the idea of *paysan*. During the trial, CP activists constructed a *paysan* with a hybrid identity of worker, international actor, and protector of consumer, environment, and democracy. Through five major discursive maneuvers, the CP deployed a series of discourses, identities, and intermediaries to promote nonproductivist agricultural policy, forge new alliances with other social movements, and frame discussion of GMOs in contexts other than risk.

In the first maneuver, the CP transformed the identity of the traditional *paysan* into that of worker-*paysan* in solidarity with other laborers. During the trial, the CP enlisted the support of leaders in the French unemployment movement and other trade unions to testify against GMOs, placing them in a context of unfair labor practices linked to globalization. In the second maneuver, it expanded the traditional identity of the *paysan*—the *homme du pays* (man of the country) who protects *le terroir* (soil and country)—to include the *paysan*'s role as an expert within contemporary consumer and environmental discourse. In this way, the traditional *paysan* became a protector of quality, a holder of local expert knowledge, or savoir faire, about French food products, culture, and environment. The CP thus disrupted the dominant GMO risk discourse by introducing competing cultural discourses on food and agricultural quality. By appropriating the discourse on quality, the CP attempted to establish itself as a site of expert knowledge, demonstrating its alliance with anti-GMO scientists and consumer and environmental groups at the trial, which in turn further legitimized farmers' claims.

The third maneuver transformed the identity of the traditional *paysan* into that of the modern citizen, deploying a discourse of democracy. Here, the CP transcended the particularistic concerns of workers, consumers, and environmentalists, invoking the idea of public interest in the GMO debate. It thus transformed GMOs into the symbol of the failure of modern democracy, appropriating recent popular discourse in France on the crisis in public representation. In so doing, the CP countered notions of the premodern, backward-looking peasant. At the trial, Claude Julien, a former president of *Le Monde Diplomatique* (a major French leftist newspaper), solidified this maneuver by stating in support of the CP: "It's they who are modern, because they want, in this democratic era, a real public debate regarding these transgenic products" (G. G. 1998). In this maneuver, the CP thus turned the discursive tables, confronting the government for recklessly authorizing the untested corn and disrespecting a rational juridical norm designed to protect citizens from public harm.

In the fourth maneuver, the CP transformed the identity of the parochial *paysan* into that of an international subject in solidarity with peasants in Europe and the Third World, as GMOs became the symbol of the disenfranchisement of peasants all over the world.[10] Invoking Third World sustainable-development and biodiversity discourses, the CP invited Vandana Shiva, an Indian activist and scientist, to testify at the trial. Shiva's presence personified the incorporation of an international struggle into the French anti-GMO movement and pointed to the alliance between the CP and an international indigenous struggle (Montel 1998).

The final maneuver introduced the identity of the "policy-making *paysan*." Countering the stereotype of the *paysan* as *passéist,* or backward-looking, the CP constructed the identity of the *paysan* as progressive subject who, through democratic agricultural policy, can produce a viable alternative to productivist agriculture. The CP promoted an *agriculture paysanne,* a detailed program calling for a protection of food quality, nature, and "the peasant way of life" (Confédération Paysanne 1995). Yet rather than propose an atavistic return to preindustrial times or invoking a bounded static identity, CP publications and literature made frequent references to "the farm of the future" that would be made possible by sustainable agricultural technologies and international *paysan* and worker solidarity. The CP's construction of the *agriculture paysanne* also disrupted the dominant French discourse of sustainable development, a discourse often deployed within critiques of GMOs, which emphasizes technical aspects of agricultural practice over social, cultural, and economic dimensions. In this maneuver, GMOs also were cast as the symbol of European agricultural policy that obliged small farmers to depend on large agrochemical companies who monopolize the production of seed and agricultural inputs (Riesel 1998).

In the summer of 1999, José Bové, René Riesel, and other members of the CP organized a series of direct actions against corporate and public institutions conducting tests of GM crops in southern France, where they both reside. Yet the incident that really brought the CP to international prominence was the demonstration organized by Bové against a local McDonald's.[11] Determined to punish Bové for the anti-GMO actions, the local judge ordered unusually high bail, which led to a three-week imprisonment. Within weeks, media coverage of this event catapulted Bové to the status of international martyr and hero in the antiglobalization movement. Since 1999, Bové has linked the issues of GMOs and McDonald's, citing them both as examples of "globalization and a decline in quality of food and life." The

French debate about GMOs has since shifted even more broadly to one about the commodification of life, the fate of the small farmer, and the global homogenization of culture by multinational capitalism.[12]

By transforming both the meaning of a trial and the identity and discourse of the traditional French *paysan*—and by linking GMOs to the more general problems of globalization and quality of life—the CP and Bové have redefined the public debate about GMOs in France. The CP struggled to shift the debate from science, health, and environmental risk to labor, culture, democracy, agrarian autonomy, and international agricultural policy. Here, agricultural biotechnology became not only a symbol of risky scientific practice, or of risky nature, but also a symbol of productivist agriculture, antidemocratic governance, and the disruption of peasant identity all over the world. And by transforming the discourse about GMOs, the CP also transformed public understanding of the French family farmer, no longer presented as a backward *passéist* but as an active promoter of democracy, public safety, and local autonomy.

KNOWLEDGE, GENES, AND PROPERTY IN RECENT ANTHROPOLOGICAL DISCOURSE

What can we learn from these cases about anthropology's contribution to the study of transnationalized genetic landscapes? Anthropologists have begun to analyze the hegemony of Western views of genes and intellectual property based on notions of possessive individualism, fully commodified social relations, and market transactions (Bush 1996; Strathern 1998). As Marilyn Strathern (1998) has pointed out, biodiversity discussions tend to reenact the division between Euro-American and other cultures. For many peasant and indigenous societies, genes and intellectual property rights are not meaningful categories. Conversely, locally meaningful categories, including blood, reciprocity, commons, and noncommodified forms of compensation, cannot be easily translated into the Western concepts genes, persons, and property. Other peasant societies do have notions of intellectual and individual property, suggesting that while Western intellectual property rights might be generally inappropriate for these societies, they could be used effectively in some cases (Cleveland and Murray 1997). Intellectual property is a potentially flexible concept, and it could perhaps be applied to instances of collective cultural property and other products of community life (Strathern 1996).

Anthropologists also have explored the role of the commons, however defined, in many peasant communities. The commons may consist of land, material resources, knowledge, ancestors, spirits, and so on, but it does not readily translate into intellectual property rights, and such rights in effect make the benefit of community innovations accrue to external capital

(Gudeman 1996). As Joan Martínez Alier (1996) puts it, the conflict between economistic reasoning and ecological reasoning that is central to biodiversity debates must be solved politically. Otherwise, conservation strategies will amount to a merchandising of biodiversity. Cases such as these lead us to question the possibility of defending a posteconomistic, ecological production rationality.

Politically, it has become possible to defend the rights of local communities to control their knowledge, territories, and economies. Anthropological debates on such questions have barely started, however. Anthropology's long-standing interest in documenting and promoting diverse understandings (here, of knowledge, property, and economy) is at stake, this time complicated by the added dimensions of globalization and technoscience. In sum, biodiversity presents anthropology with an imperative to rethink some of its long-standing interests, particularly the relation between nature and culture.

The stakes are not so different in the case of transgenic foods and agriculture. Anthropology can offer insight into newly emerging knowledges of nature that arise to contest hegemonic notions of modern science and society. As Strathern suggests, concepts such as seed patents are not recognized as legitimate categories by many peasant groups, such as the CP, who frame practices like seed saving as forms of community reciprocity, agricultural autonomy, and preservation of a *patrimoine culturel* (1998; Riesel 1998).

While much of what has been called the anthropology of risk is constituted by a psychological or cultural analysis of public risk perception, there is need for an ethnographic approach problematizing risk discourse itself as a Western notion that produces new understandings of self, society, and nature. GMO-related-risk discourse mutes competing issues of local knowledge as well as cooperative and autonomous agricultural practice that may be understood as what Enrique Leff (1993) describes as an alternative production rationality that resists the reduction of questions of land use to terms of productivism, growth, and profit. Anthropology must examine the emergence of new constructions of nature and culture that are surfacing within debates concerning agricultural biotechnology and problematizing dominant constructions of a risky "denature" and of what Neil Smith (1996) calls a socialized nature: an agricultural nature invoked by peasants within Western industrial-capitalist societies. Such constructions stand in sharp contrast to understandings of a real or pure nature existing within the "genetic wilderness" that surface within debates over biodiversity.

Biodiversity and transgenic agriculture constitute powerful networks through which concepts, policies, and ultimately cultures and ecologies are contested and negotiated. The issues advanced through these networks have a growing presence in the strategies of social movements in many parts of the world. The two movements in Colombia and France discussed here entail a cultural politics significantly mediated by ecological and technoscientific concerns. They are engaged in a laborious construction of identities that articulate old and more recent languages and concerns. Despite the negative forces opposing them, and in the climate of certain favorable ecological and cultural conjunctures, these movements might represent a real defense of social and biophysical landscapes in ways that are not dominated by the genetic essentialism and reductionism that characterize dominant trends. The movements can be seen as attempts to show that life, work, nature, and culture can be organized in ways different from those mandated by dominant models of culture and the economy.

Our two case studies also illustrate novel and troubling aspects of the intersection of globalization and genetic knowledge. The discourses of risk and property rights, couched in neoliberal ideologies of free trade, dominate the work of progressive nongovernmental organizations and social movements. As we show, however, these discourses can be transformed by social movements into discussions of food quality, nature, property, development, and democracy. French farmers and Colombian black communities alike attempt to craft nonproductivist agricultural and biodiversity policies and to frame GMOs and genetic resources issues in terms other than risk and trade-related intellectual property rights, such as those of local sustainability and cultural autonomy.

If French farmers find in transgenic agriculture another imposition of the global economy and large-scale agriculture, Colombian black activists see most approaches to bioprospecting and international negotiations on genetic resource as another attempt at extraction of resources from their territories. All these processes have entailed momentous changes in local cultures, pointing at the emergence of a veritable transformation in collective identities. Whereas French peasant identities borrow images of workers, citizenship, rationality, and expertise, Colombian black activists foster identities in terms of cultural difference, self-determination, local knowledge, and traditions of conservation. Along the way, these images and identities displace the genetic essentialism, naturalism, economism, and managerialism that characterize dominant positions. In doing so, they equally alter the terms of the debate at local, national, and international levels.

One hopes, nevertheless, that in the spaces of encounter and debate provided by the biodiversity and agricultural genetics networks, there will be found ways for academics, scientists, nongovernmental organizations, and intellectuals to reflect seriously on, and support, the alternative frameworks that, with varying degrees of explicitness and sophistication, are crafting place-based, yet transnationalized, social movements. As with the body, global capital is attempting to redesign nature by interfacing technology and production more effectively (Critical Art Ensemble 1998), and this forces social movements to separate the cultural and the natural in their strategies to oppose the inscription of a global political economy and scientific rationality into ever new biological domains. The frameworks these movements invent in doing so have much to teach us about alternative political ecologies that are continuing to be articulated in this increasingly contentious age of genetics.

NOTES

1. WRI 1994: 147; Swanson 1998: 7. Current definitions of biodiversity do not create a new object of study; they are geared toward assessing the significance of biodiversity loss to ecosystem functioning and structure, and to ascertaining the "services" ecosystems provide. See, for instance, the voluminous report by the Scientific Committee on Problems of the Environment and the United Nations Environment Program (Heywood 1995), especially the chapter by Mooney et al., and the useful review of the report in Baskin 1997.

2. In its classical formulation, the actor network theory proposed by Callon (1986a,b) and Latour (1983, 1987, 1993) was a methodology for studying the coproduction of technoscience and society. Anthropologists and feminist scholars such as Rayna Rapp, Deborah Heath, Emily Martin, and Donna Haraway have elaborated on it by adding elements from poststructuralist theory.

3. See also WRI 1992, 1994: 149–51. The debates on intellectual property rights in agriculture and in wild biodiversity developed somewhat separately but are coming together in light of advancing genetic engineering. Discussions on genetic resources and life patents started in relation to seeds and agricultural biotechnology in the early 1980s, and were generalized to include biodiversity with the Convention on Biological Diversity. For an earlier manifestation of these concerns, see the special issue of *Development Dialogue* edited by Pat Mooney in 1983, and Mooney's 1998 update. For recent statements from a progressive perspective, see the special issue of *Resurgence* (May-June 1998). A history of the genetic resources movement, written under the sponsorship of International Plant Genetic Resources Institute, is found in Pistorius 1997. We thank Nelson Alvarez of Genetic Resources Action International (GRAIN) for help with these issues, including bibliographic assistance.

4. The Malaysian-based Third World Network and Vandana Shiva's Research Foundation for Science, Technology, and Natural Resource Policy in India have both taken leading roles in the denunciation of bioimperialism and in the articulation of biodemocracy. They are joined in this effort by nongovernmental organizations in Latin America (such as Acción Ecológica in Ecuador, Semillas in Colombia, and Redes in Uruguay), Africa, Europe (GRAIN, Corner House, and Genet, among others), and Canada (Rural Advancement Foundation International, or RAFI). The Third World Network's journal, *Resurgence;* GRAIN's *Seedling;* and RAFI's communiqués are among the best sources on biodiversity and genetic resources issues from this perspective. See also Shiva 1993.

5. Ribeiro 1998; Escobar 1998. The progressive nongovernmental network complicates the geopolitical distinctions between North and South, First and Third Worlds. The Third World Network and the Barcelona-based GRAIN exemplify the

new form of transnational activism. GRAIN's goal is to counter the growing gene pool depletion arising from increasing privatization and industrial monopoly rights over biodiversity. It promotes dissemination of useful information and critical analyses on these issues, support for Third World popular movements, and capacity building for local biodiversity and genetic resource management. It maintains an active presence at key events at institutions such as the U.N. Food and Agricultural Organization, the WTO, and the Convention on Biological Diversity. GRAIN relies on a vast network of partners, particularly in Asia, Africa, and Latin America, and it is increasingly decentralizing its Barcelona office. Mary King, an anthropology graduate student at the University of Massachusetts, is currently doing research on GRAIN as an example of new practices of transnational activism.

6. The Pacific coast region of Colombia covers a vast area (about seventy thousand square kilometers) stretching from Panama to Ecuador and from the westernmost chain of the Andes to the ocean. It is one of the world's most biodiverse rain forest regions. About 60 percent of the region's 900,000 inhabitants (800,000 Afro-Colombians; about 50,000 Embera, Waunana, and other indigenous people; and about 50,000 mestizo colonists) live in the few larger towns; the rest inhabit the margins of the more than 240 rivers, most of which flow from the Andes toward the ocean. The area's black and indigenous peoples have maintained distinct material and cultural practices. For overall background on the region, see Escobar and Pedrosa 1996. For the politics of biodiversity, see Escobar 1997, 1998; and for the black movement, see Grueso, Rosero, and Escobar 1998.

7. Levidow and Carr 1997. These resistant varieties constitute what is referred to as the "first generation" of genetically modified crops. They have been criticized in France for providing economic benefits to agrochemical companies and large-scale farmers. The "second generation" of genetically modified crops is supposed to provide consumers with transgenic produce featuring improved flavor, nutrition, or pharmaceutical benefits.

8. Contradictory policies regarding different crops are related to the specific risks associated with them. The debate over Novartis's corn with *Bacillus thuringiensis* centers on the effects of antibiotic resistance genes (often used as markers in genetic engineering) on food safety and the environmental risks of transmission of genes from transgenic corn to neighboring plants. When French scientists proved that corn has no indigenous weedy relatives in France, the gene flow risk was ruled out, the corn was approved, and antibiotic resistance became the focus of concern. The debate over other genetically engineered crops, such as beets, canola, and chicory, is focused on concerns about gene flow between these crops and weedy relatives that do exist in France.

9. Until 1999, the CP was marked by two tendencies that are still debated within the CP: reformist and radical. Originally, José Bové and Réné Riesel (organizers of the Nérac action) headed up the radical wing and were responsible for much of the direct action that emerged from the southern departments. After the McDonald's incident and the union's subsequent prominence in national and international politics, however, direct action, once marginal, has become more central to the union's activities.

10. The CP's alliances with international peasant movements began in 1986, with the foundation of the Coordination Paysanne Européenne. Based in Brussels, the

association is composed of twenty-seven organizations from twelve European countries and has engaged in struggles over European and WTO agricultural policy. The CP is also part of Via Campasina, an international Third World–based coalition of seventy organizations from forty countries.

11. Having for several years demonstrated against McDonald's as "a symbol of globalization," Bové led the sabotage of a McDonald's construction site in his town of Millau in symbolic retaliation for Clinton's 1999 sanctions against Europe (punishment for the European Union's refusal to import hormone-treated beef).

12. The trial was further amplified by Internet biotechnology e-mail lists that disseminated petitions, press statements, and calls for support from the CP. An international petition was signed by one hundred organizations (from forty different countries), five thousand individuals around the world, and two hundred French organizations. The trial was even mentioned in a *New York Times* article that erroneously stated that the farmers had urinated on the corn.

REFERENCES

Baskin, Y. 1997. *The work of nature: How the diversity of life sustains us.* Washington, D.C.: Island Press.

Bush, Stephen. 1996. Prospecting the public domain. Manuscript.

Callon, M. 1986a. The sociology of an actor-network: The case of the electric vehicle. In *Mapping the dynamics of science and technology,* ed. M. Callon et al., 19–34. London: Macmillan.

———. 1986b. Some elements of a sociology of translation: Domestication of the scallops and the fishermen of St. Brieuc's Bay. In *Power, action, and belief: A new sociology of knowledge?* Sociological Review Monograph no. 32, ed. J. Law. Keele, U.K.: University of Keele.

Cleveland, D., and S. Murray. 1997. The world's crop genetic resources and the rights of indigenous farmers. *Current Anthropology* 38, no. 4:477–512.

Confédération Paysanne. 1995. *Agriculture Paysanne: Une demarche, une recherche, des practiques.* Paris: Confédération Paysanne.

———. 1997. *Technologies genetiques: Pour un moratoire sur la mise en culture et la commercialisation pour l'application du principe du precaution.* Paris: Confédération Paysanne.

Corner House. 1998. *Food? Health? Hope? Genetic engineering and world hunger.* Corner House Briefing no. 10. Dorset, U.K.: Corner House.

Critical Art Ensemble. 1998. *Flesh machine.* New York: Autonomedia.

Descola, P., and G. Pálsson, eds. 1996. *Nature and society: Anthropological perspectives.* London: Routledge.

Escobar, A. 1997. Cultural politics and biological diversity: State, capital, and social movements in the Pacific Coast of Colombia. In *Between resistance and revolution,* ed. R. Fox and O. Starn, 40–64. New Brunswick, N.J.: Rutgers University Press.

———. 1998. Whose knowledge, whose nature? Biodiversity conservation and the political ecology of social movements. *Journal of Political Ecology* 5:53–82.

Escobar, A., and A. Pedrosa, eds. 1996. *Pacífico: Desarrollo o diversidad? Estado, capital y movimientos sociales en el Pacífico Colombiano.* Bogotà: CEREC and Ecofondo.

Fortun, M., and H. Bernstein. 1998. *Muddling through: Pursuing science and truths in the twenty-first century.* Washington, D.C.: Counterpoint.
GAIA and Genetic Resources Action International (GAIA and GRAIN). 1998a. Intellectual property rights and biodiversity: The economic myths. *Global Trade and Biodiversity in Conflict,* no. 3.
———. 1998b. TRIPs versus CBD. *Global Trade and Biodiversity in Conflict,* no. 1.
Genetic Resources Action International (GRAIN). 1995. Towards a biodiversity community rights regime. Manuscript.
———. 1998. Patenting life: Progress or piracy? *Global Biodiversity* 7, no. 4:2–6.
G. G. 1998. C'est nous qui sommes modernes. *Dépêche* (4 February): 28.
Girod, J. C. 1997. Deux luttes exemplaires en France-Comté. *Campaigns Solidaires* no. 107 (April): 10.
Grueso, L., C. Rosero, and A. Escobar. 1998. The process of black community organizing in the southern Pacific Coast of Colombia. In *Cultures of politics/politics of cultures: Re-visioning Latin American social movements,* ed. S. E. Alvarez, E. Dagnino, and A. Escobar, 196–219. Boulder, Colo.: Westview Press.
Gudeman, S. 1996. Sketches, qualms, and other thoughts on intellectual property rights. In *Valuing Local Knowledge,* ed. S. Brush and D. Stabinsky, 102–21. Washington, D.C.: Island Press.
Heller, C. 2001. From risk to globalization: Discursive shifts in the French debate about GMOs. *Medical Anthropology Quarterly* 15:24–29.
Hervieu, B. 1996. *Au bonheur des campagnes (et des provinces).* Paris: Éditions de l'Aube.
Heywood, V. H., ed. 1995. *Global biodiversity assessment.* Cambridge: Cambridge University Press.
Janzen, D. 1992. A South-North perspective on science in the management, use, and economic development of biodiversity. In *Conservation of biodiversity for sustainable development,* ed. O. T. Sanlund, K. Hindar, and A. H. D. Brown, 15–26. Oslo: Scandinavian University Press.
Janzen, D., and H. Hallwachs. 1993. *All taxa biodiversity inventory.* Philadelphia: University of Pennsylvania (downloaded from the Internet Gopher, University of Pennsylvania).
Latour, B. 1983. Give me a laboratory and I will raise a world. In *Science observed,* ed. K. Knorr-Cetina and M. Mulkay, 141–70. London: Sage.
———. 1987. *Science in action: How to follow scientists and engineers through society.* Cambridge: Harvard University Press.
———. 1993. *We have never been modern.* Cambridge: Harvard University Press.
Le Déaut, J-Y. 1998. *De la connaissance des gènes à leur utilisation.* Tome 1. Paris: Assemblée Nationale, Senat.
Leff, E. 1993. Marxism and the environmental question. *Capitalism, Nature, Socialism* 4, no. 1:44–66.
Levidow, L., and S. Carr. 1997. European biotechnology regulation: Framing the risk assessment of a herbicide-tolerant crop. *Science, Technology, and Human Values* 22:4.
Marris, Claire. 1999. *Background context for public perceptions of agricultural biotechnologies in France: French national report for the Public Perceptions of Agricultural Biotechnologies in Europe Project.* Lancaster, U.K.: Lancaster University, Center for the Study of Environmental Change.

Martínez Alier, J. 1996. The merchandising of biodiversity. *Capitalism, Nature, Socialism* 7, no. 1:37–54.

Montel, M. 1998. Le vrai débat enfin lancé. *Dépêche* (4 February): 27.

Mooney, H. A., J. Lubchenko, R. Dirzo, and O. E. Sala. 1995. Biodiversity and ecosystem functioning. In *Global biodiversity assessment,* ed. V. H. Haywood, 275–452. Cambridge: Cambridge University Press.

Nelkin, D., and S. Lindee. 1995. *The DNA mystique: The gene as a cultural icon.* New York: W. H. Freeman and Company.

Pistorius, R. 1997. *Scientists, plants, and politics: A history of the Plant Genetic Resources Movement.* Rome: IPGRI.

Proceso de Comunidades Negras (PCN) and Organización Regional Embera-Waunana (OREWA). 1995. *Territorio, etnia, cultura e investigación en el Pacífico Colombiano.* Cali: Fundación Habla/Scribe.

Restrepo, E., and J. I. del Valle, eds. 1996. *Renacientes del guandal.* Bogotá: Proyecto Biopacífico–Universidad Nacional.

Ribeiro, G. L. 1998. Cybercultural politics and political activism at a distance in a transnational world. In *Cultures of politics/politics of cultures: Re-visioning Latin American social movements,* ed. S. E. Alvarez, E. Dagnino, and A. Escobar, 325–52. Boulder: Westview Press.

Riesel, R. 1998. Official statement made at the Conférence Citoyenne, Assemblée Nationale, Paris, 20 June.

Sánchez, E., and C. Leal. 1995. Elementos para una evaluación de sistemas productivos adaptativos en el Pacífico Colombiano. In *Economías de las comunides rurales en el Pacífico Colombiano,* ed. C. Leal, 73–88. Bogotá: Proyecto Biopacífico.

Shiva, V. 1993. *Monocultures of the mind.* London: Zed Books.

———. 1997. *Biopiracy.* Boston: South End Press.

Smith, N. 1996. The production of nature. In *Future/natural,* ed. G. Robertson et al., 35–53. London: Routledge.

Strathern, M. 1996. Potential property: Intellectual rights in property and persons. *Social Anthropology* 4, no. 1:17–32.

———. 1998. Cultural property and the anthropologist. Paper presented at Mt. Holyoke College, South Hadley, Mass., 8 December.

Swanson, T. 1998. *Global action for biodiversity.* London: Earthscan.

Third World Network and Research Foundation for Science, Technology, and Natural Research Policy. 1994. *Resource kit for building a movement for the protection of biodiversity and people's intellectual rights.* Kuala Lumpur: Third World Network and Research Foundation.

World Resources Institute (WRI). 1992. *Global biodiversity strategy.* Washington, D.C.: WRI.

———. 1994. *World resources, 1994–95.* Washington, D.C.: WRI.

Wright, S. 1993. The social warp of science: Writing the history of genetic engineering policy. *Science, Technology, and Human Values* 18, no. 1 (winter): 34–41.

Chapter 9

Future Imaginaries

Genome Scientists as Sociocultural Entrepreneurs

Joan H. Fujimura

Imagination is more important than knowledge.
ALBERT EINSTEIN

The image, the imagined, the imaginary—these are all terms that direct us to something critical and new in global cultural processes: the imagination as a social practice. No longer mere fantasy, no longer simple escape, no longer elite pastime, and no longer mere contemplation, the imagination has become an organized field of social practices, a form of work (in the sense of both labor and culturally organized practice), and a form of negotiation between sites of agency (individuals) and globally defined fields of possibility. The imagination is now central to all forms of agency, is itself a social fact, and is the key component of the new global order.
ARJUN APPADURAI

Imagination is a social practice deployed in the production of science and technology. Creating future imaginaries is a major part of scientists' work in the new biotechnologies that I study: genetics, artificial intelligence, and robotics research. Since these sciences are literally producing the future, I examine the social practices of imagining that form part of their work. I treat both imagining and laboratory experimentation as practices in which scientists are regularly engaged.

Science and technology have come to play increasingly important roles in defining the daily lives and bodies of people across the globe and defining the cultures and societies within which they live. The Human Genome Project, or more generally genomics, has triggered the imagination of scientists and society in ways that are not far from science fiction stories of the 1960s.

The term *genomics* refers to the new world created by molecular genetic sciences, information and computer sciences, and their institutional affiliates—the human genome projects in the United States, Japan, and Europe. This new world includes the transformation of genes into commodities in which biotechnology companies and venture capitalists have made major investments with the expectation of high profits; present and potential medical applications; and social, legal, and ethical concerns about the consequences of these technologies.

In addition to dramatically changing the production of knowledge in biology, genomics has begun to transform understandings of life, bodies, disease, health, illness, relatedness, identities, "nature," and "humanness," as well as the practical handling of related affairs. Although they occur in a global context, these transformations happen in different ways in different locations. In this essay, I focus on these specificities of the locations of scientific production.

Genomic science is simultaneously a national and a transnational enterprise. It is transnational in its flow of ideas, information, materials, protocols and practices, and people. The initiation and shape of the science was a product of national competition and collaboration among the United States, Japan, and several European nations.[1] Here I portray two Japanese scientists who present two different imaginaries for the biology and culture of the twenty-first century. Their examples illustrate three main points: First, genome scientists are imagining the future and sometimes transfiguring nature and culture through their work. They are building roads to our future and choosing where and how to build them. Second, scientific imaginings often are engaged with *other* contemporary cultural discourses. Third, these images and efforts to reinvent nature and culture and related cultural discourses must be historically situated to be understood. To appropriate another oft-used metaphor about the genome project, but with a significant one-word modification, genome scientists are writing *a* book of life. The form, content, and interpretation of this book may differ in different historical periods and locations.

JAPANESE DISCOURSES ON WESTERN TECHNOLOGY AND JAPANESE CULTURE

In twentieth-century Japanese discourses, *science-technology* and *culture* have been distinct and even contradictory terms.[2] Debates about premodernity, modernity, postmodernity, and nonmodernity in Japan are intertwined with discussions of culture and technology and of Japan's relationship to the "West."[3] The set of discourses currently called Nihonjin-ron, or Japanese cultural uniqueness, represents Japanese culture as existing in its purest form in the period before Japan's modernization. Although technology and science were being imported from the West in the mid–nineteenth century, those technologies and sciences were said to have been carefully translated through Japanese culture so as to make the foreign "native." "Japanese culture" here is represented as the source of a firm, unshakable self-knowledge, and technology as the foreign, Western object that required translation through culture.

The modern period is often represented as a time when serious conflicts between modernization and Japanese culture arose. Debates about importation of Western science and ambivalence about modernization began in the late 1800s (Harootunian 1970; Najita 1989; Pyle 1969).[4] During the Meiji era (1868–1912), often marked as the time when Japan was transformed into a modern industrialized nation-state, Japanese intellectuals began to criticize the modernization of Japan and the attendant introduction of Western knowledge and technologies, and this criticism persisted. The critics argued that technological progress through increasingly efficient production came at a price: it took precedence over unique Japanese aesthetic and cultural forms. *Modernity* and *the West* are conflated here. Japanese novelists of the time wrote essays and novels that expressed their concerns over this loss of Japanese culture to modernization and the West. The writer Natsume Soseki (1911, cited in Najita 1989: 11) argued that the Western system of knowledge and production was overwhelming Japanese culture and was producing a pervasive "nervous exhaustion" in place of social well-being and happiness.[5] His conclusion was that Western technology had produced a crippled personality and a crisis of culture in Japan. The writer Tanizaki Junichiro (1933, cited in Najita 1989: 12) argued similarly that the laws and epistemologies accompanying Western technology had "distorted the ethical and aesthetic sensitivities of the Japanese" and produced a form of self-colonization. He argued that the Japanese "should be self-consciously identifying with culture as an internalized space of resistance."

After World War II, this earlier discourse was modified to argue that Japan was the first Asian nation to modernize, because Japanese culture was compatible with technological development and industrialization. For example, the sociologist Robert Bellah (1985 [1957]) argued that Japan's early modernization was aided by "authentic Japanese values" like group harmony and loyalty, and individual and collective achievement.[6] A description of Japan's vertical stratification system, by the sociologist Nakane Chie (1970), and a description of *amae* (psychological dependence in relationships) by psychologist Doi Takeo (1973, 1986), were used similarly to articulate a social organization well suited for modernization. These values and patterns purportedly made it possible for the Japanese to be effective in the organized processes of high-growth economics.

In contrast to these social scientists, cultural nationalist writers like Mishima Yukio (1969) argued that Japanese culture was epistemologically and ethically different from the politics and technologies of modernization. He invoked the prewar discourse's notion of Japanese culture as a pure, authentic sphere to criticize the new consumer culture of high-growth economics and its accompanying bureaucracies (Ivy 1995; Pincus 1996).

In the postmodern period, again the main question has been whether or not rapid modernization and economic growth have actually helped Japanese social and cultural life. Have they instead created a homogenized and harried consumer culture with few intellectual and social benefits?[7]

Controversial writers on civilization theory in Japan carry these criticisms one step further.[8] They challenge the scientific authority of Euro-American nineteenth-century positivist philosophies and their claims to universal relevance. They note that the West has had the special historical privilege of being able to appear at once unique and universal. Writers like the economic historian Kawakatsu Heita "define post-enlightenment western thought as the product of a particular society and era, and open up the possibility that the traditions of other societies may contain the seeds of new theories to fill the gap left by the withering of old certainties" (Morris-Suzuki 1995: 760). The cultural traditions of other societies can then be the source of "better" civilizations and societies.[9]

However, theories of Japanese cultural uniqueness have not gone unchallenged. Most recently both Japanese and American writers in anthropology, history, and literature have begun to question and complicate these early representations of Japanese culture and the Japanese self.[10] Their studies interrogate how what is assumed to be a coherent, unified Japanese culture has been and is being constituted, reconstituted, deconstructed, and challenged through history. They critique the earlier Japanese literature on the Japanese self as yet another example of Orientalism, where Japanese writers created *themselves* as the Other in response to Western culture.[11] In contrast to both premodern and modern discourses, these histories, ethnographies, and cultural studies discuss the located and contingent courses of events, actions, and practices rather than static generalizations about Japanese social, cultural, and identity categories.[12] These historical discourses of Japanese cultural uniqueness form part of the context within and against which images of the West, the East, and Japan, of science and culture, are being constituted and reconstituted in the imagination and entrepreneurship of genomic scientists in Japan.

THE 1990S: THE JAPANESE HUMAN GENOME PROJECT AND GENOME INFORMATICS

In addition to physical and genetic linkage mapping, structural analysis, cDNA cloning and mapping, and sequencing, there has been a separate program set up to address computational aspects of biology in both the first and second five-year plans (1991–1996, 1996–2001) of the Japanese genome project.[13] The project called "Genome Informatics" has received a substantial portion of human genome funds, in part because of the efforts of a major architect of the Human Genome Project by Monbusho (the Ministry of Education, Sport, and Culture).[14] Professor Suhara (a pseudonym), a leading molecular biologist in Japan, was instrumental in making the Japanese Human Genome Project a reality in the late 1980s.[15] Suhara was influential in convincing the ministry to fund this research and, as head of the project in its first five-year period, in constructing the project in its initial and present format. Both before and during his directorship of the Japanese genome project, Suhara was involved in organizing the coordinated efforts of the Japanese, American, and European genome projects to map the human genome. He spent several years as a postdoctoral student in the United States and speaks excellent English, the de facto official language of genomics.

Suhara envisioned the genome project as more than the investigation of human genes. For him, it also represented the "installation" of a new science for twenty-first-century Japan. In his view, Japanese biological laboratories were not well integrated with scientific information networks in the United States and Europe. One of Suhara's primary missions was to create an infrastructure of "thick lines" of communication, both literally and figuratively. He accomplished this in the first phase of the genome project (1991–1996) by building into the funding and institutional structure of the genome project a major commitment to genome informatics, that is, biology using computer technology.[16] Suhara believed that traditional wet lab biology would decrease in importance in comparison to the emerging field of computational biology.[17] While many of his colleagues in Japan disagreed, he managed to persuade Monbusho officials to create a separate institutional and budgetary structure to fund genome informatics and computational genome analysis.[18]

Suhara's rationale was based on his view of Japan's noncompetitive position in the biological sciences. At the beginning of the 1970s, when molecular biology was becoming firmly established as the "new biology" in the United States and Europe, Japanese biologists were not interested. In Suhara's estimation, this indifference had left Japan behind in the field. He worried that the situation would repeat itself in genome informatics. Thus, in 1990, Suhara planned and argued for Japanese genomics to focus more resources on genome informatics, a field in which Japanese and American bioinformaticians were at about the same level of investment, expertise, and experience relative to their national budgets and personnel. Suhara believed Japan could make its mark in bioinformatics, in part because of its competitive position in the computing and information sciences.

Reinventing the East and the West via Genomics

Informatics includes techniques for comparing genetic and protein sequences and thus provides methods for comparing genomes of different species. These comparisons became Suhara's tool for promoting the genome project to the Japanese public and the basis of his imagined transformation of the West by genomic knowledge. His vision is apparent in this assessment of the significance of the Human Genome Project:

> Too much stress has been placed on human dignity [in the West]. It's much easier for us [in Japan] to accept [man's place in nature] because we have not been brought up under the influence of Christianity. Most Japanese are either Buddhist or Shinto, and they have a much wider view of all living things. They don't put man as the representative of God to be placed above all the other living things.[19] This attitude is very firmly imprinted during our childhood.
>
> [The Japanese have a] much cooler concept of man. We look at man as one [among other] living creatures. By slowly changing the concept of life, I think . . . our attitudes toward technology [and toward] making use of the Human Genome Project will be slowly changed, particularly in Asian countries where the majority of people are not living under the influence of Christianity or [Islam], but under the influence of Buddhism or Confucius *[sic]* or Shinto.
>
> Everybody's bound to the contract [with one God] in the Christian community. You don't have to change this [Christian] social contract. But you do have to get better views on what man is by taking the flow of information from the Human Genome Project and extend[ing] the thought on evolution to man, [e.g., the idea] that a man is a result of a process of nature, has very close ties with other living things, and has to live together [with them] on earth. Culture plays the most important role in accepting evolution and the life of man among other lives.

In Suhara's narrative, culture is a set of values imprinted on us in our early years that then governs how we act in the world. For him, religious differences between Eastern and Western cultures explain why the Christian West values humans above other animals.[20] Suhara uses his notions of East and West as a basis for criticizing Western actions and attitudes and promoting his view of Asian values and attitudes, which he in turn uses to promote genomics research in Japan.

Suhara appeals to a view of the Japanese as sharing a common culture steeped in a Buddhism and a Shinto that are radically different from Christianity. But in contrast to Nihonjin-ron claims that the principles of science and freedom in modern industrial civilization have led to negative social and spiritual consequences, Suhara emphasizes the harmonious effects of science. He contends that a science that seems to represent the epitome of modernist interventions into nature—that is, genomics and its accompanying manipulations—will produce knowledge of the harmony and relatedness of all living animal species and, especially, of humans with other animals. This relatedness fits well with both Buddhist and Shinto views as he understands them.[21]

Unlike Japanese critics of the Meiji adoption of Western technologies, Suhara does not equate science and technology with the West. Therefore, the adoption of molecular biology technologies is not an adoption of Western cultural views. Indeed, in his view, these technologies are resisted in the West precisely because they (or at least their rhetorics) threaten to impose evolutionary biological knowledge over what he calls "cultural knowledge" or "the Christian social contract."[22]

In Suhara's future imaginary, once this truth is known, the West will eventually have to change its concept of humans and other animals. He envisions a turn, or return, to nature or immanence in the more animistic sense of a Buddhism that has been heavily influenced by Shinto.[23] Although Suhara is interested in the potential medical payoffs and products proclaimed by many American technocrats as *their* reason for promoting genomics, his imaginary engages more with the potential transformation of Western cultural values.

In order to realize this imaginary, Suhara had to persuade the Japanese public to accept and support genomics. In this effort, his narrative subverts the tropes of both Nihonjin-ron Orientalism and Western Orientalism, where tradition and religion are attributes of a premodern Orient, and science and modernity are attributed to a modern Occident. In Suhara's rhetoric, there is no Western or Eastern science. Instead, there is only science. It is not science that is the problem in the West; it is culture. Western culture does not allow it to benefit from the fruits of its own technologies. Indeed, culture and science are in direct conflict and contradiction in the West.

This rhetoric was useful to Suhara's advocacy efforts. Cognizant that it takes work to prepare "the community" for the introduction of new technologies, Suhara spoke to, and with, various Japanese public groups before and during his tenure as director of the genome project's first five years. As he describes it,

> Actually, I have spent four or five years working very hard in [arranging for] the Genome Project to be acceptable in the community. . . . It's important. You have to be prepared before the community or society raises its hand to ask questions. What are the implications? What good will it bring about? What bad will it create? You must be prepared beforehand. Our experience is still that, if we fail in doing this, none of the scientific activities will get real support from the community.

Suhara prepared this strategy after learning from the experience of those who had previously tried to promote organ transplants in Japan, which he says is one of the worst examples of improper introduction of new technologies: "In organ transplantation, we are twenty years behind the world, because they were not welcomed here at the time. When people began asking questions and brought up some problems, the medical people were not able to answer them." The anthropologist Margaret Lock (2001) has pointed to cultural factors—the fact that in Japan death is defined as the death of the heart, not of the brain—that have slowed the acceptance of organ transplantation in Japan. In contrast to Lock's cultural explanation, Suhara insists that "the medical community failed in preparing, at least in educating the community and preparing for the questions and the protests against having these activities." For Suhara, acceptance of, or resistance to, organ transplantation is changeable. It does not occupy his pantheon of cultural values that are "firmly imprinted during our childhood."

Suhara also prepared scientists to deal with opposition to the project:

> I have spent some time in preparing scientific communities [for] the questions and the oppositions from the communities to the introduction of DNA research. There has been much reluctance and fears, just like [Jeremy] Rifkin's work has [generated] in the States, in the community. Some people are still trying hard [to generate these fears], but after [the] spread of [information], particularly for the high school kids, the attitude has become more and more a minority.

Public education was one of Suhara's missions and a significant part of his work as director of the human genome project.[24] He lectured to the public, to other scientific communities, and to the press about genomics, often in the face of community and media criticisms.

More important for my argument, Suhara's linking of his cultural imaginary to his efforts at public education was strategic. When arguing for funding from the government ministry, Suhara promoted the medical and pharmaceutical (e.g., drug design) benefits of the project as well as the development of information infrastructures. But he was aware that describing scientific benefits was not as useful when addressing public concerns about the social and ethical implications of the project. For these audiences, Suhara spoke to how genomics would change our understandings of life: "People in general are not so much interested in the forthcoming change in basic science or [the] setting up of the infrastructure of the scientific community. They are more interested in life."

Anthropologists, historians, and literary scholars writing about invented traditions have argued that such traditions have been invented and manipulated toward particular ends (e.g., Hobsbawm and Ranger 1983). Recent scholarship has demonstrated that many prominent Japanese "traditions" were in fact created during the late nineteenth and early twentieth centuries as Japan underwent the arduous, and at times traumatic, experience of modernization. For example, the historian Takashi Fujitani (1996) informs us that the system of emperor worship, with its many spectacles and "traditional" displays in modern Japan, was not a holdover from feudal times but instead a very modern invention of monarchy in the late nineteenth and early twentieth centuries by a particular class of people. This invention served both to produce a national culture over and against the many different ways of living that had existed in Japan until that time and to promote the interests of a few people: a ruling aristocracy and its retainers. Monolithic notions of Japanese culture as constituted by imperial traditions here served the purposes of particular people and not others.

Suhara's campaigns wielded notions of Shinto and Buddhism that have been similarly rendered in Japanese society as traditionally Japanese. In a fashion similar to the building of an infrastructure for the imperial system (e.g., the spectacles of emperor worship) and the production of the nation-state, Suhara worked to lay the foundations for public acceptance of genome research, technologies, and therapeutics. He saw it as his task as the first director of the Japanese human genome project to educate and prepare the public for this new future. Using Nihonjin-ron-like rhetorics and invented traditions, Suhara promoted very modern ends. This cultural imaginary contested the idea that the new science was bad for Japan while it simultaneously ratified Nihonjin-ron claims of cultural uniqueness.

Contrary to Nihonjin-ron discourse, Suhara's imaginary inferred a congruence, and even a necessary bond, between modern science and traditional Japanese cultural values. In Suhara's rhetoric, these traditional Asian values are critical to the success of science, in which science and Asian values will be confederates in the transformation of the world. Japan and Asia will become the leaders of a modernity based on religious and scientific understandings that show the Christian West to be premodern, irrational, and "traditional." Ironically, this transformation will happen because of molecular genetic technologies first developed in the West. The cutting-edge technologies that will, in Suhara's eyes, be instrumental to this transformation were developed in a culture that will change as a result of the cutting-edge knowledge it has produced. Genomic knowledge will make Western culture consistent with Asian culture and sensibilities. In Suhara's future imaginary, "Christianity will have difficulties in changing the concept [of humankind] in the near future when we know about the basic structure of the human genome; but still, that time will come." Suhara's vision of Japanese culture transforming the world has some parallels with the philosopher Ueyama Shumpei's vision of Japanese and Asian cultures as the sources of the antidote for the ills of civilization.

Although Suhara's notion of culture appears at first to be undertheorized, he uses his notion in a rhetorical strategy of reversal to weave a representation of the West as the place of tradition and the East as the forefront of a scientific modernity. Suhara is promoting modernity through nostalgia. That is, he uses tropes like Japanese traditions of Buddhism, Shinto, and their attendant views of nature to promote science and modernity. Modernity and tradition are not binaries in Suhara's vision. Instead, they are concomitant productions. They can exist together; each can even create the grounds for the existence of the other.[25] Suhara's rhetoric uses these "authentic" productions to promote the emblem of modernity—that is, science.

KITANO HIROAKI, THE GLOBAL SCIENTIST-ENTREPRENEUR

The nostalgic literature, or Nihonjin-ron discourses of cultural purity and uniqueness, contradict the realities of multiple border crossings and transnational interchanges in people, trade, education, fashion, music, and technoscience. Indeed, they may even be a conservative nationalist reaction to these transnational interchanges. While Suhara promotes interchange, he uses a cultural nationalist rhetoric that resonates with Nihonjin-ron and civilizationalist discourses.[26] In contrast, Kitano Hiroaki—a physicist turned computer scientist, robotics entrepreneur, and systems biologist—is a prime exemplar of border crossing, yet his transnational enterprises are also useful to the nationalist agendas of Japanese government bureaucrats.

A caption on the cover of the January 13, 1999, issue of *Japan Newsweek* reads, "The 21st Century's 100 Leaders: These are the stars who are opening the New Era in Politics, Business, and Art" (Seiki no riidaa hyakunin: Seiji, Business, Art shinjidai o kirihiraku shuyaku wa karera da). Among the thirty or so faces of political and media stars on the cover, a few spots to the right of Cameron Diaz and George W. Bush, is Kitano Hiroaki. In summer 1999, Kitano received an award from the Science and Technology Agency for his work on robotics. The governors of two Japanese prefectures were at the awards ceremony to speak about their RoboFesta events, inspired by Kitano's RoboCup. They were sponsoring RoboFesta to try to reverse the drop in interest in science among Japanese schoolchildren. In October 2000, one of Kitano's robotics designs was installed in the Venetian Biennaire Exhibition (2000), and it won the Prix Ars Electronica (Austria, 2000). In November 2000, under the auspices of the Science and Technology Corporation, Kitano organized the first international conference on systems biology. In February 2001, the editor of the Japanese business magazine *President* commissioned a writer to follow Kitano and write a portrait article on his various projects around the world. In June 2001, Kitano appeared on a Japanese television game show, where he talked about his lab's robotics research. Miniature dolls modeled on the lab's humanoid walking robot, PINO, went on sale in August 2001 at KiddyLand, the most famous toy store in Japan.

Who is Kitano Hiroaki and how has his vision of the future captured the attention of the popular media and the Japanese government? He was educated in Japan in particle physics. He worked in software engineering at NEC Corporation, a Japanese electronics company, and then moved to Carnegie Mellon University in Pittsburgh to conduct research in artificial intelligence. In 1993, he moved back to Japan to work at Sony Corporation on virtual reality modeling language and entertainment robotics (including Sony's AIBO; see below). In 1999 he established his own research institute. Meanwhile, he has created an international organization, RoboCup, to organize matches between soccer-playing robots in the interest of developing artificial intelligence.

Kitano is still officially employed by Sony but spends most of his days at his own not-(yet)-for-profit research institute in the cosmopolitan Harajuku section of Tokyo. In the middle of Tokyo, Kitano has built an institute that is combining research on advanced robotics, artificial intelligence, and systems biology. The institute is located on Omotesando Street, also home to the chic fashion houses of Issey Miyake, Yooji Yamamoto, and Rei Kawakubo, but its members and laboratories are located in other institutions both in Japan and across the Pacific, at the California Institute of Technology in Pasadena.

The Kitano Symbiotic Systems Project was funded by the Japan Science and Technology Corporation under the auspices of the Exploratory Research for Advanced Technology (ERATO) program. Kitano wrote not a single grant proposal for this project. Instead, ERATO officials, who had followed his work for some time, offered him a grant to conduct whatever research interested him. ERATO annually selects several scientists to receive their substantial awards. The agency's unofficial goal is to show that Japanese science can be innovative, in response both to intimations in American and European research institutions that Japanese science is merely mimetic and to direct statements that Japan exploits the innovations of the West to produce lucrative technologies—that it does not fund enough fundamental research. ERATO scientists also seem to be charismatic, creative, and imaginative people with an international orientation.

Kitano's research focuses on the designing and modeling of symbiotic systems—simulated biological and intelligence systems. On his web page at Sony, the description of his personal project states that the unified theme of his research is the "emergence and evolution of intelligence."

> Diverse approaches must be taken to tackle this grand problem. As a basic researcher, I am focusing on computational aspects of the evolution of neurogenesis and morphogenesis. Research on high-level intelligence is based on the genetic supervision theory and active perception, so that phenomena such as emotion and selective attention can be incorporated. In the long run, these issues will be integrated as "Symbiotic Systems Theory." A robust real-time translation of closed-caption and entertainment applications are expected fruits of these basic researches.

The following statement by the agency funding Kitano's research, and which appears on his web page, summarizes Kitano's project:

> Biology, unlike physics, has yet to find a unifying way to deal with its diverse subject matter. Hence, it has so far been impossible to use the great power of mathematical simulations to help overpower the inherent complexity and to gain greater predictive power.

> Kitano . . . is trying to recapture the essence of "complexity" and "symbiosis." "Complexity" is used concerning phenomena involving a very large number of elements interacting in a very non-linear complex way; "symbiosis" is well-known through work in cellular evolution and the Gaia hypothesis, that Earth is a self-regulating cybernetic system in which all of the many species interact to maintain homeostasis (Japan Science and Technology Corporation and ERATO 1999: 6)

Kitano's rhetoric combines the cultural capital that physics and mathematics have accrued as a result of their high positions in the hierarchy of scientific disciplines, with the literal capital that computer science holds in Japanese government, academia, and private industry, as well as with the cultural capital that the Gaia hypothesis holds in international popular culture. The Gaia theme is compatible with the Shinto theme of the harmony of all living things.[27] But by using Gaia rather than Shinto in his rhetoric, Kitano avoids Japanese cultural nationalism. It has public appeal without being nationalistic.

Kitano's rhetoric has worked because the funding agencies are based in government ministries. These ministries have been mandated to increase Japan's international presence in the world. In response to accusations from outside Japan that it has not been a sufficiently responsible international citizen, the Japanese government has instituted a campaign to internationalize at all levels of society.[28] As noted earlier, Kitano's rhetoric convinced these bureaucrats of the usefulness of his project because of its international appearance and appeal and because of its novelty.

There are two parts to Kitano's strategies for tackling the grand problem of "the emergence and evolution of intelligence." The first part is to model development in organisms using techniques from artificial intelligence and computer science research. The second is to use the results of molecular developmental biology and molecular neuroscience research to develop novel methods for building intelligent robotics systems.

Systems Biology

Using bioinformation databases, computer modeling and simulation, artificial intelligence tools, and complexity theory, Kitano is attempting to model organic developmental systems. He wants to establish methodologies and techniques that enable scientists to understand (1) the structure of the biological system, such as gene-metabolic-signal transduction networks and physical structures, (2) the dynamics of such systems, (3) methods to control systems, and (4) methods to design and modify systems for desired properties (Kitano 2002: 1662).

Many computational researchers moving into biology believe that the field has become too complex for biologists to handle. They argue that the mass of information now stored in databases primarily in the United States, Europe, and Japan must be analyzed by people experienced with masses of complex data. One proposed approach is to build and use simulations to understand the dynamical properties of biological systems—for example, developmental systems. This approach may yield answers to questions such as: How does the body develop from its original single cell? How do cells differentiate into lung and heart and muscle cells when they all have the same DNA in their nuclei? Another approach is to develop software to mine the data for nuggets of knowledge that can be used to produce pharmaceuticals and other therapeutics. The first approach could be used in the long run to produce therapeutics, while the hope of the latter is to develop products in the short term. Kitano is interested in the former.

Kitano's interest in developmental biology derives from his interest in developing models of intelligent systems: "To develop artificial intelligence, you need to know about the actual reality of the brain. You need to know how the brain evolved, how it functions, how it grows. I was particularly interested in evolution and growth, so my primary interest was in developmental biology."[29]

The Kitano project combines work in computer simulations, dynamic systems, and molecular biology. Kitano states, "It aims to establish a new comprehensive methodology, systems biology, which emphasizes an understanding of a biological system as a system." Using systems theory, Kitano's group is creating software that simulates a developmental process in order to identify the dynamics of genes and proteins. He observes, "Ideally, it should be possible to predict a certain gene X having a specific location and function [using the simulation model]. If the result is not consistent with reality, this method also includes automatic hypothesis generation and a mathematical model to suggest the shortest research path to identify the gene. This process shows great promise to obtain a better basic understanding of biology while providing improved predictive and preventive medicine."

Kitano's group is currently working on three projects, including studies of human cell aging, developmental systems (eye, wing, leg, segmentation) in *Drosophila,* and embryogenesis and neural systems of the worm *Caenorhabditis elegans.* According to Kitano and his colleagues, "The goal of this research program is to establish a new paradigm of biological research. . . . We propose that computer simulation which models mechanisms of biological processes should be used together with actual biological experiments, instead of using abstract mathematics describing average behavior of the system, so that results of simulation[s] (which are virtual experiments), can be verified by tangible experiments" (Kitano et al. 1997: 275). The "virtual cell laboratory" is creating detailed simulations of intracellular genetic interactions and metabolic cascades. It involves modeling and analyzing aging, differentiation, and the cell-cycle regulation of human cells.

More specifically, in one experiment, the researchers are trying to model aging by first translating it into a problem of cell senescence. Kitano and the microbiologist Imai Shin-Ichiro worked together to study cell aging by building a probability model that used computer simulations of the cell aging process. Even though they did not believe that cellular processes are random, their model is based on random generation of cellular events. The researchers attempted to match the results from the computer model with results of wet lab experiments. For example, they produced graphed curves of life spans of cells produced by their computer algorithm and compared these to curves produced in the wet lab experiments. Their comparison showed that their simulated curves matched the wet lab experimental results better than curves produced by other models.[30] More important, their results suggested to them an innovation in aging research: the idea that cellular senescence involves two independent regulatory processes (Kitano and Imai 1998).

Kitano and Imai jointly created this model of aging. In 1996, they had produced interesting simulation results that identified a particular gene as critical to aging and had sent a paper reporting these simulation results to a well-regarded biology journal. The journal editor returned the paper, saying that they needed wet lab proof that the gene was important in aging processes. Imai's laboratory did the experimental work of isolating and testing the gene, and they then resubmitted the paper for publication. The journal editor then said that, since they had located the relevant gene, the theoretical and simulation work was irrelevant.

This journal's response was not unusual. Until recently, theoretical or computational biologists have had a difficult time convincing wet lab scientists that their work is worthy of regard, funds, and publication. While neural network research has received some support from psychologists and molecular biologists like Francis Crick, theoretical biology generally has not been held in high regard since the advent of molecular biology.[31] Some researchers in the field call this "the other two-culture problem." Dry lab versus wet lab competitions in biology usually have been won by wet lab molecular biologists, in part because wet lab biologists have controlled the means of production in biology. But computational biology slowly has managed to make its presence felt in recent years. Biologists have been willing to have computer scientists and mathematicians help them figure out how to map and locate genes and how to search for similarities, but they have been loath to grant computational biology status equal to that of molecular biology. But the other side has its own version of elitism. For the past several years, I have studied computational biologists doing what they call data extraction or data mining. Some of them have argued that their methods allow them to produce better, unbiased knowledge than the molecular biologists who, according to their computational rivals, work in such great detail with one gene or protein or part of a protein that they cannot see the forest for the trees.

Kitano decided to bypass the battle in order to get his work done. He enrolled several wet and dry lab scientists in his project to study developmental pathways and to design software systems to assist in this process. However, most of the project's early articles were published in artificial intelligence journals and in biology journals like *Experimental Gerontology,* which the researchers consider less prominent than *Cell, Nature,* and *Science.*

Another hurdle for Kitano's project was, in his view, the present state of biology in Japan. Like Suhara, Kitano recognizes the lead that the United States has held in molecular biology and the competitive position that Japan has held in computer science. Japan has excelled in the production of robotics, imaging techniques, cybernetics, artificial intelligence, and simulation technologies like virtual reality.[32] But rather than engage in a competitive response to this situation as Suhara did, Kitano takes advantage of the situation by organizing research teams that are transinstitutional and transnational. He says,

> If U.S. research is more advanced, we can create a team with American researchers. It's not completely conflict-free, but we can team up with the top American biologists, because what they are doing and what we are doing are usually complementary. It's not competitive. So if we come up with a very good model, they can take advantage of that, if we make a team.

Kitano's team currently includes computer scientists, biologists, and engineers from several institutions in Japan and California. Kitano has built a new laboratory at the California Institute of Technology that incorporates molecular biology, computing, artificial intelligence, and robotics research in one physical laboratory.

Artificial Intelligence

The second part of Kitano's imaginary includes a plan to use systems biology to reanimate artificial intelligence and robotics research:

> Current research is aiming at the development of novel methods for building intelligent robotics systems, inspired by the results of molecular developmental biology and molecular neuroscience research. Symbiotic intelligence . . . incorporates a new type of robotics system having many degrees of freedom and multimodal sensory inputs. The underlying idea is that the richness of inputs and outputs to the system, along with the coevolving complexity of the environment, is the key to the emergence of intelligence. As many sensory inputs as possible, as well as many actuators, are being combined to allow smooth motion and then [are] integrated into a functional system. The brain

> is an immense system with heterogeneous elements that interact specifically with other elements. It is surprising how such a system can create coherent and simple behaviors which can be building blocks for complex behavioral sequences, and actually [can] assemble such behaviors to exhibit complex but consistent behavior.

Aiding this symbiotic systems design for both biology and artificial intelligence is Kitano's global venture, RoboCup, a trademarked nonprofit organization registered in Switzerland in 1997. RoboCup organizes and coordinates a World Cup series of soccer matches played annually by teams of robots (both real and simulated) from thirty-five countries. Kitano has also worked to develop the AIBO, a robotic pet, developed by Sony. The AIBO plays soccer, kicks and heads the ball, rolls over, and stands up after being tackled. At the Third Robot World Cup, AIBO dogs played in three-member teams. Sony has also claimed that the AIBO has instincts and emotion, and that it can autonomously act on these capabilities. But RoboCup and the AIBO are not just fun and games. They are part of an effort to use the collaboration and competition between robots and robot teams to study and then improve the interactive capabilities of robotics systems. Kitano has also proposed developing robotic systems for use in large-scale disaster rescue operations. The 1995 Kobe earthquake demonstrated the vulnerability of human rescue systems. Rescue personnel themselves were victims of the disaster and unable to carry out their responsibilities. Rescue robots that can be controlled from afar, and computerized systems for simulating possible disaster and rescue scenarios, are the serious work of and rationale for these robots, while play and entertainment are what animates them. Work and play produce a powerful synthesis. For Kitano, these robotic and artificial intelligence systems also provide an immediate practical benefit for his systems biology project: he can attract top robotics engineers and designers to work on developing artificial intelligence and engineering systems to assist in designing his biological systems.

Future Possibilities?

Is the goal of systems biology just knowledge and understanding? No, says Kitano: "Overall, the project aims to obtain a breakthrough in the methodology for understanding, controlling, and creating biological systems." Kitano imagines cloning human organs from human cells in a precisely controlled manufacturing system. Simulation technologies, control technologies, and instrumentation could be used to clone organs customized for each person: manufactured life. Projects in the United States aimed at cloning human organs from stem cells have yet to devise the precise technologies for growing organs from an individual's cell or DNA. Kitano's idea is to simulate such growth and development before ever moving into the production stage. As noted above, organ donation is still a rare event in Japan. When transplantation has occurred, the organs have come from outside Japan—which has also produced criticism from both within and outside the country. In this climate, cloning organs to order is a provocative offer, despite the uneasiness that many people might have about cloning.

Long-range imaginings are part of Kitano's rhetoric. Systems biology will allow humans to intervene in the natural workings of bodies to prevent problematic outcomes and to build organs that can be transplanted when the natural ones are frail or damaged.

Kitano's visions are not universally shared. Researchers in computational biology criticize his particular framing of the project, although they like his use of the name *systems biology*. And it is not clear that his visions can be realized. Nevertheless, he has created a demand, audience, and resources for himself, his visions, and his work.

CONCLUSION

These instances of technosocial design for the future serve as examples of the social practice of imagination. These future imaginaries are distinct from fantasy, especially in the sense that fantasy refers to "thoughts disconnected from projects and actions" (Appadurai 1996: 7). Indeed, the imaginaries presented here are visions of future possibilities around which scientific practices and communities are organized. They are collective enterprises and not simply an individual's dreams. They are the products and producers of networks of humans that include cultural intellectuals and writers, government bureaucrats and politicians, molecular biologists, artificial intelligence researchers, robotics engineers and designers, business executives and long-range planners and the public. These networks also can be said to include technologies—nonhuman actors, in Bruno Latour's words—without which scientific imaginaries cannot be made into projects and actions.

Technosocial imagining is serious work done by serious people. The work of the two scientists I discuss here has led to enterprises that have enrolled and engaged many people, funds, and government agencies, and much public and consumer interest. Through their work we see new possible futures in the making.

But are such visions merely hyperbole, the rhetorical strategies of persuasion that lead to nothing but a waste of good resources? Should we not wait to see what actually happens before we study such projects? I argue that if we wait for the future to become the past, we leave the design of the future to others. *Especially* if one does not support the possibilities being imagined today, it is critical to study them. I am arguing for a sociology of the future.

In the design of the future, rhetorical strategies of persuasion are in part a requirement of imagination as a social practice because the future is being imagined as *a new possibility*, as something that does not already have a constituency. Hyperbole is part of the very process of persuading people to support that new possibility and, thus, of convincing them to change their present ways. For example, when Kitano projects into the future his vision of what systems biology, robotics, and artificial intelligence could become, he relies on an imagined vision of future possibilities to win people's support. Kitano's imaginary is a world where simulated biological and intelligence systems help us understand and emulate the emergence and evolution of intelligence, which in turn can be used to solve complex problems, build intelligent robotics systems, produce disaster rescue systems, clone human organs for transplantation, and otherwise provide improved predictive and preventive medicine. Kitano's vision provides both animated machines and manufactured life. To achieve his goals, Kitano must create cultural capital literally from the imagination. Hyperbole can be seen as a means to that end.

Future imaginings also have to be located in their present contexts. Appadurai's concept of "globally defined fields of possibility" refers to the fact that possible futures never before imagined by people in one locality are now available to them because of the influence and reach of global media. But these fields of possibility are also produced and consumed within particular contexts, particular locations.

For example, Suhara uses Nihonjin-ron and civilizationalist discourses in Japan to produce an image of genomics that will reshape Western culture to be more in line with Eastern cultural values. Genomics will transform the assumptions of the "Christian West." Suhara has turned cultural nationalist arguments around and used them to attempt to convince the Japanese public that new genomic technologies are more congruent and even harmonious with traditional Japanese and Asian cultural values than they are with Western cultures.

Suhara uses nationalism to promote a transnational, even postnationalist, project. Kitano's supporters use an internationalist agenda to promote a national agenda, while Kitano uses their nationalist agendas to promote his transnational project. Contradictions and complications abound—and can become resources.

Both of these scientists are imagining what they might create with DNA, amino acids, and other biological information collected in databases around the world. And by doing so, they and their colleagues are participating not merely in the practice of science but also in redesigning culture and society. Genomic scientists are building maps of genomes, national and transnational identities, notions of culture, new institutions, and future realities. Although one could argue that genomic maps, cloned organs, and visions of nature are first-order products, and that notions of national identities, culture, and institutions are by-products, these aspects of nature and culture are inseparable. Social and cultural organization may not be first-order objects of the everyday practice of scientists, but they are clearly tools manipulated in their efforts to produce genomes and other such scientific knowledge and medical technologies. Both their manipulation of these tools and their specifically scientific products have consequences for the constitution of society and culture. It is critical to pay attention to the practices of scientists as social actors and to the future worlds they are imagining. Politicians, political philosophers, sociologists, and anthropologists may mark off their territories, and may designate themselves as the makers and keepers of society and culture, but science and technology have already demonstrated their powers in making and remaking culture and society.

NOTES

This paper has benefited from many readers and audiences, of whom I can list only a few. I thank my colleagues at the Wenner-Gren International Symposium on "Anthropology in the Age of Genetics," my colleagues in the School of Social Science at the Institute for Advanced Study, and the Princeton University and University of Pennsylvania Joint Workshop in History of Science who heard and read different versions of this paper. For their detailed comments, I also thank Adam Ashforth, Troy Duster, Clifford Geertz, Ken George, Alan Goodman, Marta Hanson, Deborah Heath, Deborah Keates, Claire Kim, Dorothy Ko, Rob Kohler, Roddey Reid, Danilyn Rutherford, Joan W. Scott, and Sharon Traweek. I am grateful to the respondents who took the time and effort to talk with me about their work and allowed me to observe their activities. Finally, I thank the Social Science Research Council, the Abe Foundation, the Henry R. Luce Foundation, and the Institute for Advanced Study for their support of this research and writing.

1. See Cook-Deegan 1994 and Fujimura 2000 for stories of the international competition that led to the development of the American Human Genome Initiative.

2. The two terms are complex and paradoxical, and their interrelations are intertwined. Morris-Suzuki (1995) discusses "some of the paradoxes of the concept [civilization] by examining its evolution in the work of a few Japanese scholars who have reflected particularly deeply on the meaning of 'culture' as a framework for analysis."

3. These debates assumed that there was and is an essential, bounded Japanese culture that would be violated by "Western" ideas. This discourse also assumed a desire to maintain Japanese culture as it had been conceived collectively by these intellectuals. In contrast, recent work in anthropology, history, and cultural studies has demonstrated that culture is not synonymous with nation-state. For a discussion of these recent arguments, see Gupta and Ferguson 1992.

4. However, the discourse of certainty in the "inviolate" Japanese culture continued in the language of Japanese exceptionalism until circa 1910.

5. Names of Japanese authors living and writing in Japan are given in this essay with family name first.

6. Bellah partially recanted on this claim in the revised version of his book *Tokugawa Religion* (1985).

7. Miyoshi and Harootunian (1989) point out that this question applies to Europe and America as well.

8. For Western versions of this kind of civilization theory, see Huntington 1996.

9. The philosopher and cultural theorist Ueyama Shumpei has similarly argued that modern industrial civilization, based on the principles of science and freedom, has produced great benefits for humanity but has also produced negative social and spiritual consequences—"poisons" to which it is necessary to seek an antidote. Ueyama credited Western culture for the origin of industrialization and argued that Japan, with its roots in Chinese civilization, was, therefore, a more promising source for this antidote (Ueyama 1990, as discussed in Morris-Suzuki 1995: 740). For critiques of Kawakatsu's and Ueyama's work, see Fujii 1998 and Morris-Suzuki 1995.

10. These include Befu 2001; Field 1993; Fujii 1993; Fujitani 1996, 1998; Ivy 1995; Kondo 1990, 1997; Morris-Suzuki 1995; Ohnuki-Tierney 1993; Sakai 1997; Tanaka 1993; Traweek 1992; Vlastos 1998; Yoneyama 1995.

11. For example, immersed in the discourse of Japanese cultural uniqueness, authors like Doi (1973) have created an image of a Japanese self as interdependent in contradistinction to the image in the West of a self that is individualist and independent. Both images are cultural productions, and the image of Japanese interdependence was viewed as one of Japanese Orientalizing themselves. For critiques on Nihonjin-ron literature, see especially Kondo 1990; Ivy 1995; and Sakai 1997; see Said 1979 on Orientalism.

12. Some historians and anthropologists use these histories and anthropologies of Japan to question the organization of area studies in the American academy (e.g., Harootunian and Sakai 1999).

13. The acronym cDNA refers to DNA in the genome that actually codes for genes in more complex organisms such as humans. The cDNA for a particular gene appears as split fragments in the genome. In the laboratory, cDNA can be produced from mRNA (messenger RNA). Scientists are interested in cDNA because it allows them to study the expression of human genes in tissues.

14. The Japanese genome projects have been undertaken on a scale much smaller than those in the United States; the research was organized into five different projects, each belonging to a different government ministry. Japanese ministries are organized in a top-down fashion, and each ministry is highly competitive with other ministries, although this traditional division of labor recently has begun breaking down. (For example, several agencies now fund different aspects of the research being conducted at the Monbusho-funded Human Genome Center at the University of Tokyo.) Aside from Monbusho, the ministries involved in genomics research include the Science and Technology Agency, the Ministry of Agriculture, Koseisho (the Ministry of Health and Human Welfare), and the Ministry of International Trade and Industry. Monbusho and the Science and Technology Agency have recently been combined to form Monkasho (the Ministry for Education, Science, and Culture). Although Monbusho's genome project was the largest, with researchers and facilities in universities throughout Japan, other agencies have been expanding their projects. In the face of recent financial crises, the Japanese government has also decided to boost the economy by putting more money into the development of biotechnology. As a result of this decision, competition between ministries is beginning to take second place to coordinating the expertise for developing biotechnol-

ogy and competing with the biotechnology industries of the United States and Europe.

15. Professor Suhara's comments throughout the essay are from several interviews I conducted with him, the first in summer 1994, the last in summer 1999.

16. Genome informatics, or bioinformatics, is the study of biological information at all scales, from the study of submolecular functional groups and bond lengths to that of molecules, especially DNA and proteins. Bioinformatics researchers aspire to include other kinds of information on organisms and environments, but most of the information collected in databases to date pertains to the molecular level.

17. "Wet lab" biology refers to the research done on living biological materials, in contrast to "dry lab" research, which uses computational tools to examine and manipulate biological information.

18. Walter Gilbert, a molecular biologist at Harvard University and winner of the Nobel Prize, was thinking along the same lines at that time. In 1991, he published an article in the journal *Nature* titled "Toward a Paradigm Shift in Biology," which was subtitled "Molecular Biology Is Dead—Long Live Molecular Biology." In it, he argued that molecular biology would have to reinvent itself as an information science to have relevance in the future.

19. This representation ignores, among other things, the exploitation of forests and marine life around the world by Japanese companies.

20. In contrast, the discipline of anthropology has been entangled in controversies over its central organizing concept, culture. The notion of culture has been criticized, articulated, rearticulated, rejected, defended, and embraced. Suhara elides these agonistic struggles.

21. I qualify this statement because there are many different versions of both religions in many different countries and throughout history.

22. Suhara refers not only to creationism but also to other Christian views.

23. See Ketelaar (1990) on the conflicts between Buddhism and Shinto in Japan. Through this process, Buddhism transformed into a more immanent view of spirituality.

24. See Fujimura 1996 (chapter 7) on articulation work. Scientific work takes many different forms, including public education. Experiments at the laboratory bench are just one kind of work practice.

25. Indeed, for Foucault, tradition was an effect of the discourse of modernity.

26. See n. 8 on civilization theory, which promotes a view of the world as being divided into several major civilizations that are incommensurable.

27. Lovelock's Gaia hypothesis describes the earth as a complex, self-regulating cybernetic system. However, New Agers have understood this to mean a Goddess-guided system, and this spiritual definition has traveled far. Kitano subscribes to the Lovelock definition but benefits from the wide appeal of the term.

28. James Fujii (1998) and Marilyn Ivy (1995) discuss the Japanese state-sponsored *kokusaika* (internationalization) as an effort to domesticate the foreign, not as an actual opening of Japanese industry and education to the outside. Jennifer Robertson argues that internationalization is both a product of and central to the ongoing (since the Meiji period) formation of a Japanese national cultural identity (1998: 129).

29. Kitano Hiroaki, interview by author, 19 July 1999, 54. Except where otherwise attributed, all quotes by Kitano are from this interview.

30. Other projects include the attempt to model the genetic and enzyme cascades of yeast. Kitano says they "will [also] study the development of neural systems in order to understand how very large complex systems can be evolved and developed." In another project, Kitano and his colleagues are building a detailed simulation of the embryogenesis and neural system of *C. elegans,* with a focus on "a detailed model of a gene regulatory network for cell fate determination."

31. Francis Crick and James Watson won the Nobel Prize for proposing the double-helix structure of DNA.

32. Indeed, the translation of this technological advantage into commercial success during the 1980s produced both the trade wars and the praise and criticisms of Japan that accompanied them. See Morley and Robins (1995) for the effect of these wars on discussions of Japanese culture.

REFERENCES

Appadurai, A. 1996. *Modernity at large: Cultural dimensions of globalization.* Minneapolis: University of Minnesota Press.

Befu, H. 2001. *Hegemony of homogeneity: An anthropological analysis of Nihonjinron.* Honolulu: TransPacific Press.

Bellah, R. 1985 [1957]. *Tokugawa religion: The cultural roots of modern Japan.* New York: Free Press.

Cook-Deegan, R. 1994. *The gene wars: Science, politics, and the human genome.* New York: W. W. Norton.

Doi, T. 1973. *The anatomy of dependence.* Tokyo: Kodansha.

———. 1986. *The anatomy of self.* Tokyo: Kodansha.

Field, N. 1993. *In the realm of a dying emperor: Japan at century's end.* New York: Vintage Books.

Fujii, J. A. 1998. Internationalizing Japan: Rebellion in Kirikiri and the International Center for Japanese Studies. *Journal of Intercultural Studies* 19:149–69.

Fujimura, J. H. 1996. *Crafting science: A socio-history of the quest for the genetics of cancer.* Cambridge: Harvard University Press.

———. 2000. Transnational genomics in Japan: Transgressing the boundary between the "modern/West" and the "pre-modern/East." In *Cultural studies of science, technology, and medicine,* ed. R. Reid and S. Traweek, 71–92. New York: Routledge.

Fujitani, T. 1996. *Splendid monarchy: Power and pageantry in modern Japan.* Berkeley: University of California Press.

———. 1998. Minshushi as critique of orientalist knowledges. *Positions* 6, no. 2:303–22.

Gilbert, W. 1991. Toward a paradigm shift in biology: Molecular biology is dead—long live molecular biology. *Nature* 349:99.

Gupta, A., and J. Ferguson. 1992. Beyond culture: Space, identity, and the politics of difference. *Cultural Anthropology* 7:6–23.

Harootunian, H. D. 1970. *Toward restoration: The growth of political consciousness in Tokugawa Japan.* Berkeley: University of California Press.

Harootunian, H., and N. Sakai. 1999. Dialogue: Japan studies and cultural studies. *Positions* 7, no. 2:593–647.

Hobsbawm, E., and T. Ranger. 1983. *The Invention of tradition.* Cambridge: Cambridge University Press.

Huntington, S. P. 1996. *The clash of civilizations and the remaking of world order.* New York: Simon and Schuster.

Ivy, M. 1995. *Discourses of the vanishing: Modernity, phantasm, Japan.* Chicago: University of Chicago Press.

Japan Science and Technology Corporation and Exploratory Research for Advanced Technology (ERATO). 1999. *Japan Science and Technology Corporation and ERATO program bulletin.* Tokyo: Japan Science and Technology Corporation.

Ketelaar, J. E. 1990. *Of heretics and martyrs in Meiji Japan: Buddhism and its persecution.* Princeton: Princeton University Press.

Kitano, H. 2002. Systems biology: A brief overview. *Science* 295:1662–64.

Kitano, H., S. Hamahashi, J. Kitazawa, K. Tako, and S-I. Imai. 1997. The virtual biology laboratories: A new approach to computational biology. *Proceedings of the Fourth European Conference on Artificial Life,* 274–83.

Kitano, H., and S-I. Imai. 1998. The two-process model of cellular aging. *Experimental Gerontology* 33:381–91.

Kondo, D. K. 1990. *Crafting selves: Power, gender, and discourses of identity in a Japanese workplace.* Chicago: University of Chicago Press.

———. 1997. *About face: Performing race in fashion and theater.* New York: Routledge.

Lock, M. 2001. *Twice dead: Organ transplants and the reinvention of death.* Berkeley: University of California Press.

Lovelock, J. 1998. *Ages of Gaia: A biography of our living earth.* New York: W. W. Norton.

Mishima, Y. 1969. *Bunka oei ron* [Discussion on the defense of culture]. Tokyo: n.p.

Miyoshi, M., and H. D. Harootunian. 1989. Introduction to *Postmodernism and Japan,* ed. M. Miyoshi and H. D. Harootunian, vii–xx. Durham, N.C.: Duke University Press.

———, eds. 1989. *Postmodernism and Japan.* Durham, N.C.: Duke University Press.

Morley, D., and K. Robins. 1995. *Spaces of identity: Global media, electronic landscapes, and cultural boundaries.* New York: Routledge.

Morris-Suzuki, T. 1995. The invention and reinvention of "Japanese culture." *Journal of Asian Studies* 54, no. 3:759–81.

Najita, T. 1989. On culture and technology in postmodern Japan. In *Postmodernism and Japan,* ed. M. Miyoshi and H. D. Harootunian, 3–20. Durham, N.C.: Duke University Press.

Nakane, C. 1970. *Japanese society.* Berkeley: University of California Press.

Natsume, S. 1965 [1911]. Gendai Nihon no kaika [The enlightenment of modern Japan]. In *Han kindai no shiso,* ed. Fukuda Tsuneari, 53–72. Gendai Nihon shiso taikei 31. Tokyo: n.p.

Ohnuki-Tierney, E. 1993. *Rice as self: Japanese identities through time.* Princeton: Princeton University Press.

Pincus, L. 1996. *Authenticating culture in imperial Japan: Kuki Shuzo and the rise of national aesthetics.* Berkeley: University of California Press.

Pyle, K. B. 1969. *The new generation in Meiji Japan: Problems of cultural identity, 1885–1895.* Stanford: Stanford University Press.

Robertson, J. 1998. It takes a village: Internationalization and nostalgia in postwar

Japan. In *Mirror of modernity: Invented traditions of modern Japan,* ed. S. Vlastos, 110–29. Berkeley: University of California Press.

Said, E. 1979. *Orientalism.* New York: Vintage.

Sakai, N. 1997. *Translation and subjectivity: On "Japan" and cultural nationalism.* Minneapolis: University of Minnesota Press.

Tanaka, S. 1993. *Japan's orient: Rendering pasts into history.* Berkeley: University of California Press.

Tanizaki, J. 1965 [1933]. In'ei raisan [In praise of shadows]. In *Han kindai no shiso,* ed. Fukuda Tsuneari, 114–46. Gendai Nihon shiso taikei 31. Tokyo: n.p.

Traweek, S. 1992. Border crossings: Narrative strategies in science studies and among physicists in Tsukuba Science City, Japan. In *Science as Practice and Culture,* ed. A. Pickering, 429–65. Chicago: University of Chicago Press.

Vlastos, S., ed. 1998. *Mirror of modernity: Invented traditions of modern Japan.* Berkeley: University of California Press.

Yoneyama, L. 1995. Memory matters—Hiroshima's Korean Atom Bomb Memorial and the politics of ethnicity. *Public Culture* 7, no. 3:499–527.

Chapter 10

Reflections and Prospects for Anthropological Genetics in South Africa

Himla Soodyall

On June 16, 1999, day six of the Wenner-Gren International Symposium that led to this volume, a significant event was taking place in my home country: the inauguration of President Thabo Mbeki—the second president elected in democratic South Africa. Although I was not at home to be part of the celebration, I watched some of the proceedings on television. Several issues that President Mbeki raised in his inauguration speech delivered at Pretoria, South Africa, struck me as crucial to both the discussions at our symposium and the challenges facing academics in the "new" South Africa:

> We will also work to rediscover and claim the African heritage, for the benefit especially of our young generations. From South Africa to Ethiopia lie strewn ancient fossils that, in their stillness, speak still of the African origins of all humanity. Recorded history and the material things that time left behind also speak of Africa's historic contribution to the universe of philosophy, the natural sciences, human settlement and organisation and the creative arts.
>
> Being certain that not always were we the children of the abyss, we will do what we have to do to achieve our own Renaissance. We trust that what we will do will not only better our own condition as a people, but will also make a contribution, however small, to the success of Africa's Renaissance, towards the identification of the century ahead of us as the African Century.[1]

Mbeki stresses two points: African heritage and the African Renaissance. The African Renaissance theme, which he has promoted since 1996, aims to advance the "rebirth and renewal of our continent" and emphasizes the role Africans have played in the history of the world. The African Renaissance seeks to build a new world, "one of democracy, peace and stability, sustainable development and a better life for the people, non-racism and non-sexism, equality among the nations, and a just and democratic system of international governance" (Mbeki 1999). For two days in September 1998, 470 delegates from all over Africa gathered in Johannesburg to deliberate a spectrum of issues (Makgoba et al. 1999). These included: Who are the Africans? Where do they come from? What is their history and where are they going? Has Africa a history of scientific and technological culture? How can Africa best harness and exploit its natural and indigenous resources and its human diversity for its own benefit?

Africa has been an astonishing crucible for the earth's history for the last 2 billion years. At an international scientific congress held in South Africa in 1998, Philip Tobias reached the conclusion that "nearly everything of note or consequence started in Africa. . . . Africa is the home of the first eukaryotes, the first mammals, the first hominids, the first marked enlargement of the brain, the first signs of spoken language, the oldest evidence of stone tools, the oldest testimony of the mastery and control of fire" (Diop 1999).

Few Africans, however, realize the magnitude of Africa's contributions to the development of the world. The African Renaissance conference was conceived as "part of a small contribution towards a larger process of our history, our consciousness, our roots and our realities" (Makgoba et al. 1999).

While there are a number of important social issues that have to be addressed in the African context, a major challenge to all Africans is to restore African pride, culture, and identity. Mbeki reminds us,

> None of us can estimate or measure with any certainty the impact that centuries of the denial of our humanity and contempt for the colour black by many around the world have had on ourselves as Africans. But clearly it cannot be that successive periods of slavery, colonialism and neo-colonialism, and the continuing marginalisation of our continent could not have had an effect on our psyche and therefore our ability to take our destiny into our own hands." (1999)

These views are echoed in the words of one conference delegate, Sémou Pathé Guéye:

> No future and no renaissance can be envisaged with peoples who are psychologically defeated and have lost their confidence in themselves and their ability to change their own situation according to their own needs and aspirations. We therefore have to restore the self-confidence of Africans, their pride and the historical internal dynamics of their cultures, by recalling the original contribution of Africa, the continent where human history began, to the process of world civilisation. (1999)

This essay is an exercise in exploring the "original contribution of Africa." Molecular anthropology offers another perspective on the rich history of the peoples of Africa, making use of the tools of molecular biology to identify changes in the genomes of living people and reconstruct the evolutionary history of these changes to a point of common ancestry in the past. Research conducted in my laboratory has embraced the new technology to examine the genetic affinities of southern African populations (Khoisan, Bantu-speakers, Khoisan-speaking Negroids, "Coloureds" and seaborne immigrants) with other sub-Saharan African populations. In addition, genetic information has to be used in conjunction with data gleaned from the studies of historians, anthropologists, archaeologists, linguists, and palaeontologists before any meaningful conclusions can be reached. Of course, as my opening suggests, this essay is not only about the value of molecular anthropology in understanding the history of southern African populations, but is also about some of my experiences as a researcher in South Africa in the midst of an emerging African Renaissance.

THE REVOLUTION IN GENETICS AND MAPPING RELATEDNESS

Before the "new genetics," scholars interested in the peoples of southern Africa made use of a number of disciplines to reconstruct their history, including linguistics, archaeology, physical anthropology, cultural anthropology, history, and paleoanthropology. The genetic era in South Africa began within a few years of the demonstration by Hirschfeld and Hirschfeld (1919) that the frequencies of the genes of the ABO blood group system varied among populations. The Hirschfelds' pioneering paper demonstrated that the three common genes of the ABO system occurred in all populations even though their frequencies differed from one population to the next (Jenkins 1988). A. E. Mourant (1961) pointed out the importance of this study, "not simply as making the discovery of one particular anthropological character but as being the first application to anthropology of a totally new method, the study of gene distribution: since there was no necessary distinction between the individuals of one population and of another, the populations themselves became the units of study."

South African researchers like Harvey Pirie (1921) and Adrianus Pijper (1930, 1932, 1935) embraced the anthropology and exploited the value of blood groups in "a search for the real past of this vast continent" (Dart 1951). As additional blood typing systems, like Rhesus, MNSs, Kell, and Duffy, were discovered in the 1940s and 1950s, they were incorporated into studies to better understand genetic diversity in southern African populations (Zoutendyk et al. 1953, 1955).

Trefor Jenkins, working at the South African Institute for Medical Genetics and the University of the Witwatersrand, adopted these approaches. He recalls:

> Although my studies on the peoples of southern Africa began, I suppose, as early as 1960 when I was a medical officer at Wankie (now Kwange) in Zim-

> babwe (then southern Rhodesia), they were at that stage rather amorphous and without a clear objective. I had the good fortune to be encouraged by Desmond Clark . . . and I had the even greater good fortune to find myself three years later on the staff of Philip Tobias's department. (1988)

Jenkins had the foresight to recognize the value of the molecular technology as it became available in South Africa in the mid-1980s. He started the first molecular genetics laboratory in the country, coupling human genetics with anthropology.

The dramatic technological advances of the past fifteen to twenty years have produced powerful tools for present-day scientists, including geneticists. These advances have revolutionized my own understanding of genetic variation in human populations. In 1987, Rebecca Cann, Mark Stoneking, and the late Allan Wilson claimed that mitochondrial DNA (mtDNA) found in modern humans originated in an African ancestor who lived about two hundred thousand years ago (Cann et al. 1987). These researchers advanced a method of high-resolution DNA mapping in which they digested mtDNA with a number of different restriction enzymes (proteins isolated from bacteria, which recognize specific sequences of DNA and then cut DNA at specific nucleotide positions) and then examined the products of digestion to map the different patterns of variation in mtDNA.

The discovery of the polymerase chain reaction (PCR) brought another revolution to the study of DNA variation. It allowed researchers to amplify mtDNA from small amounts of starting material taken from hair, bone, teeth, cheek swabs, and blood spots, among other sources. This technique, coupled with automated methods for DNA sequencing, is now the method of choice for DNA analysis.

Now chip technology and robotics are enhancing our capabilities of deciphering the human genome. Electronic media are valuable sources of information that can be accessed by anyone around the globe. We cannot fight the technological era but must embrace it and make it part of our daily activities without losing sight of what is humane and what is right.

When I joined the Department of Human Genetics as a medical scientist in March 1987, the first challenge I faced was that of purifying mtDNA. Jenkins was keen to introduce DNA markers like mtDNA and Y chromosome markers to the ongoing studies on genetic variation in local populations to supplement studies based on serological typing that made use of blood groups and serum proteins. Having been successful at purifying mtDNA and perfecting the technique for screening for variation using restriction enzyme mapping, I demonstrated that there was a high level of mtDNA variation in local populations. This sparked my interest in human population genetics and my quest to reconstruct the history of the peoples of southern and sub-Saharan Africa.

FEARS, CRITICISMS, AND CONCERNS

Many of the objections to molecular anthropology are analogous to the ethical, legal, and social criticisms of the Human Genome Diversity Project (HGDP). Concerns include:

1. The ethics of doing biomedical research in developing countries and the problem of insuring the protection of human subjects (for example, it is difficult to ascertain what "informed consent" means in different cultures—that is, to know if an individual is voluntarily participating in the sample-collection process);
2. The legal issues that spring from the possible commercial value of the project's samples or results; and
3. The social and political issues surrounding the possible misuse or misinterpretation of the information generated. (Collins and Galas 1993)

These are extremely important concerns that must be addressed. But it is also important to notice the benefits of anthropological genetic research in South Africa.

The concept of race has over the years helped to reinforce human prejudice—nowhere, of course, more visibly than in South Africa (Jenkins 1988). Tobias points out that, during the apartheid era,

> the entire life and destiny of everyone in South Africa was determined by an artificial, arbitrary and totally unscientific system of race classification.... Every citizen's supposed race was determined, usually by ludicrous and insulting means, such as pencils being placed in the hair of a subject! ... Once labelled ... each person was subjected to a series of laws governing where she or he was permitted to live, what jobs might be held or not held, what schools and universities might be attended or not attended, what entertainment, hotel and restaurant facilities might be enjoyed or not enjoyed, with whom marriage and sexual relations were permitted or forbidden. (1998)

He reminds us:

> It shocks one today to realise that some scholars from abroad—and also in South Africa [—] considered the Khoisan, and especially the San, to have been less than human or even on a side-branch from the rest of humanity. The Swedish naturalist, Linnaeus, an obsessional classifier of living and dead things, first placed human beings among the Primates in the Animal Kingdom. While he put modern humans in the species *Homo sapiens,* he classified the Khoikhoi as a separate species, which he deemed to be on a side-branch of human evolution. That was in the 18th century. But the idea that the Khoisan were a completely different species of creature persisted, astonishingly enough, into the 20th century. (1998)

Anthropological and genetic research has helped resolve several myths of the peoples of Africa. Tobias dismissed the myth advanced by the human geneticist F. Lenz of Göttingen—that "neither African pygmies nor Bushmen interbreed with Negroes or with Europeans"—by demonstrating "fertile unions between San and Whites," thus dispelling scientifically "what was possibly the last surviving remnant of the school of thought that wished to place the Khoisan in a separate category from the rest of mankind" (Tobias 1954, 1998).

Another myth asserted that the Khoisan were not Africans, and that they had reached the continent from Asia. In 1668, O. Dapper referred to the Khoisan as "yellowish Javanese." From early European visitors, there was reference to the "Chinese Hottentots," a name that the Europeans gave to the Gonaqua in the eastern Cape Province (Stow 1905). The supposed linkage of the Khoisan with Asians was also supported by Robert Broom (1923, 1941) on the basis of physical features and one set of blood groups. Later, Raymond Dart (1952, 1954) opined that the "alien racial features" had been transmitted to the Khoisan by Asians during a period of active Oriental contacts with the East African seaboard (Tobias 1998). Subsequent genetic research has demonstrated that the genetic makeup of the Khoisan relates them more closely to the peoples of Africa than to any other people (Weiner and Zoutendyk 1959; Weiner et al. 1964; Barnicot et al. 1959; Nurse and Jenkins 1977; Nurse et al. 1985; Soodyall and Jenkins 1992, 1998).

I am optimistic that our research can play a major role in contributing to the African Renaissance and in restoring pride and a sense of identity to every South African. In the words of Sydney Brenner, "Molecular genetics has become so directed toward medical problems and the needs of pharmaceutical companies that most people do not recognise that the most challenging intellectual problem of all time, the reconstruction of our biological past, can be tackled with some hope of success" (1998).

THE HISTORY OF SOUTHERN AFRICAN POPULATIONS

Historical, linguistic, anthropological, and archaeological studies together confirm that the twentieth-century people who often were referred to collectively as the Khoisan are the aboriginal inhabitants of southern Africa. The term *Köisan* (later *Khoisan*) was coined by Leonard Schultze (1928) in his biometric study of "Hottentot" and "Bushman" populations, intended for use as a biological label. However, Isaac Schapera (1930) popularized the term *Khoisan* as a cultural and linguistic label. The term was derived by combining *Khoi* (part of the term *Khoikhoi,* used by the Khoikhoi, or "Hottentots," to refer to themselves) with *San,* the term the Khoikhoi used to refer to the hunter-gatherers, or "Bushmen." Although the San and Khoikhoi are not easily distinguishable in physical appearance, certain cultural and linguistic differences are apparent. Whereas the San are classically hunter-gatherers and keep no domesticated animals, the Khoikhoi were, until recently, pastoralists who herded large flocks of sheep and cattle (Vedder 1928; Ehret 1982a,b).

Khoisan history is also closely connected with the history of people of the later Stone Age. San art, tools, and other remains of the hunter-gatherer lifestyle date as far back as 20,000 years ago, but their social structure is most clearly traceable during the Holocene, the last 10,000 years. Only during the latter period has there been consistent evidence of deliberate burial of the dead by Stone Age people in South Africa found in shell middens along the coast and in rock shelters along the western and eastern Cape and in the Cape Folded Mountains. The herder way of life associated with people who spoke Khoe languages (Khoikhoi) extends back about 2,000 years in South Africa (Deacon and Deacon 1999).

Southern Africa received three major waves of immigration in the last two millennia; the first, which arrived within the last 2,000 years, was made up of peoples speaking Bantu languages; the second, within the last 350 years, was composed of seaborne European immigrants; and the third, within the past 100–120 years, was composed of indentured laborers from India and the Malay Archipelago.

The migration of Bantu-speakers to southern Africa was the result of the "Bantu expansion." The Bantu expansion hypothesis was formulated by linguists in the late 1880s. It was hypothesized that Bantu languages originated in West Africa in the region of the boundary between present-day Cameroon and Nigeria approximately 3,000–5,000 years ago (Guthrie 1962). One wave of migration, associated with western Bantu culture, is thought to have arisen in the region of the Cameroon grassland between 1600 and 700 B.C., before spreading to parts of west-central Africa and southwestern Africa (Vansina 1984). Another wave of migration, hypothesized as having taken place around 1,000 B.C. from the Bantu homeland in West Africa, resulted in the spread of Bantu-speakers to central and eastern Africa, giving rise to the eastern Bantu family of languages (Guthrie 1962). Some of these Bantu-speakers eventually migrated into southern Africa, and these routes have been mapped using Early and Late Iron Age technologies from the archaeological record (Ehret 1982a,b; Huffman 1980, 1982, 1989; Phillipson 1977, 1985). A branch of eastern Bantu-speakers who subsequently migrated into southern Africa, possibly as recently as about 1,000 years ago, gave rise to the southeastern Bantu-speaking groups—the predominant language group in South Africa.

Prehistoric and recent contact between Khoisan and indigenous non-Khoisan populations has resulted in several cultural and linguistic exchanges between them (de Almeida 1965; Ehret 1982a,b; Westphal 1963). For example, the Dama, who are current residents of Namibia, are biologically

Negroid but, through the course of history, have acquired the language spoken by the Nama, presumably a consequence of their enslavement by the Nama (Nurse et al. 1976). Also, the Kwengo of southern Angola and the Kavango region of Namibia, referred to by the Portuguese anthropologist Antonio de Almeida (1965) as the "Black Bushmen," are biologically Negroid but have acquired the language and culture of the Sekele, or "Yellow Bushmen" as de Almeida prefers to call them. Furthermore, the click sound characteristic of Khoisan languages is a trait borrowed by some Bantu-speaking groups.

GENETIC AFFINITIES AND ORIGINS OF SOUTHERN AFRICAN POPULATIONS

Differences in gene frequencies among the various people of southern Africa, and indeed other sub-Saharan African populations, shed light on the interrelationships and origins of southern African populations. Classical genetic markers like the immunoglobin allotype $Gm^{1,13}$ haplotype, the Duffy Fy^a, and acid phosphatase P^r alleles have been found in the San and Bantu-speakers but not in central and West Africans. These markers have been used to estimate the amount of gene flow from the San to different Bantu-speaking chiefdoms (Jenkins et al. 1970; Jenkins 1972). These studies confirmed that the migrations of Bantu-speakers into southern Africa resulted in varying degrees of admixture between the incoming Bantu-speakers and the resident Khoisan.

Gender specific markers (mtDNA and Y chromosome) are particularly useful in shedding light on the way in which admixture occurred. Since mtDNA is maternally inherited in the absence of recombination, phylogenetic relationships between mtDNA types reflect the maternal genealogical relationships between individuals sampled. Several mtDNA polymorphisms have been used to trace population ancestry. The region between the genes that code for cytochrome oxidase II and the transfer RNA for lysine normally contains two copies of a 9 base pair (BP) repeat. The loss of one copy of the repeat is referred to as the 9-bp deletion. The deletion is not found in Khoisan peoples and is rare or absent in West and southwestern African populations. However, the deletion does occur in the Aka and Mbuti groups in central Africa and in southeastern Bantu-speakers in southern Africa (Soodyall et al. 1996). The frequency and the distribution of the 9-bp deletion can be explained by the migration of Bantu-speakers due to the Bantu expansion.

MtDNA control-region sequence data reveal that certain polymorphisms found at high frequencies in the Khoisan are found in some Bantu-speaking chiefdoms (Soodyall and Jenkins 1998) and in the enigmatic Dama and Kwengo groups at frequencies of 21.1 percent and 58.7 percent, respectively. These frequencies represent the proportion of female gene flow from the

Khoisan into these groups. Also, certain mtDNA polymorphisms found commonly among Bantu-speakers have been introduced into the Khoisan through gene flow (Soodyall 1993; Soodyall and Jenkins 1992, 1993). Approximately 26 percent of mtDNA types found in the Nama (Khoikhoi), 41.3 percent in the Kwengo, and 78.9 percent in the Dama are derived from Bantu-speakers.

We also dated a Khoisan-specific cluster of mtDNA types, using the mtDNA mutation rate of one mutation every 20,180 years calibrated by Peter Foster and colleagues (1996), to 120,000 years ago. Elizabeth Watson and colleagues (1997) previously estimated that the most recent common ancestor lived between 111,000 and 148,000 years ago. The present data, therefore, provide the first direct evidence that the most ancient mtDNA signatures, dating back to about 120,000 years ago, have been retained in contemporary Khoisan individuals. These data suggest that modern humans could have originated in southern Africa instead of East Africa as is commonly suggested.

The nonrecombining portion of the human Y chromosome has become an important tool for population and evolutionary studies. Its exclusive paternal inheritance and lack of recombination with other chromosomes makes it an attractive genome in tracking the history of mutational events recorded in the Y chromosomes of present-day males to previous generations and, ultimately, to a point of common origin or coalescence. Using nine diallelic polymorphic sites on the Y chromosome, Michael Hammer and colleagues (1998) resolved into ten haplotypes the Y chromosomes found in 1,544 males drawn from a worldwide sample. By comparing the variants found in humans with primates at each locus, they established the root of the Y chromosome haplotypes and showed how all ten haplotypes were related evolutionarily within a network.

Haplotype 1A, established as the ancestral haplotype in the network, was found exclusively in sub-Saharan African populations, but was restricted in its distribution among the populations examined: it was present in 20 percent of Khoisan individuals, in 3 percent of West Africans, in 4 percent of southeastern Bantu-speakers, and in 5 percent of the Dama. More recently, Rosaria Scozzari and colleagues (1999) found that haplotype 1A was also present in 22 percent of Ethiopians, in 4 percent of West Africans, in 2.5 percent of Cameroonians, in 28 percent of the !Kung (Sekele), and in 11.5 percent of Khwe (Kwengo) from southern Africa. Haplotype 1A is estimated to have arisen approximately 145,000 years ago (Hammer et al. 1997), but further resolution of this haplotype using faster evolving short tandem repeat (STR) markers reveals that STR haplotypes in the San are unrelated to those found in other Africans (Scozzari et al. 1999). Thus, Y chromosomes associated with haplotype 1A in San and Ethiopians have diverged substantially since their time of origin. In addition, Y chromosome studies lend further support to the results of mtDNA studies showing that southern African Khoisan populations claim the greatest antiquity in Africa.

GENETIC DIVERSITY AND DISEASE

Medical genetics in South Africa has benefited enormously from studies centering on genetic diversity and disease. Largely through the efforts of Tobias (1962), who argued that medical genetics ought to be included in the curriculum for medical students, two human genetics departments were created in South Africa by 1975 (Jenkins 1990). The interests of Trefor Jenkins—appointed chair of the Department of Human Genetics at the University of the Witwatersrand in 1975—in population genetics and hematology has resulted in a better understanding of the frequencies and distribution of a number of diseases prevalent in the region.

As Jenkins points out in a review of medical genetics in South Africa (Jenkins 1990), the founder effect has played a significant role in the high frequencies of a number of diseases in the Afrikaans population, for example, porphyria variegata and familial hypercholesterolemia. The demonstration of significant, albeit low, frequencies of certain mutant alleles associated with diseases such as cystic fibrosis, galactosemia, and albinism, extending over vast areas of sub-Saharan Africa (Stevens et al. 1997; Manga et al. 1999; Padoa et al. 1999), indicates migrations and assimilation of formerly small populations in which these mutations had attained high frequencies as "private" polymorphisms. These observations further attest to the biological unity of Bantu-speakers—a consequence of the Bantu expansion.

The populations being studied suffer from diseases shaped by both environmental and genetic forces. These must be understood in tandem. As Ernst Mayr (1961) reminds us, to fully understand disease, we must first understand the population's history and the evolutionary factors that contribute to variation at the gene level. Genetic studies incorporating these two ideas, together with the application of technologies currently available and those that might be developed in the future, undoubtedly would help strengthen medical science and improve the primary health care infrastructure, to the eventual benefit of the whole population (Jenkins 1995).

ANTHROPOLOGY IN THE AGE OF GENETICS: PRACTICE, DISCOURSE, AND CRITICISM

One of the major criticisms of molecular anthropology deals with the collection of blood or other tissue samples. Over the past several years, I have confronted such issues in my fieldwork in different regions of sub-Saharan Africa and Madagascar.

For several years I worked on material stored in freezers in the department that had been collected by Trefor Jenkins and colleagues before I had an opportunity to conduct research outside the confines of the laboratory. I accompanied a colleague, who had had limited fieldwork experience, to Schmidsdrift, situated approximately one hundred kilometers north of Kimberly in northern Cape Province, to collect blood samples from San men who were members of the South African Defence Force (SADF). The San soldiers came from southern Angola and had previously been based at the Omega Military Base in the Caprivi Strip close to the Angola-Nambia border. At the time of Nambia's independence in 1990, they were relocated to Schmidsdrift. The soldiers and their families, consisting of about four thousand Sekele and one thousand Kwengo, were living under canvas in facilities provided by the SADF. After obtaining permission from the commander at the base, and with the help of a medical officer attached to the unit, we set up our workstation in the clinic within the settlement.

At an informal gathering, the medical officer and my colleague explained the purpose of our research. Thereafter, several individuals volunteered to contribute a blood sample for our studies. We collected about one hundred samples from the two groups of San at the base. Although at the time I felt comfortable that the participants had voluntarily donated their blood samples, I wondered later if, for soldiers, a request from a ranked medical officer in the SADF could have been interpreted by the soldiers as a command.

A few years later I learned that some researchers from a neighboring university had also visited the San at Schmidsdrift and the soldiers were resampled. I am sure that these scientists would have followed the same procedures we undertook in obtaining the necessary permission to collect samples from the soldiers. Two things bother me about this: The first is that the authorities at the camp could have informed these scientists of our earlier visit and could have suggested that the researchers meet with us to share samples. Second, the blood samples collected this time were sent by the local scientist to a laboratory in the United States, where they were transformed into cell lines, despite the fact that the local researcher who played "postman" knew about the interest of our department at the University of the Witwatersrand and my research in particular.

There are a few lessons to be learned from this incident and what scientists in developing countries experience in general. First, if South Africa had some kind of coordinated human genome project, such exploitation might be less likely to occur. Trained field-workers could, with the help of anthropologists, keep accurate records of information about the populations that would contribute to a more robust analysis of the generated genetic data. Second, the pressures for collaboration by South African scientists with overseas researchers often result in samples being sent away for testing—with perhaps coauthorship or an acknowledgement for their troubles. If coauthorship is offered, often the overseas "collaborator" writes the manuscript without even consulting the South African researcher. Sending samples to be tested overseas precludes opportunities for development of technology locally and, hence, poorer training of African scientists. If overseas scientists were seriously interested in collaborations with researchers in the developing world, with mutual benefit, then there would be some hope of training local scientists, thereby improving the numbers, and level of training, of scientists in developing countries. Another problem I have faced personally is that sometimes researchers from abroad request samples for one purpose and then cannot resist the temptation to use the material for other purposes without obtaining the appropriate permission to do so.

Despite these trying circumstances, the lesson I appreciate most from my fieldwork experiences is the value of a DNA sample. Having worked in areas characterized by abject poverty, inadequate facilities, and stressful living conditions, I know that a DNA sample is much more than just a chemical you can retrieve from the freezer. It represents a person, his or her community, his or her history.

One of the highlights of my career has been to work with Trefor Jenkins in the field. His compassion and concern for the people among whom we work is striking. He has the ability to joke with both the young and the old, making them feel comfortable almost immediately. It does not bother him if, after days of chatting with people, they do not want to donate a sample of blood for our studies. His emphasis is on completely voluntary participation; and he will not turn away any person who volunteers to donate a blood sample irrespective of his or her ethnic background. I am sure that these humanitarian attributes have contributed to the success Professor Jenkins has had over the years in collecting blood samples wherever he has worked. I aspire to possess these humanitarian values and attitudes toward research throughout my own career.

CONCLUSIONS

The growing understanding of human genetic diversity emphasizes the point that, if we are to understand the causes of disease, including complex disorders, we must study all humankind. We must understand the evolutionary histories of allelic variants for normal genetic markers at candidate loci, since, if a locus really has genetic variation influencing susceptibility to a complex trait, that variation also has an evolutionary history and will be tied to the history of the adjacent normal DNA sequence variation. Our understanding of human population histories also relates to how readily a finding in one ethnic group or geographical region will generalize to other populations: only some generalizations are valid, and the causes common in one group may not be the common causes of the same apparent disorder in another group.

We cannot fully understand the global picture of variation at any locus without a better picture of the variation in Africa. Almost all polymorphisms arose in Africa, and only a small fraction of some of these variants spread to other parts of the world when humans left Africa to populate the rest of the world. While most genetic studies have focused on disorders affecting populations in the developed world, little work has been conducted in sub-Saharan African populations to elucidate the genetic factors implicated in the etiology and pathogenesis of disease. We can learn more about diseases by understanding the evolution of normal and disease-related genes and by examining the geographic patterning of the genetic differences among living populations, in particular, sub-Saharan Africans. By examining various ethnic groups in southern Africa and comparing their genetic structure with other sub-Saharan African populations, it would be possible to identify the source(s) of genes in South African populations and to reconstruct the history of southern African populations. These data would be invaluable in assisting national health care programs to initiate appropriate measures to deal with diseases that occur at different frequencies among ethnically classified groups.

My research has focused on the value of genetic studies in reconstructing the history of the people of Africa. The integration of genetic studies with other studies like linguistics, archaeology, anthropology, and history ought to enhance our understanding of the history of African people. Before I can achieve my research objectives, I must overcome a number of challenges. The ultimate challenge for me, and indeed all other researchers, is not to lose sight of the fact that all research and technological advancement ought to be directed toward the betterment of all humankind. Moreover, we must not overlook the rights of individuals and the risks involved and the humanitarian aspect of research.

NOTES

I feel particularly privileged to have had the opportunity of participating in the Wenner-Gren International Symposium, and I wish to extend my gratitude to the organizers for inviting me. I am most grateful to my colleagues Dr. Bharti Morar and Professor Trefor Jenkins for their contributions in making this research possible and for their critical evaluation of and comments on this manuscript. My colleagues join me in extending our gratitude to all volunteers who donated blood samples and to Dr. Thomy de Ravel for his assistance with fieldwork. This research was supported by the National Health Laboratory Service (formerly known as the South African Institute for Medical Research), the South African Medical Research Council, and the University of the Witwatersrand.

1. Found on the Internet at Polity, a policy and law on-line news source,

http://www.polity.org.za/govdocs/speeches/1999/sp0616.html (accessed on 2 December 2002).

REFERENCES

Barnicot, N. A., J. P. Garlick, R. Singer, and J. S. Weiner. 1959. Haptoglobin and transferrin variants in Bushmen and some other South African peoples. *Nature* 184:2042.

Brenner, S. 1998. The impact of society on science. *Science* 282:1411–12.

Broom, R. 1923. Contribution to the craniology of the yellow-skinned races of South Africa. *Journal of the Royal Anthropological Institute* 53:132–49.

———. 1941. Bushmen, Koranas, and Hottentots. *Annals of the Transvaal Museum* 20:217–49.

Cann, R. L., M. Stoneking, and A. C. Wilson. 1987. Mitochondrial DNA and human evolution. *Nature* 325:31–36.

Collins, F., and D. Galas. 1993. A five-year plan for the United States Human Genome Program. *Science* 262:43–46.

Dart, R. A. 1951. African serological patterns and human migrations. Paper presented at the South African Archaeological Society Meeting, Cape Town, South Africa.

———. 1952. A Hottentot from Hong Kong. *South African Journal of Medical Sciences* 17:117–42.

———. 1954. *The Oriental horizons of Africa.* Johannesburg: Witwatersrand University Press.

Deacon, H. J., and J. Deacon. 1999. *Human beginnings in South Africa: Uncovering the secrets of the Stone Age.* Cape Town: David Philip, Publishers.

de Almeida, A. 1965. *Bushmen and other Non-Bantu peoples of Angola.* Johannesburg: Witwatersrand University Press.

Diop, D. 1999. Africa: Mankind's past and future. In *African renaissance: The new struggle,* ed. M. W. Makgoba, 3–9. Johannesburg: Mafube and Tafelberg, Publishers.

Ehret, C. 1982a. The first spread of food production to southern Africa. In *The archaeological and linguistic reconstruction of African history,* ed. C. Ehret and M. Posnansky, 158–81. Berkeley: University of California Press.

———. 1982b. Linguistic inferences about early Bantu history. In *The archaeological and linguistic reconstruction of African history,* eds. C. Ehret and M. Posnansky, 57–65. Berkeley: University of California Press.

Foster, P., R. Harding, A. Torroni, and H-J. Bandelt. 1996. Origin and evolution of Native American mtDNA variation: A reappraisal. *American Journal of Human Genetics* 59:935–45.

Guéye, S. P. 1999. African renaissance as an historical challenge. In *African renaissance: The new struggle,* ed. M. W. Makgoba, 243–65. Johannesburg: Mafube and Tafelberg, Publishers.

Guthrie, M. 1962. Some developments in the prehistory of the Bantu origins. *Journal of African History* 3:273–82.

Hammer, M. F., T. Karafet, A. Rasanayagam, E. T. Wood, T. K. Altheide, T. Jenkins, R. C. Griffiths, A. R. Templeton, and S. L. Zegura. 1998. Out of Africa and back again: Nested cladistic analysis of human Y chromosome variation. *Molecular Biology and Evolution* 15:427–41.

Hammer, M. F., A. B. Spurdle, T. Karafet, M. R. Bonner, E. T. Wood, A. Novelletto, P. Malaspina, R. J. Mitchell, S. Horai, T. Jenkins, and S. L. Zeruga. 1997. The geographic distribution of human Y chromosome variation. *American Journal of Human Genetics* 145:787–805.

Hirschfeld, L., and H. I. Hirschfeld. 1919. Serological differences between the blood of different races: The result of researches on the Macedonian front. *Lancet* 2:675–79.

Huffman, T. N. 1980. Ceramics, classification, and Iron Age entities. *African Studies* 39:123–74.

———. 1982. Archaeology and ethnohistory of the African Iron Age. *Annual Review of Archaeology* 11:133–50.

———. 1989. Ceramics, settlements, and Late Iron Age migrations. *African Archaeological Reviews* 7:155–82.

Jenkins, T. 1972. Genetic polymorphisms of man in southern Africa. M.D. thesis, University of London.

———. 1988. *The peoples of southern Africa: Studies in diversity and disease.* Raymond Dart Lectures, no 24. Johannesburg: Witwatersrand University Press.

———. 1990. Medical genetics in South Africa. *Journal of Medical Genetics* 27:760–79.

———. 1995. Genetic diseases in the tropics. In *Tropical pathology,* ed. G. Seifert and W. Doerr, 61–123. Berlin: Springer-Verlag.

Jenkins, T., A. Zoutendyk, and A. G. Steinberg. 1970. Gammaglobulin groups (Gm and Inv) of various southern African populations. *American Journal of Physical Anthropology* 32:197–218.

Makgoba, M. W., T. Shope, and T. Mazwai. 1999. Introduction to *African renaissance: The new struggle,* ed. M. W. Makgoba, i–xii. Johannesburg: Mafube and Tafelberg, Publishers.

Manga, N., T. Jenkins, H. Jackson, D. A. Whittaker, and A. B. Lane. 1999. The molecular basis of transferase galactosaemia in South Africa. *Journal of Inherited Metabolic Disease* 22:37–42.

Mayr, E. 1961. Cause and effect in biology. *Science* 134:1501–6.

Mbeki, T. 1999. Prologue to *African renaissance: The new struggle,* ed. M. W. Makgoba, xiii–xxi. Johannesburg: Mafube and Tafelberg, Publishers.

Mourant, A. E. 1961. Evolution, genetics, and anthropology. *Journal of the Royal Anthropological Institute* 91:151–65.

Nurse, G. T., and T. Jenkins. 1977. *Health and the hunter-gatherer: Biomedical studies on the hunting and gathering populations of southern Africa.* Basel: S. Karger.

Nurse, G. T., A. B. Lane, and T. Jenkins. 1976. Sero-genetic studies on the Dama of South West Africa. *Annals of Human Biology* 3:33–50.

Nurse, G., J. S. Weiner, and T. Jenkins. 1985. *The peoples of southern Africa and their affinities.* Oxford: Clarendon Press.

Padoa, A., A. Goldman, T. Jenkins, and M. Ramsay. 1999. Cystic fibrosis carrier frequencies in populations of African origin. *Journal of Medical Genetics* 36:41–44.

Phillipson, D. W. 1977. The spread of the Bantu language. *Scientific American* 236:106–14.

———. 1985. *African archaeology.* Cambridge: Cambridge University Press.

Pijper, A. 1930. The blood groups of the Bantu. *Transactions of the Royal Society of South Africa* 18:311–15.

———. 1932. Blood groups of Bushmen. *South African Medical Journal* 6:35–37.

———. 1935. Blood groups in Hottentots. *South African Medical Journal* 2:192–95.

Pirie, J. H. H. 1921. Blood testing preliminary to transfusion, with a note on the group distribution among SA natives. *Medical Journal of South Africa* 16:109–12.

Schapera, I. 1930. *The Khoisan peoples of South Africa.* London: Routledge and Kegan Paul.

Schultze, L. 1928. *Zur Kenntnis des Körpers der Hottentotten und Buschmänner,* 147–227. Jena: Gustav Fischer.

Scozzari, R., F. Cruciani, P. Santolamazza, P. Malaspina, A. Torroni, D. Sellitto, B. Arredi, G. Destro-Bisol, G. De Stefano, O. Rickards, C. Martinez-Labargo, D. Modiano, G. Biondi, P. Moral, A. Olckers, D. Wallace, and A. Novelletto. 1999. Combined use of biallelic and microsatellite Y-chromosome polymorphisms to infer affinities among African populations. *American Journal of Human Genetics* 65:829–46.

Soodyall, H. 1993. Mitochondrial DNA variation in southern African populations. Ph.D. diss., University of the Witwatersrand.

Soodyall, H., and T. Jenkins. 1992. Mitochondrial DNA polymorphisms in Khoisan populations from southern Africa. *Annals of Human Genetics* 56:315–24.

———. 1993. Mitochondrial DNA polymorphisms in Negroid populations from Namibia: New light on the origins of Dama, Herero, and Ambo. *Annals of Human Biology* 20:477–85.

———. 1998. Khoisan prehistory: The evidence of genes. In *The Proceedings of the Khoisan Identities and Cultural Heritage Conference,* ed. A. Bank, 374–82. Cape Town: Institute for Historical Research and INFOSOURCE.

Soodyall, H., L. Vigilant, A. V. Hill, M. Stoneking, and T. Jenkins. 1996. Mitochondrial DNA control region sequence variation suggests multiple independent origins of an "Asian-specific" 9-bp deletion in sub-Saharan Africans. *American Journal of Human Genetics* 58:595–608.

Stevens, G., M. Ramsay, and T. Jenkins. 1997. Oculocutaneous albinism (OCA2) in sub-Saharan Africa: Distribution of the common 2.7-kb P gene deletion mutation. *Human Genetics* 99:523–27.

Stow, G. W. 1905. *The native races of South Africa.* London: Sonnenschein.

Tobias, P. V. 1954. On a Bushman-European hybrid family. *Man* 54:179–82.

———. 1962. Some perspectives in human genetics. *Leech* 32:76–83.

———. 1998. Myths and misunderstandings about Khoisan identities and status. In *The Proceedings of the Khoisan Identities and Cultural Heritage Conference,* ed. A. Bank, 19–29. Cape Town: Institute for Historical Research and INFOSOURCE.

Vansina, J. 1984. Western Bantu expansion. *Journal of African History* 25:129–45.

Vedder, H. 1928. The Nama. In *The native tribes of South West Africa,* ed. C. H. L. Hahn, 109–49. Cape Town: Cape Times.

Watson, E., P. Forster, M. Richards, and H-J. Bandelt. 1997. Mitochondrial footprints of human expansions in Africa. *American Journal of Human Genetics* 61:691–704.

Weiner, J. S., G. A. Harrison, R. Singer, R. Harris, and W. Jopp. 1964. Skin colour in southern Africa. *Human Biology* 36:294–307.

Weiner, J. S., and A. Zoutendyk. 1959. Blood group investigation on Kalahari Bushmen. *Nature* 183:843–44.
Westphal, E. O. J. 1963. The linguistic prehistory of southern Africa: Bush, Kwadi, Hottentot, and Bantu linguistic relationships. *Africa* 39:237–365.
Zoutendyk, A., A. C. Kopec, and A. E. Mourant. 1953. The blood groups of the Bushman. *American Journal of Physical Anthropology* 11:361–68.
———. 1955. The blood groups of the Hottentots. *American Journal of Physical Anthropology* 13:691–98.

SECTION B
Race and Human Variation

Chapter 11

The Genetics of African Americans

Implications for Disease Gene Mapping and Identity

Rick Kittles and Charmaine Royal

Although the history of Africans in the Americas predates institutional slavery, it is marked by the brutal period of kidnapping and mass transport of millions of indigenous Africans during the transatlantic slave trade from approximately 1619 to 1850. Thus, the vast majority of contemporary African Americans are descendants of enslaved Africans. Human identity is usually defined in relation to familial, cultural, and genetic ancestry. However, because enslavement has obliterated this history for the vast majority of African Americans it is even more critical to find other ways to trace and understand their ancestry.

Shipping and trade documents provide some insight into the ethnic and geographic origin of enslaved Africans. They were kidnapped primarily from eight coastal regions, ranging from Senegal south through the Cape of Good Hope and north along eastern Africa to Cape Delgado (Curtin 1975). The eight major regions were Senegambia (Gambia and Senegal), Sierra Leone (Guinea, Sierra Leone, and parts of Liberia), the Windward Coast (Ivory Coast and Liberia), the Gold Coast (Ghana west of the Volta River), the Bight of Benin (between the Volta and Benin Rivers), the Bight of Biafra (east of the Benin River to Gabon), central Africa (Gabon, Congo, and Angola), and the southern coast of Africa (from the Cape of Good Hope to Cape Delgado, including the island of Madagascar).

Patterns of enslavement and acquisition significantly influenced the ethnic and geographic ancestry of African Americans. For instance, as plantation agriculture developed in the United States, plantation owners became more particular in selecting specific groups of Africans for their labor forces (Holloway 1990; Thompson 1987). The owners of rice plantations in the Carolinas and Louisiana requested enslaved Africans from the Senegambian region, and tobacco planters in Maryland and Virginia requested Gold

Coast Africans (Jackson 1997; Pollitzer 1994). Many of these regional geographic patterns may still be observed among contemporary African Americans.

While some relatively isolated African American communities, such as the Gullah off the coast of South Carolina, may still resemble a particular African population, the vast majority of the African American gene pool is highly heterogeneous because of the intermixture of various indigenous African populations and gene flow from non-Africans. Very little genetic information is available on the population of enslaved Africans in early America, which is why the genetic analysis of archeological sites such as the African Burial Ground in New York City is of immense interest.

During 1991 and 1992, human remains were discovered in lower Manhattan in an eighteenth-century burial ground of enslaved Africans. The United States General Services Administration was preparing to build a federal office tower on the site at Broadway and Duane Streets. Although historical maps indicated that the site had been a "Negroes Burying Ground," the agency did not anticipate the storm of controversy that arose after excavations began (La Roche and Blakey 1997). The cemetery dates from around 1712 until 1794, a period when New York was a major slave port, and is the resting place for more than twenty thousand Africans. The genetics of the African Burial Ground population may provide insights into the development of the African American gene pool. During the burial ground's period of use, Africans in New York included those born in the colony, those born in other North American colonies, and those brought from the Caribbean and directly from West and central Africa.

African American human biology and disease profiles have been significantly shaped by periods of intermixture, which created high heterogeneity, and selective pressures emanating from the unique and particularly adverse social, economic, and political conditions in the United States (Jackson 1993). All these factors might contribute to the high incidence of diseases with a significant genetic component, such as type 2 diabetes, asthma, hereditary cancers (prostate, breast, and lung), and hypertension.

THE NEW YORK AFRICAN BURIAL GROUND

Construction of the federal building on the African Burial Ground was eventually halted in response to outrage and activism by the descendant community. After much negotiation, the remains of 408 human beings were disinterred from a portion of the burial ground and sent to the Cobb Laboratory at Howard University in Washington, D.C. With community involvement, a research program was proposed to address three major research questions: What was the physical quality of life in eighteenth-century New York City? What can the site reveal about the biological and cultural transition from African to African American identity? And what are the origins of the population?

Skeletal analyses of the burial ground population have revealed evidence of nutritional stresses, possible treponemal disease, and anemias. Polymorphic DNA loci defining β-hemoglobin S (HbS) haplotypes for sickle cell disease and polymerase chain reaction (PCR) markers for the various treponemal infections (pinta, yaws, and syphilis) will add to the skeletal data and help scientists working on the African Burial Ground to determine the physical quality of life for the enslaved Africans in New York.

The second research question will be much more difficult than the other two for genetics to answer. The biological transition of African to African American is marked by the transition from environmental stresses in Africa to those in the Americas, and, to a lesser extent, by the incorporation of non-African alleles into the African American gene pool. The American environment imposed new selective pressures on the Africans. These selective pressures may have favored certain genes while eliminating others. This evolutionary hypothesis has been a controversial explanation for the high incidence of diseases such as hypertension in African Americans (Wilson and Grim 1992; see discussion in Cooper et al. 1999).

Genetic markers called population specific alleles, or more accurately, population *associated* alleles, differ in frequency between Africans (mainly West and central Africans) and Europeans, and will be useful in determining the genetic changes of the African population. We will also examine all the genetic data to see if any burial patterns are evident within the cemetery. Specific analyses will test for evidence of spatial patterns of burials that correlate with genetic lineages or possible ethnic groups.

An assessment of mitochondrial DNA (mtDNA) sequence data from the skeletal remains of forty-eight people has revealed a high level of genetic diversity; 95 percent of the sequences were unique (Kittles et al. 1999). All sequences are closely related to those of West Africans. However, due to the limited sampling of indigenous African populations and recent gene flow within Africa, the exact geographic or ethnic origin of many of the samples is difficult to determine. The identification and sampling of indigenous African populations for genetic studies historically has been highly problematic and categorical (Keita and Kittles 1997). However, more recent work has attempted to obviate racial thinking (see Kittles and Keita 1999) and fill large unsampled geographic gaps within Africa (Watson et al. 1996, 1997; Tishkoff et al. 1996, 1998; Kittles et al. 2001).

Although there has been limited and sporadic sampling of African populations for genetic studies, studies have observed at least three mtDNA haplogroups in African populations, L1, L2, and L3. All three haplogroups have been observed in the African Burial Ground population. The L1 haplogroup is a group of isolated mtDNA lineages common in central and southern

Africa. We have observed this lineage in 10 percent of the samples. The L1 group of haplotypes appears to be older but less frequent than the other two haplogroups. Not surprisingly, the L1 haplogroup is observed in the least sampled area of Africa, so we expect that it may be more common than reported.

Haplogroup L2 is common among the Niger-Kordofanian speakers in the Senegambia and Gold Coast regions of Africa. L2 haplotypes may represent the descendants of migrants of Bantu speakers into western Africa. L2 haplotypes represent 66 percent of the African Burial Ground population. The third haplogroup, L3, is an interesting group of haplotypes. L3 haplotypes are quite common in eastern Africa and the Cape Horn region. The L3 haplogroup is closely related to an mtDNA haplotype common in European populations (the Cambridge Reference Sequence, the reference sequence used to compare other human sequences; Anderson et al. 1981). A subgroup of related mtDNA haplotypes appears to be specific to East Africa and may represent a common ancestral sequence for most of Europe and Eurasia (Kittles and Keita 1999).

The most obvious signature of migration among African populations is the dramatic eastern-to-western Africa cline of mitochondrial DNA haplogroup L3a frequencies (Watson et al. 1997). Although the L3 group is more common in East Africa, it is also observed at an appreciable frequency in West Africa (mainly, but not restricted to, speakers of Afro-Asiatic languages). The L3 haplogroup is present in 21 percent of the burial ground population. This may be explained by the fact that many of the enslaved Africans came from inland areas of western Africa, such as northern Nigeria, Ghana, and southern Niger. One surprising example of this is burial 101, an adult male about thirty years of age. This man possessed an L3 mtDNA haplotype, and his coffin lid contained a heart-shaped design identified by African art historians as the Adinka symbol Sankofa. This symbol is commonly used among the Akan peoples of Ghana and the Ivory Coast. Other cultural artifacts, such as beads and shells, were found in graves throughout the burial ground and provide independent confirmation of ancestry for this population.

ASSESSING GENETIC DIVERSITY OF AFRICAN AMERICANS

African Americans represent a recent yet highly heterogeneous and regionally diverse macroethnic group (Jackson 1997). The majority of African Americans resemble West and central Africans with respect to autosomal, mtDNA, and Y chromosome markers. Interestingly, Y chromosome markers (paternally inherited) reveal a genetic distance between African Americans and indigenous Africans that is slightly larger than what autosomal mtDNA markers show. The Y chromosome analysis of male descendants of Thomas Jefferson, who was rumored to have fathered a child with Sally Hemings, provides a recent example of the "contribution" of European men to the African American gene pool. With the help of PCR and sophisticated data analyses, these oral histories have been verified by genetics (E. A. Foster et al. 1998). The Sally Hemings story is not an isolated incident. We have recently examined the Y chromosome and mtDNA variation in African American males from Washington, D.C., and Columbia, South Carolina, and estimated that 28 to 30 percent of the Y chromosomes observed were of European origin (Doura and Kittles 2002). This pattern of variation was not observed for the maternally inherited mtDNA marker in the two populations. European mtDNA was found in less than 1 percent of the two African American populations.

Specific Y chromosome and mtDNA markers provide direct information on paternal and maternal lineages because they are not affected by recombination. This is not the case for autosomal markers. In order to examine lineage, or the "gene history" of autosomal markers, it may be important to examine closely linked loci. A set of polymorphic, linked alleles inherited as a unit is considered a haplotype. When the occurrence of pairs of specific alleles at different loci on the same haplotype is not independent, the deviation from the independence is termed *linkage disequilibrium* (LD). This deviation has been useful in gene mapping efforts. LD can usually be found in populations for genes that are tightly linked (that is, have a short genetic distance), and can be generated by mutation, selection, or admixture of populations with different allele frequencies. Generally, disequilibrium depends on population size, time (number of generations), and distance between genetic markers. Normally, the greater the distance between markers, the faster the decay of disequilibrium.

In terms of the African American population, the extent of gene flow between various African American communities and specific non-African groups is strongly correlated with geographic region of residence (Jackson 1997; Parra et al. 1998). A 1998 study by Parra and colleagues on admixture examined ten populations of African descent in the Americas for nine genetic markers with alleles that were either population specific or showed frequency differences (>45 percent) between Africans from West and central Africa, and western Europeans (1998). European genetic ancestry ranged from 6.8 percent in Jamaica to 29 percent in Seattle, Washington. The results also suggested a northern-southern and eastern-western clinal pattern of non-African gene flow into African American communities. One striking example of regional differences in admixture is revealed by the higher frequencies of HbS in African Americans from Charleston, South Carolina, and the Gullah Islands than in African Americans in other parts of the United States (Pollitzer 1958; Bowman and Murray 1990). Pollitzer (1958) proposed that African Americans in Charleston were less admixed than other African American communities. This claim was later confirmed by Parra's study

(1998), which showed that African Americans in Charleston had the lowest level of European admixture of all populations studied, with the exception of that in Jamaica. A recent follow-up study examined admixture proportions in six different African American samples from South Carolina, taken from the Gullah speakers living in coastal South Carolina; four different counties in the "Lowcountry" (Berkeley, Charleston, Colleton, and Dorchester); and Columbia, the state capital, located in central South Carolina (Parra et al. 2001). The results of the study indicated, in accordance with previous historical, cultural, and anthropological evidence, very low levels of European admixture in the Gullah population (3.5 percent). The proportion of European alleles was higher in the Lowcountry (ranging between 9.7 percent and 13.8 percent), and the highest levels were observed in the more cosmopolitan city of Columbia (17.2 percent).

SIGNIFICANCE OF THE AFRICAN AMERICAN POPULATION FOR GENE MAPPING

A genetic consequence of the unique population history of African Americans is increased linkage disequilibrium. Much of the disequilibrium may not actually be due to genetic linkage but represents artifacts of divergent allele frequencies in the parental populations. However, we expect that linked loci will also show significant disequilibrium in the African American population.

The analysis of LD between marker and disease loci has proven to be a powerful tool for positional cloning of disease genes (Hastbacka et al. 1992; see de la Chapelle 1993; Jorde 1995). When a disease or trait manifests variation between populations, admixed populations provide a population-based approach to evaluating the relative importance of genetic factors (Chakraborty and Weiss 1998). A variety of statistical genetic methods for disease studies exploit the LD created by admixture. These include the transmission disequilibrium test (Ewens and Spielman 1995; McKeigue 1997) and mapping by admixture linkage disequilibrium (Stephens et al. 1994; Briscoe et al. 1994). An important assumption of many of these methods is that the ancestry of alleles at each locus can be assigned to one of the two founding populations. The assignment of alleles to parent populations is problematic at times; however, as more informative genetic markers are found and more individuals and populations sampled, the statistical power to assign alleles increases.

ETHICAL CONSIDERATIONS IN THE AGE OF GENETICS

The involvement of African Americans in genetics research has several ethical, legal, and social implications that must receive ample consideration by researchers seeking to involve this population in such studies. In formulating hypotheses and conceptualizing study designs, it is essential that the ultimate goal be improvement in the health and well-being of the community. As previously stated, African Americans suffer disproportionately from several common and complex diseases. The ethical principles of beneficence and justice demand that genetics research on African Americans focus primarily on elucidating the genetic and environmental components of these diseases, thus facilitating early detection, effective treatment, and ultimately, prevention. The increasing interest in, and misrepresentation of, information on genetic factors purported to influence intelligence, crime, and other sociobehavioral traits has (justifiably) generated suspicion and fear that genetics research has become just another tool for perpetuating racism and lending credence to the notion that African Americans are inferior (Dula 1994; King 1997). This oversimplification of the determinants of such complex traits consistently leads to the generation of spurious explanations for disparities (real or perceived) and the ills of society, ignoring the more serious moral, political, and social contributors (Blakey 1998).

Population genetics studies should be aimed at identifying gene-based differences and similarities within and among populations, with the hope of providing researchers with a better understanding of their biomedical significance and a greater appreciation for the diversity that contributes to the uniqueness of our species. In view of the stigma already associated with the "minority" status, any negative trait associated with African Americans or other nonwhite groups will likely be emphasized in public discussions (King 1992; Nickens 1996). Researchers should endeavor to ensure that their publications, both technical and general, do not foster stigmatization and discrimination. They should also be prepared to actively oppose misrepresentation or misinterpretation of their findings.

Also critical to minimizing harm to the population is the involvement of African Americans in all aspects of the research process (Blakey 1997; Jackson 1997). This not only increases the likelihood that the research agenda will correlate with priorities of the community but also raises sensitivity to African American history and culture. The United States Public Health Service Syphilis Study (formerly known as the Tuskegee Syphilis Study) has had disastrous effects on the willingness of African Americans to participate in biomedical research (Cox 1998; Bonner and Miles 1997; Shavers-Hornaday et al. 1997; Talone 1998).[1] However, the general mistrust within the community dates back to the 1800s, when brutal "experiments" were conducted on enslaved Africans (Blakey 1987; Dula 1994; Gamble 1993). This history of medical experimentation, as an outgrowth of societal racism, has obliged genetic and other biomedical researchers to expend extra effort in gaining the trust of the African American community in order to procure their participation in studies, thereby increasing the capacity of this population to benefit from scientific research.

One model for genetics research on the African American community is the African American Hereditary Prostate Cancer (AAHPC) Study (Royal et al. 2000). The AAHPC Study Network was established in 1997 to examine the genetics of hereditary prostate cancer in African Americans. This research collaboration involves investigators at Howard University (the coordinating center for the collaboration), the National Human Genome Research Institute, the National Institutes of Health, and six collaborative recruitment centers across the United States. The network comprises urologists, radiation oncologists, molecular biologists, geneticists, nurses, epidemiologists, data managers, and statistical geneticists. Because the majority of investigators are African American, the project offers a unique opportunity for the involvement of African Americans in human genetics research.

As project coordinator for the AAHPC study, I (R. A. K.) visited all six collaborative recruitment sites. On a site visit to Chicago, I met a seventy-eight-year-old African American man dying of prostate cancer. Four of his first-degree relatives had also been affected by the disease. I sat with him in the radiation oncologist's office, trying to convince him that not only he but also his three brothers affected with prostate cancer and three unaffected family members should participate in our study. I was reminded of my grandfather, who had recently died of prostate cancer. I felt simultaneous compassion for this dying man and excitement at finding a family who fit the stringent criteria for study participation.

As I was preparing to deal with questions from his family about the study, he abruptly inquired, "What will this project do for me? I probably will not be living to see next year." The answer was that the study would do nothing for him personally. But I reminded him about his family and said that, ultimately, finding predisposing genes for prostate cancer might lead to the development of better drugs and treatment and genetic screening tests. Fortunately, he understood the importance of this type of research for future generations.

Furnishing adequate information about the research goals, as well as their relevance and anticipated value to the study community, is one of the first steps in obtaining informed consent. This is arguably the tenet of biomedical ethics that is most difficult to implement in genetic (or other) research. Informed consent for genetic studies requires a paradigm shift from focusing on minimal physical risks to focusing on the more detrimental psychosocial risks, many of which may still be unknown; indeed, *informed consent* in genetics is something of a misnomer. As a result, researchers must disclose as fully as possible the known benefits (without overpromising), risks (physical and psychosocial), and limitations of their studies to potential participants and must ensure that participation is truly voluntary.

The guarantee of confidentiality is also vital in genetics research involving human participants. The general principles governing privacy of individual and familial information obtained through genetics research have been well articulated (OPRR 1993). However, there has been much concern and discussion about the storage and use of anonymous samples that may be linked to groups identifiable by common ancestry (M. W. Foster et al. 1998; NBAC 1999; Wadman 1998). Until the implementation of specific guidelines protecting groups from collective risks regarding stored samples, and even thereafter, the history of exploitation of African Americans necessitates the involvement of African American scientists and others respectful of the community to minimize the likelihood of abuse.

Genetic tests are usually developed soon after genes have been isolated, and this discussion would be incomplete without an examination of the implications of genetic testing and screening in African Americans. The widespread discrimination in the 1970s against African Americans with sickle cell trait has undoubtedly set the tone for the current spectrum of concerns regarding genetic testing and screening in this population. One of the leading causes of the sickle cell screening fiasco was the pervasive repetition of inaccurate information (Bowman 1992; Murray 1997). The natural history and symptomatology of the disease, the distinction between disease and trait, the interpretation of test results, the prevalence of the disease, and the value of testing were widely misrepresented among health care providers and the general public, as well as among state and federal agencies. The current rapid advances in genetic mapping and sequencing technology will likewise culminate in a period of incomplete knowledge, creating more opportunities for the propagation of misinformation and its adverse effects, especially for the more challenging, complex disorders, many of which are prevalent in African Americans.

Two of the most pressing concerns regarding genetics research are insurance and employment discrimination (Hudson et al. 1995; Rothenberg et al. 1997). Individuals whose genetic makeup shows them to be at risk for certain heritable conditions are also at risk of being denied employment, promotion, and health insurance; being charged higher insurance premiums; and having genetic information disclosed to third parties. Fear of genetic discrimination could create additional public health problems by making individuals reluctant to participate in genetics research, share genetic information with health care providers or family members, and utilize available preventive and treatment services (Hudson et al. 1995).

A significant body of literature indicates that African Americans have diminished access to health care in general and use it less (Schensul and Guest 1994; Russell and Jewell 1992; and Nickens 1996). These barriers may further limit access to and use of genetic services. Approximately 25 percent of African Americans lack health insurance, and many who do have insurance are likely to be insured through the Medicaid program, which provides very limited coverage (Nickens 1996). Therefore, genetic services may be beyond the financial reach of many.

Structural barriers include inconvenient locations and the significant underrepresentation of African Americans among providers of genetic services. In addition, diverse and remote communities are generally unaware of genetic risks and available services, a problem which may be attributed to limited outreach by institutions. The delivery of culturally relevant and nondiscriminatory genetic services to African Americans may also be hampered by the fact that only 1.1 percent of members of the United States–based American Society of Human Genetics—medical geneticists, genetic counselors, cytogeneticists, and so on—are African American (Mittman 1998). Even though an increase in the number of African Americans providing genetic services may not, by itself, guarantee the provision of culturally appropriate and equitable health care, it will undoubtedly ease cross-cultural communication and contribute to empowerment of the community (King 1992).

Spirituality and the concept of time also appear to influence African Americans' utilization of genetic services (Hughes et al. 1996). For many, spirituality as an integral facet of their culture encourages reliance on faith in God and the power of prayer in all aspects of life (Pinderhughes 1982; Russell and Jewell 1992). Consequently, some may reject genetic services for fear of intervening in divine destiny. With regard to time orientation, Wade Nobles (1991) noted that African philosophy emphasizes a focus on the past and present rather than the future. Na'im Akbar (1991) illustrates African Americans' treatment of time as a possible outgrowth of this philosophy: "The Black Psychology time focus is on the recent past of the African American experience and the present conditions of oppression and its multifarious manifestations. The future is not considered as relevant." Consequently, the predictive and probabilistic nature of genetics may discourage some African Americans from taking advantage of certain genetic services. Finally, another plausible and possibly overriding determinant is African Americans' inherent mistrust of the entire health care system. More studies are needed to explore these perspectives.

If certain genetic diseases occur more frequently among nonwhite groups, then group members are at increased risk for stigmatization and discrimination. Another, even more disturbing possibility is that, because of their association with a "minority" group, these diseases might receive lower priority in funding research for treatment and prevention (King 1992; Nickens 1996). Unfortunately, there is evidence to suggest that such inequitable allocation of resources has indeed occurred. In a comparison of funding levels for cystic fibrosis and sickle cell disease, Herbert Nickens (1996) points out that, in spite of the differences in disease frequency (1 in 2,500 white

Americans is born with cystic fibrosis; 1 in approximately 600 black Americans is born with sickle cell disease), and despite the fact that in the United States sickle cell disease is 1.5 times as common as cystic fibrosis, the National Institutes of Health allocated about $18 million for sickle cell research in fiscal year 1992 while budgeting $46 million for cystic fibrosis. Such discrepancy in funding is a likely contributor to the nearly half-century hiatus between identification of the cause of sickle cell disease and effective therapeutic options (hydroxyurea use and bone marrow transplantation). Despite their limitations, these advances now offer some promise to the many sufferers of this life-threatening illness. Considering that sickle cell disease is a single gene disorder, and that it was the first human disease to be understood at the molecular level, it is inevitable that there will be even greater challenges in attempting to address the more common complex diseases.

CONCLUSION

Despite the power and promise of the emerging genetic technology that permits researchers to assess origins, determine genetic affinities, and map susceptibility genes for diseases and complex traits, there are various factors that may determine the opportunities for African Americans to reap the anticipated benefits of anthropological interests and improved diagnosis, treatment, and prevention of disease. Genetics researchers, health care providers, anthropologists, policy makers, and the African American community have a unique opportunity to create a model for conducting biomedical research on the African American population. Together they can help develop and implement better mechanisms to protect African Americans from the potentially harmful effects of genetics research and to maximize the improved health outcomes that may be made possible by advances in human genetics.

NOTES

We thank Drs. Georgia Dunston, Fatimah Jackson, S. O. Y. Keita, and Michael Blakey for helpful discussions. We also thank Dale Young, Debra Parish-Gause, Gay Morris, Ebony Bookman, Matt Neapolitano, Nadeje Sylvester, and the National Human Genome Center at Howard University for technical support. This work was supported in part by the National Institutes of Health (G12-RR-03048 and N01-HG-75418).

1. This forty-year (1932–1972) study involved 399 African American men with syphilis, who were enrolled by investigators interested in learning more about the course of the disease. The men were denied treatment for syphilis even after penicillin became available in the 1940s. Deception, lack of informed consent, and injustice in the selection of participants have always contributed to the relatively low participation of African Americans in studies. Low participation became even more serious when information about the Tuskegee study became well-known in the 1980s.

REFERENCES

Akbar, N. 1991. The evolution of human psychology for African Americans. In *Black psychology,* ed. R. L. Jones, 99–123. 3d ed. Berkeley: Cobb and Henry.

Anderson, S., A. Bankier, B. Barrell, M. de Bruijn, A. Coulson, J. Drouin, I. Eperon, D. Nierlich, B. Roe, and I. Young. 1981. Sequence and organization of the human mitochondrial genome. *Nature* 290:457–65.

Blakey, M. L. 1987. Skull doctors: Intrinsic social and political bias in American physical anthropology, with special reference to the work of Ales Hrdlicka. *Critique of Anthropology* 7:7–35.

———. 1997. Past is present: Comments on "In the realm of politics: Prospects for public participation in African-American plantation archaeology." *Historical Archaeology* 31, no. 3:140–45.

———. 1998. Beyond European enlightenment: Toward a critical and humanistic human biology. In *Building a new biocultural synthesis: Political-economic perspectives on human biology,* ed. A. H. Goodman and T. L. Leatherman, 379–405. Ann Arbor: University of Michigan Press.

Bonner, G. J., and T. P. Miles. 1997. Participation of African Americans in clinical research. *Neuroepidemiology* 16:281–84.

Bowman, J. E. 1992. The plight of poor African Americans: Public policy on sickle hemoglobins and AIDS. In *African American perspectives on biomedical ethics,* ed. H. E. Flack and E. D. Pellegrino, 173–87. Washington, D.C.: Georgetown University Press.

Bowman, J. E., and R. F. Murray. 1990. *Genetic variation and disorders in peoples of African origin.* Baltimore, M.D.: Johns Hopkins University Press.

Briscoe, D., J. C. Stephens, and S. J. O'Brien. 1994. Linkage disequilibrium in admixed populations: Applications in gene mapping. *Journal of Heredity* 85:59–63.

Chakraborty, R., and K. M. Weiss. 1988. Admixture as a tool for finding linked genes and detecting that difference from allelic association between loci. *Proceedings of the National Academy of Sciences, USA* 85:9119–23.

Cooper, R. S., C. N. Rotimi, and R. Ward. 1999. The puzzle of hypertension in African Americans. *Scientific American* (February): 56–63.

Cox, J. D. 1998. Paternalism, informed consent, and Tuskegee. *International Journal of Radiation Oncology Biology Physics* 40, no. 1:1–2.

Curtin, P. D. 1975. *The Atlantic slave trade: A census.* Madison: University of Wisconsin Press.

de la Chapelle, A. 1993. Disease gene mapping in isolated human populations: The example of Finland. *Journal of Medical Genetics* 30:857–65.

Doura, M., and R. Kittles. 2002. Paternal lineages of African Americans in South Carolina. *American Journal of Human Genetics* 71, no. 4:178.

Dula, A. 1994. African American suspicion of the healthcare system is justified: What do we do about it? *Cambridge Quarterly of Healthcare Ethics* 3:347–57.

Ewens, W. J., and R. S. Spielman. 1995. The transmission/disequilibrium test: History, subdivision, and admixture. *American Journal of Human Genetics* 57:455–64.

Foster, E. A., M. Jobling, P. Taylor, P. Donnelly, P. de Kniff, R. Mieremet, T. Zerjal, and C. Tyler-Smith. 1998. Jefferson fathered slave's last child. *Nature* 396:27–28.

Foster, M. W., D. Bernstein, and T. H. Carter. 1998. A model agreement for genetic research in socially identifiable populations. *American Journal of Human Genetics* 63:696–702.

Gamble, V. 1993. A legacy of distrust: African Americans and medical research. *American Journal of Preventive Medicine* 9, no. 6 suppl.:35–38.

Handt, O., M. Krings, R. H. Ward, and S. Paabo. 1996. The retrieval of ancient human DNA sequences. *American Journal of Human Genetics* 59:368–76.

Hastbacka, J., A. de la Chapelle, I. Kaitila, P. Sistonen, A. Weaver, and E. Lander. 1992. Linkage disequilibrium mapping in isolated founder populations: Diastrophic dysplasia in Finland. *Nature Genetics* 2:204–11.

Hedrick, P. W. 1987. Gametic disequilibrium measures: Proceed with caution. *Genetics* 117:331–41.

Holloway, J. E. 1990. *Africanisms in American culture.* Bloomington: Indiana University Press.

Hudson, K. L., K. H. Rothenberg, L. B. Andrews, M. J. E. Kahn, and F. S. Collins. 1995. Genetic discrimination and health insurance: An urgent need for reform. *Science* 270:391–93.

Hughes, C., C. Lerman, and E. Lustbader. 1996. Ethnic differences in risk perception among women at increased risk for breast cancer. *Breast Cancer Research and Treatment* 40:25–35.

Jackson, F. L. 1993. Evolutionary and political economic influences on biological diversity in African Americans. *Journal of Black Studies* 23, no. 4:539–60.

———. 1997. Concerns and priorities in genetic studies: Insights from recent African American biohistory. *Seton Hall Law Review* 27, no. 3:951–70.

Jorde, L. B. 1995. Linkage disequilibrium as a gene mapping tool. *American Journal of Human Genetics* 56:11–14.

Keita, S. O. Y., and R. A. Kittles. 1997. The persistence of racial thinking and the myth of racial divergence. *American Anthropologist* 99, no. 3:534–44.

King, P. A. 1992. The past as prologue: Race, class, and gene discrimination. In *Gene mapping: Using law and ethics as guides,* ed. G. J. Annas and S. Elias, 94–111. Oxford: Oxford University Press.

———. 1997. The dilemma of difference. In *Plain talk about the Human Genome Project: A Tuskegee University conference on its promise and perils . . . and matters of race,* ed. E. Smith and W. Sapp, 75–81. Tuskegee, Ala.: Tuskegee University.

Kittles, R., and S. O. Y. Keita. 1999. Interpreting African genetic diversity. *African Archaeological Review* 16, no. 2:87–91.

Kittles, R., G. Morris, M. George, G. Dunston, M. Mack. F. L. C. Jackson, S. O. Y. Keita, and M. Blakey. 1999. Genetic variation and affinities in the New York burial ground of enslaved Africans. *American Journal of Physical Anthropology* 28, suppl. 28:16.

Kittles, R., D. Young, S. Weinrich, J. Hudson, G. Aryropoulos, F. Ukoli, L. Adams-Campbell, and G. Dunston. 2001. Extent of linkage disequilibrium between the androgen receptor gene CAG and GGG repeats in human populations: Implications for prostate cancer risk. *Human Genetics* 109:253–61.

La Roche, C. J., and M. Blakey. 1997. Seizing intellectual power: The dialogue at the New York African Burial Ground. *Historical Archaeology* 31, no. 3:84–106.

Lewontin, R. C. 1964. The interaction of selection and linkage. I. General considerations: Heterotic models. *Genetics* 49:49–67.

———. 1988. On measures of gametic disequilibrium. *Genetics* 120:849–52.

McKeigue, P. 1997. Mapping genes underlying ethnic differences in disease risk by linkage disequilibrium in recently admixed populations. *American Journal of Human Genetics* 60:118–96.

Mittman, I. S. 1998. Genetic education to diverse communities employing a community empowerment model. *Community Genetics* 1, no. 3:160–65.

Murray, R. F., Jr. 1997. The ethics of predictive genetic screening: Are the benefits worth the risks? In *Plain talk about the Human Genome Project: A Tuskegee University conference on its promise and perils . . . and matters of race,* ed. E. Smith and W. Sapp, 139–50. Tuskegee, Ala.: Tuskegee University.

National Bioethics Advisory Commission (NBAC). 1999. Research involving human biological materials: Ethical issues and policy guidance. Available at http://www.georgetown.edu/research/nrcbl/nbac/pubs.html.

Nei, M. 1987. *Molecular evolutionary genetics.* New York: Columbia University Press.

Nickens, H. 1996. Health services for minority populations. In *The Human Genome Project and the future of health care,* ed. T. H. Murray, M. A. Rothstein, and R. F. Murray Jr., 58–78. Bloomington: Indiana University Press.

Nobles, W. W. 1991. African philosophy: Foundations for black psychology. In *Black psychology,* ed. R. L. Jones, 47–63. 3rd ed. Berkeley: Cobb and Henry.

Office of Protection from Research Risks, Department of Health and Human Services (OPRR). 1993. Human genetic research. In *Protecting human research subjects—institutional review board guidebook,* chap. H. Bethesda, Md.: OPRR.

Parra, E. J., R. Kittles, G. Argyropoulos, K. Hiester, C. Bonilla, N. Sylvester, D. Parrish-Gause, W. Garvey, M. Kamboh, L. Jin, R. Ferrell, W. Pollitzer, P. McKeigue, and M. Shriver. 2001. Ancestral proportions and admixture dynamics in geographically defined African Americans living in South Carolina. *American Journal of Physical Anthropology* 114:18–29.

Parra, E. J., A. Marcini, J. Akey, J. Martinson, M. A. Batzer, R. Cooper, T. Forrester, D. B. Allison, R. Deka, R. E. Ferrell, and M. D. Shriver. 1998. Estimating African admixture proportions by use of population-specific alleles. *American Journal of Human Genetics* 63:1839–51.

Pinderhughes, E. B. 1982. Family functioning of Afro-Americans. *Social Work* 27:91–96.

Pollitzer, W. S. 1958. The Negroes of Charleston (S.C.): A study of hemoglobin types, serology, and morphology. *American Journal of Physical Anthropology* 16:241–63.

———. 1994. Ethnicity and human biology. *American Journal of Human Biology* 3, no. 6:6.

Rothenberg, K., B. Fuller, M. Rothstein, T. Duster, M. Kahn, R. Cunningham, B. Fine, K. Hudson, M-C. King, P. Murphy, G. Swergold, and F. Collins. 1997. Genetic information and the workplace: Legislative approaches and policy challenges. *Science* 275:1755–57.

Royal, C., A. Boffoe-Bonnie, R. Kittles, et al. 2000. Recruitment experience in the first phase of the African American Hereditary Prostate Cancer (AAHPC) Study. *Annals Epidemiology* 10:S68-S77.

Russell, K., and N. Jewell. 1992. Cultural impact of health-care access: Challenges for improving the health of African Americans. *Journal of Community Health Nursing* 9, no. 3:161–69.

Schensul, J. J., and B. H. Guest. 1994. Ethics, ethnicity, and health care reform. In *It just ain't fair: The ethics of health care for African Americans,* ed. A. Dula and S. Goering, 24–40. Westport, Conn.: Praeger Publishers.

Shavers-Hornaday, V. L., C. F. Lynch, L. F. Burmeister, and J. C. Torner. 1997. Why are African Americans under-represented in medical research studies? Impediments to participation. *Ethnicity and Health* 2, no. 1–2:31–45.

Shriver, M. D., M. W. Smith, L. Jin, A. Marcini, J. Akey, R. Deka, and R. E. Ferrell. 1997. Ethnic-affiliation estimation by use of population-specific DNA markers. *American Journal of Human Genetics* 60:957–64.

Stephens, J. C., D. Briscoe, and S. J. O'Brien. 1994. Mapping by admixture linkage disequilibrium in human populations: Limits and guidelines. *American Journal of Human Genetics* 55:809–24.

Stoneking, M. 1995. Ancient DNA: How do you know when you have it and what can you do with it? *American Journal of Human Genetics* 57:1259–62.

Talone, P. 1998. Establishing trust after Tuskegee. *International Journal of Radiation Oncology Biology Physics* 40, no. 1:3–4.

Thompson, V. B. 1987. *The making of the African diaspora in the Americas, 1441–1900.* Boston: Addison-Wesley.

Tishkoff, S., et al. 1996. Global patterns of linkage disequilibrium at the CD4 locus and modern human origins. *Science* 271:1380–87.

———. 1998. A global haplotype analysis of the mytonic dystrophy locus: Implications for the evolution of modern humans and for the origin of mytonic dystrophy mutations. *American Journal of Human Genetics* 62:1389–1402.

Wadman, M. 1998. "Group debate" urged for gene studies. *Nature* 391:314.

Watson, E., K. Bauer, R. Aman, G. Weiss, A. von Haeseler, and S. Paabo. 1996. MtDNA sequence diversity in Africa. *American Journal of Human Genetics* 59:437–44.

Watson, E., et al. 1997. Mitochondrial footprints of human expansions in Africa. *American Journal of Human Genetics* 61:693–704.

Wilson, T., and C. Grim. 1992. The possible relationship between the transatlantic slave trade and hypertension in blacks today. In *The Atlantic slave trade: Effects on economies, societies, and people in Africa, the Americas, and Europe,* ed. J. Inikori and S. Engerman, 339–59. Durham, N.C.: Duke University Press.

Chapter 12

Human Races in the Context of Recent Human Evolution

A Molecular Genetic Perspective

Alan R. Templeton

This chapter examines the significance of human "racial" diversity, using the same types of genetic diversity measurements, criteria, and analytical procedures applied to other life on this planet. This is not to say that humans are not a unique species—we certainly are—but it does acknowledge the fact that our genetic diversity is subject to the same evolutionary forces that shape diversity in all life. Moreover, modern molecular genetics provides comparable means of screening for genetic variation in virtually all living species. Consequently, the amount, pattern, and significance of genetic diversity within any species can now be evaluated with common molecular measurements, analytical techniques, and interpretive criteria. This was not true when genetic variation was primarily observed and monitored through morphological variation within a species. Because most species have distinct morphologies, it was not always possible to have comparable measures of morphological variation; interpretive criteria were often subjective, and their application varied dramatically with the species being studied. In particular, the interpretative criteria applied to our own species' variation often were, and still are, inconsistent with how humans analyze and interpret diversity in other species. This unique manner of interpreting human populational differences has powerful implications culturally, politically, and economically. From a social perspective, race is a real factor in human interactions (see Troy Duster, chapter 13, this volume). From a biological perspective, a valid scientific understanding and interpretation of human genetic diversity must use the same criteria that have been applied to genetic diversity in nonhuman species.

GENETIC DIVERSITY WITHIN AND AMONG LOCAL POPULATIONS

All concepts of race ultimately are based on patterns of genetic diversity within and among the local breeding populations of a species. The existence of genetic differences among populations is a necessary *but not sufficient* condition for any of the modern concepts of race. Hence, before addressing race directly, I discuss the problem of measuring genetic differences among populations relative to genetic diversity within populations.

The basic units of genetic diversity are alleles, alternative forms of the same gene at a locus. Even an individual can display genetic diversity when the gene from the mother is of a different allelic type than the gene from the father, a phenomenon called heterozygosity. If both copies of the gene are of the same allelic form, the individual is said to be homozygous. The combined state of both gene copies defines the individual's genotype. Genetic diversity also exists among individuals within a single population because different individuals can have different genotypes. The genetic diversity found within a local population can also be characterized by noting all the different types of alleles shared by the individuals and the frequencies with which those alleles occur. Finally, genetic diversity can exist between two populations if some alleles are found in one population but not in the other, or if the same alleles occur in both populations but at different frequencies.

There are many ways of quantifying genetic diversity at these various levels (individual, local population, between populations). The mathematical details of this quantification are not important for the arguments I raise in this chapter, but it is important to measure genetic diversity both within and among populations in a manner that can be applied to all species consistently. Of particular importance for the concept of race is the amount of genetic diversity that exists among individuals *within* a population relative to that which exists *among* populations. Many evolutionary forces affect the balance between these two types of genetic diversity, but four are particularly important. The first is mutation, the ultimate creator of all genetic diversity. When a new allele is created by a mutation, it obviously exists initially in only one local population. Therefore, mutation is both a source of genetic diversity within a local population and a source of genetic diversity among populations.

The second evolutionary force is genetic drift (random sampling error). The laws of Mendelian inheritance are probabilities, not certainties, so by chance alone alleles can change their frequency in a finite population or even be lost altogether. Genetic drift, therefore, causes genetic diversity within local populations to decrease, but, at the same time, also causes different local populations to become genetically differentiated—that is, to have different alleles or different allele frequencies.

The third evolutionary force that affects this balance is gene flow, or genetic exchange—the interbreeding of individuals who were born in different local populations. Gene flow causes genetic interchange between populations, thereby giving any local population access to alleles created by mutations originating elsewhere. This infusion of new alleles by gene flow augments genetic diversity within local populations. Gene flow also causes alleles to be shared by multiple local populations and causes convergence to a common allele frequency, thereby diminishing genetic diversity among local populations. When there is no gene flow between two populations, such populations are said to be *genetically isolated,* and the populations themselves are said to be *isolates.* Because isolated populations are not subject to the homogenizing force resulting from genetic interchange, mutation and genetic drift ensure that isolates will become increasingly different genetically over time following the "split" (that is, the time at which all gene flow ceased). *Admixture* occurs if genetic interchange resumes between former isolates. The definitions given above for all the words or phrases in italics are the standard definitions as used in the literature on nonhuman species. These words are frequently used in a nonstandard fashion in the human literature.

The fourth evolutionary force considered here is natural selection. If several local populations are adapting to the same environment, natural selection can be a powerful homogenizing force that maintains the same alleles at the same frequencies for those genes involved in adapting to the common environment. On the other hand, if different local populations live in different environments, natural selection will accentuate genetic differentiation among the local populations for those genes involved in the adaptations to the differing environments.

Although natural selection can be a powerful force for either genetically homogenizing or differentiating populations, depending on the nature of the environment, selection primarily influences only those genes directly involved in the adaptation. In contrast, all genes are subject to genetic drift, gene flow, and mutation. Hence, when there are many different genes, the idiosyncratic associations with selection are often averaged out, and the overall genetic diversity patterns are determined primarily by mutation, genetic drift, and gene flow. Natural selection is important only for specific traits and their underlying genes, and these traits can and often do show patterns of genetic diversity that are inconsistent with the overall pattern of genetic diversity. Moreover, different selected traits that represent adaptations to different environmental variables can also be inconsistent with one another in their diversity patterns. As a consequence, traits under natural selection are not regarded in the literature on nonhuman species as reliable indicators of racial status (Futuyma 1986).

Note that genetic diversity within local populations is decreased by drift but increased by gene flow, whereas genetic differentiation among populations is increased by drift but decreased by gene flow. As a consequence, the relative amounts of genetic diversity within and among local populations often reflect a dynamic balance between genetic drift and gene flow. One common method of quantifying the balance between the two types of genetic diversity is the use of the F statistics of Sewall Wright (1969). F statistics are used to partition a species' genetic diversity into two components: F_{is}, the portion due to differences among individuals within a local population (called subpopulations by Wright, hence the subscript "is," which refers to individuals within subpopulations), and F_{st}, the portion due to differences among subpopulations relative to the total collection of all subpopulations. F_{st} is the more relevant measure of diversity for the question at hand, as F_{st} directly gives the proportion of genetic variation due to differences among subpopulations (or "racial" differences, in the present context). F_{st} and related statistics range from 0 (no differences among populations in either the alleles they collectively share or in the frequencies of those alleles) to 1 (different local populations do not share any alleles in common and all individuals within local populations are genetically identical).

An alternative method to F_{st} for measuring the extent of genetic differentiation specifically between two populations is to convert the allele and the allele frequency differences, or both, between two populations into a genetic distance. There are several measures of genetic distance available, and sometimes the biological conclusions depend on the precise measure chosen (Perez-Lezaun et al. 1997). However, I ignore this problem here because the relative distances among the major human "races" appear robust to differing genetic distance measures (Cavalli-Sforza 1997). For the purposes of this chapter, a genetic distance is simply a number that measures the extent of genetic differentiation between two populations in terms of the alleles that are unique to each population and the extent to which shared alleles have different frequencies.

DEFINITIONS OF RACE

The word *race* is rarely used in the modern, nonhuman evolutionary literature because its meaning is so ambiguous. When it *is* used, it is generally as a synonym for *subspecies* (Futuyma 1986: 107–9), but this concept also has no precise definition. All concepts of a subspecies are based on genetic differences between populations living in different geographic areas; but these differences alone are insufficient to define a subspecies because genetic surveys usually reveal so much variation that some combination of characters distinguishes virtually every population from all others (Futuyma 1986). As a consequence, if genetic differentiation alone were required to recognize a subspecies or race, then every local population would become a race, making the category of race superfluous. Indeed, evolutionary biologists have long made this argument, and many feel that the entire concept of race or subspecies should be completely abandoned (e.g., Wilson and Brown 1953).

Although many current evolutionary biologists find much merit in abandoning the concept of race, there has been a recent resurgence in the use of the concepts of subspecies or races in the nonhuman evolutionary literature. This resurgence has come primarily from the conservation biology community, and it results from political as well as scientific considerations. The primary legal vehicle for conservation policy in the United States (a country that has a large influence on global science) is the Endangered Species Act. This act not only allows for a species to be declared threatened or endangered but also requires preservation of vertebrate subspecies (Pennock and Dimmick 1997). (The restriction to vertebrate subspecies reveals a strong bias in this central piece of biodiversity legislation.) However, since the act does not provide a definition of *subspecies,* much effort has been initiated within the conservation biology community to provide such a definition. The critique of E. O. Wilson and W. L. Brown (1953) is still regarded as valid by most evolutionary biologists; so to prevent the term *subspecies* from being a synonym for *local population,* it is necessary to add further conditions beyond mere genetic differentiation among populations in order to recognize a race or subspecies. Three main additional criteria are applied: (1) a quantitative threshold of genetic differentiation among populations, (2) a genetic differentiation marking the qualitative state of being an isolate or distinct evolutionary lineage, and (3) genetic differentiation for special "racial" traits. Of these criteria, only the first two are used in the modern nonhuman evolutionary literature, particularly the conservation biology literature. The third solution—recognition of special "racial" traits—has no legitimacy in the literature on nonhuman species, and toward the end of this chapter I illustrate why defining special "racial" traits in humans is misleading in the recognition of subspecies.

Concerning the first solution, a standard quantitative threshold used in the literature on nonhuman species is a threshold of about .25 to .30 using F_{st} or related measures (Smith et al. 1997). Do humans exceed this threshold? Much of the anthropological literature portrays humans as a remarkably diverse, polytypic species that contains highly differentiated subtypes. For example, Milford Wolpoff and Rachel Caspari (1997), who clearly acknowledge that gene flow occurs among human populations, and that human populations are not isolates, nevertheless state that "our species is unusual and difficult to model because it is polytypic," and that "the human pattern . . . of a widespread polytypic species with many different ecological niches . . . is a very rare one." However, the human F_{st}, measured on a global basis, is about .15, a cutoff that has proven remarkably stable across studies using very different methods to assay molecular genetic variation (Lewontin 1972; Nei and Roychoudhury 1974, 1982; Barbujani et al. 1997). The value of .15 is well below the recognized threshold for recognizing subspecies in nonhuman organisms. Indeed, it is hard to find any widespread species that shows so little genetic differentiation among its populations as humans, even when considering only large-bodied mammals with strong dispersal capabilities (Templeton 1998a). Moreover, polytypic species with subspecies well-defined by this criterion are not rare or unusual (e.g., Shaffer and McKnight 1996); rather, it is the genetic homogeneity among human populations that is rare and unusual in the animal kingdom. Hence, the existence of human races cannot be demonstrated by using this quantitative threshold definition of race.

A major criticism of the threshold definition of race is that the threshold is arbitrary. There is no clear or objective rationale for choosing an F_{st} threshold value of .25 or .30 versus .20 or .35. To avoid this flaw, the modern evolutionary perspective defines a subspecies (the second criterion noted above) as a distinct evolutionary lineage or isolate within a species (Shaffer and McKnight 1996). This definition requires that a subspecies be genetically differentiated as a result of barriers to genetic exchange that have persisted for a period of time sufficient to have created detectable genetic consequences; that is, the subspecies must be composed of isolates having not only historical continuity but also current genetic differentiation. The best traits for identifying subspecies are simply those with the best phylogenetic resolution. In this regard, advances in molecular genetics have greatly augmented our ability to resolve genetic variation and provide the best current resolution of recent evolutionary histories (Avise 1994). This has permitted the identification of evolutionary lineages in an objective, explicit fashion (Templeton 1994, 1998c; Templeton et al. 1995). Note that being an identifiable evolutionary lineage is a qualitative state. This lineage definition of race thereby avoids an arbitrary quantitative threshold. Not surprisingly, the lineage definition of race or subspecies has come to dominate the recent nonhuman evolutionary literature and has become the de facto definition of a subspecies in much of conservation biology (Amato and Gatesy 1994; Brownlow 1996; Legge et al. 1996; Miththapala et al. 1996; Pennock and Dimmick 1997; Vogler 1994). In the next section, I examine the issue of human races as distinct evolutionary lineages within a species. I address this issue with molecular genetic data and through the application of the same, explicit criteria used for the analyses of nonhuman organisms. A more detailed analysis of this issue appears in Templeton 1998a.

HUMAN RACES AS DISTINCT EVOLUTIONARY LINEAGES

The two dominant models of recent human evolution during the last half of this century are the candelabra and trellis models (figure 12.1). Both models accept the evolutionary origin of the genus *Homo* in Africa and the spread of *Homo erectus* out of Africa a million years ago or more. Candelabra models posit that the major Old World geographic groups (Europeans, sub-

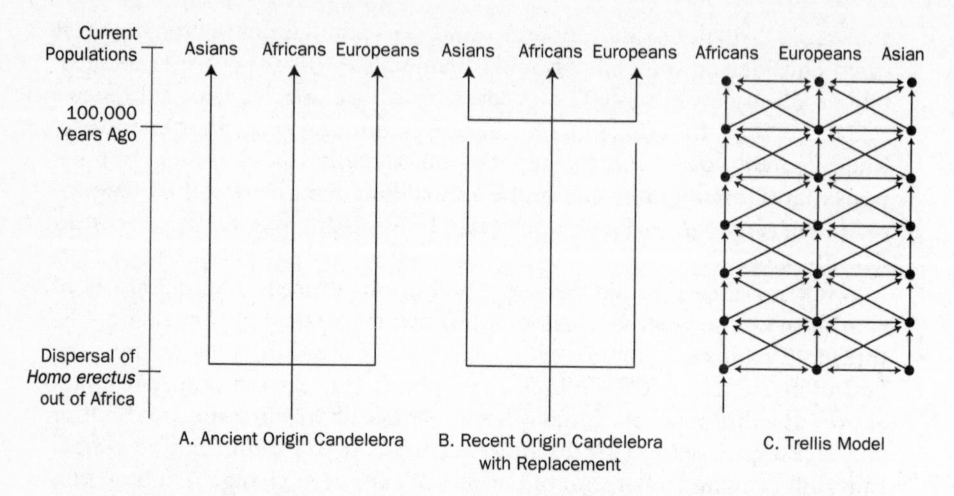

Figure 12.1 Models of recent human evolution. Part A illustrates the ancient origin candelabra model, under which the major human races split at the time of dispersal of *Homo erectus* out of Africa and independently evolved into their modern forms. Part B illustrates the recent origin candelabra model with replacement. After leaving Africa, modern humans split into separate lineages that independently acquired their modern racial variation. Part C shows the trellis model of recent human evolution. Under this hypothesis, *Homo erectus* dispersed out of Africa and established populations in Africa and southern Eurasia, as indicated by the circles. Double-headed arrows indicate gene flow among contemporaneous populations, and single-headed arrows indicate lines of genetic descent.

Saharan Africans, and Asians) split from one another and since have behaved as isolates with nearly independent evolutionary histories (but perhaps with some recent admixture). Therefore, the evolutionary relationships among Africans, Europeans, and Asians can be portrayed as an evolutionary tree—in this case with the topology of a candelabra (figs. 12.1a and 12.1b), although most candelabra models actually regard the African/non-African split as being older than the Asian/European split. The major human geographic populations are portrayed as the branches on this candelabra and are therefore valid races under the evolutionary lineage definition. The ancient origin candelabra model (figure 12.1a) regards the split between the major races as occurring with the spread of *Homo erectus* followed by independent evolution of each race into its modern form. This version has been thoroughly discredited and has no serious advocates today. However, a recent origin candelabra model known as the out-of-Africa replacement hypothesis (figure 12.1b) has become widely accepted. Under this model, anatomically modern humans evolved first in Africa. Next, a small group of these anatomically modern humans split off from the African population and colonized Eurasia about one hundred thousand years ago, driving the *Homo erectus* populations to complete genetic extinction everywhere (the "replacement" part of the hypothesis). The ancient and recent candelabra models differ only in their temporal placement of the ancestral node; they share the same tree topology that portrays Africans, Europeans, and Asians as distinct branches on an evolutionary tree. It is this branching *topology* that defines "races" under the evolutionary lineage definition, and not the *time* since the common ancestral population. Recall that the lineage definition of race is based on a qualitative feature (being a historical lineage or branch) and not on a quantitative threshold (such as the length of the branch). Hence, human "races" are valid under the out-of-Africa replacement model.

In contrast to the candelabra models, the trellis model (figure 12.1c) posits that *Homo erectus* populations had the ability to disperse not only out of Africa but also back in, resulting in recurrent genetic interchange among Old World human populations (Weidenreich 1946; Lasker and Crews 1996; Wolpoff and Caspari 1997). Under the trellis model, anatomically modern traits could evolve anywhere in the range of *Homo erectus* (which includes Africa) and subsequently spread throughout all of humanity by selection and gene flow. Hence, an African origin for anatomically modern humans is compatible with either model. The two models do differ in their interpretation of interpopulational genetic differences. Under the candelabra models, "races" are valid subspecies; but under the trellis model there was no separation of humanity into evolutionary lineages and, hence, human "races" are not valid subspecies, by the lineage definition.

The recent human genetic literature contains many papers that claim to support the out-of-Africa replacement hypothesis. However, much of this support is in the form of data that reject the ancient origin candelabra model but are equally consistent with both the recent origin and trellis models of human evolution (Templeton 1997). Hence, I will deal only with data sets and analyses that have the potential for discriminating between the recent origin candelabra model and the trellis model.

Because the F_{st} statistic is incapable of discriminating between these two models (Templeton 1998b), much of the literature on recent human origins focuses instead on genetic distance measures between pairs of existing human populations. Pairwise, genetic distances in turn can be converted into an evolutionary tree of populations by various computer algorithms. An evolutionary tree is simply a branching diagram that represents the evolutionary relationships among the members of the tree. Although one normally thinks of evolutionary trees as referring to species, an evolutionary tree can be diagrammed for any biological entity that displays clear ancestral/descendant relationships through genetic continuity. Thus, evolutionary trees can be estimated for the various alleles at a gene locus (a gene tree) if there has been little or no internal recombination within the gene. When internal recombination occurs, a new allele can be created from the parts of two different alleles, so that different sections of the new allele can have different evolutionary histories. In this case, the alleles themselves cannot be used as members of a legitimate evolutionary tree, although it may be possible to use smaller sections of the gene to define an evolutionary tree (Templeton and Sing 1993). Similarly, an evolutionary tree can be estimated for populations *if the populations are isolates.* The analogue of recombination in this case is gene flow or admixture. If genetic interchange occurs among members of the populations, then populations cannot be legitimately diagrammed as an evolutionary tree.

The computer programs that generate population trees from pairwise genetic distances do so regardless of whether populations are legitimate ancestral/descendant units. Assuming for the moment that human populations are true isolates, figure 12.2a shows such a population tree estimated from pairwise genetic distance data (Cavalli-Sforza et al. 1996). This and most other human genetic distance trees have the deepest divergence between Africans and non-Africans and interpret this divergence as accumulated genetic distance since the populations split. This split is commonly estimated to have occurred around one hundred thousand years ago (Cavalli-Sforza et al. 1996; Cavalli-Sforza 1997; Nei and Takezaki 1996). All this seems to validate the existence of human races. However, nonzero genetic distances can also arise and persist between populations due to the dynamic balance of genetic drift and recurrent gene flow. As shown by Montgomery Slatkin (1991), recurrent gene flow results in an average divergence time of gene lineages between populations even when no population-level split occurred. Therefore, an apparent time of genetic divergence does not necessarily imply a time of actual population splitting—or any population split at all! Without a split, human "races" are not truly races under the modern phylogenetic definition.

Fortunately, these two interpretations of genetic distance can be distinguished. If human "races" truly represent branches on an evolutionary tree, then the resulting genetic distances should satisfy several constraints. For example, under the evolutionary tree model, all non-African human populations split from the Africans at the same time, and therefore all genetic distances between African and non-African populations have the same expected value (figure 12.2a). When genetic distances instead reflect the amount of gene flow, the constraints of "treeness" are no longer applicable. Because gene flow is commonly restricted by geographic distance (Wright 1943), gene flow models are expected to yield a strong, positive relationship between geographic distance and genetic distance. Figure 12.2b places the populations on a two-dimensional plot in a manner that reflects their genetic distances from one another while otherwise attempting to minimize

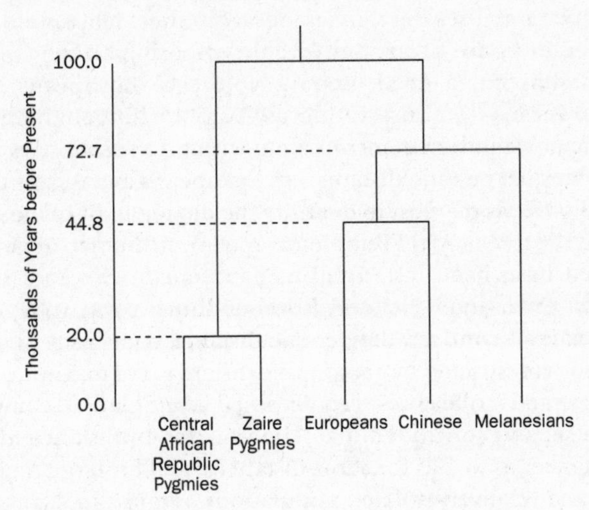

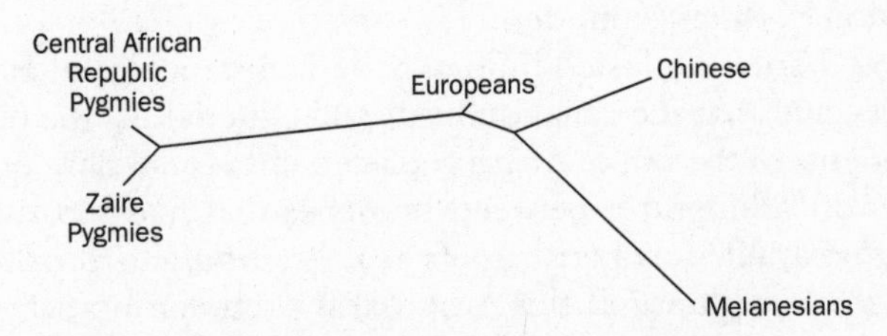

Figure 12.2 Portrayals of human genetic distances. Part A shows an evolutionary tree of human populations as estimated from the data given by Bowcock et al. (1991). Human population evolution is depicted as a series of splits, and the numbers on the left indicate the estimated times of divergence in thousands of years. This figure is redrawn from figure 2.4.4 by Cavalli-Sforza et al. (1996: 91). Part B shows the same data drawn with the neighbor-joining method, but without the constraints of a tree. This figure is redrawn from figure 2.4.5 by Cavalli-Sforza et al. (1996: 91).

the total sum of branch lengths (formally, a neighbor-joining dendrogram). Figure 12.2b uses the same genetic distance data employed to generate the tree in figure 12.2a, but without imposing all the constraints of treeness (Cavalli-Sforza et al. 1996). Note that Europeans fall *between* Africans and Asians—an observation consistent with geographic location under a gene flow model but inconsistent with the evolutionary tree model prediction of equal genetic distances of Europeans and Asians to Africans.

Statistical procedures exist to quantify the degree of fit of the genetic distance data to the tree model (Templeton 1998a). *All* human genetic distance data sets that have been tested fail to fit treeness (Bowcock et al. 1991; Cavalli-Sforza et al. 1996; Nei and Roychoudhury 1974, 1982; Templeton 1998a). In marked contrast, the genetic distance data easily fit a recurrent gene flow model restricted by geographic distance. For example, Luigi Luca Cavalli-Sforza and colleagues (1996: 124) assembled a comprehensive human data set and concluded that "the isolation-by-distance models hold for long distances as well as for short distances, and for large regions as well as for small and relatively isolated populations." Figure 12.3, a redrawing of a figure from Cavalli-Sforza and colleagues 1996, illustrates the excellent fit of the isolation-by-distance model.

Given that there is no tested human genetic distance data set consistent with treeness, and that the isolation-by-distance model fits the human data well, proponents of the out-of-Africa replacement model have postulated a complex set of "admixtures between branches that had separated a long time before" (Cavalli-Sforza et al. 1996: 19). According to Terrell and Stewart (1996), the key phrase in this proposal is "between branches that had separated a long time before." This phrase implies that human races have been valid biological categories for much of human evolutionary history, with recent admixture obscuring the long-standing racial status of human populations. Proponents of the out-of-Africa replacement model attempt to reconcile the genetic distance data by using an admixture model that mimics some of the effects (and the good fit) of recurrent gene flow. By invoking admixture events as needed, they can still treat human "races" as separate evolutionary lineages, but now with the qualification that the "races" were purer in the past—the paradigm of the "primitive isolate" (Terrell and Stewart 1996). However, even advocates of the races-as-evolutionary-branches model acknowledge that these postulated admixture events are "extremely specific" and "unrealistic" (Bowcock et al. 1991: 841). In contrast, the isolation-by-distance model fits the human data well and requires only that humans have tended to mate primarily with others born nearby but often outside their own natal group (Lasker and Crews 1996; Santos et al. 1997).

The hypothesis of admixture can be tested directly. For periods when isolation truly existed, the previous isolates should show many allele frequency differences. For periods when they came into contact again, admixture

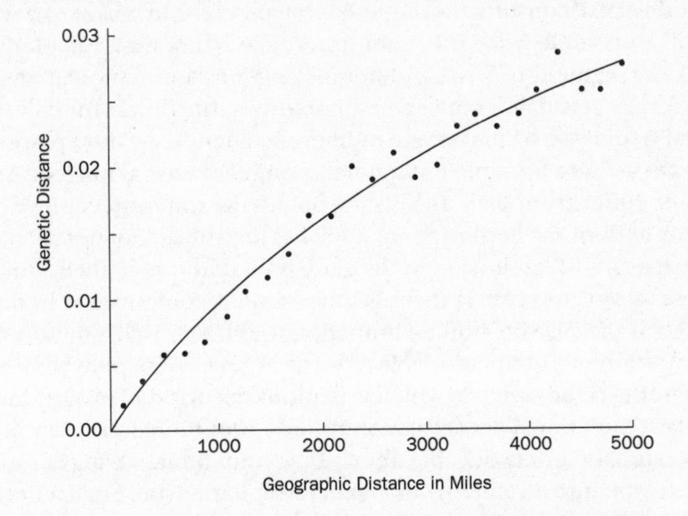

Figure 12.3 Genetic distance and isolation by geographic distance. The global human genetic distances (the ordinate) are plotted against geographic distance in miles (the abscissa). The circles indicate the observed values, and the curved line is the theoretical expectation under an isolation-by-distance model. This figure is redrawn from figure 2.9.2 by Cavalli-Sforza et al. (1996: 123).

should result in a genetic cline, a continuous shift from one allele frequency to another over geographic space. Such genetic clines should be set up simultaneously for all differentiated loci, thereby resulting in a strong geographic concordance in the clines for all genetic systems. In contrast, isolation by distance may result in geographic concordance for systems under similar selective regimes (Endler 1977; King and Lawson 1995), but otherwise no concordance should be expected. The lack of concordance in the geographic distribution of different genetic traits in humans has been thoroughly and extensively documented and has been one of the primary traditional arguments against the validity of human races (Cavalli-Sforza et al. 1996; Futuyma 1986). This lack of concordance across genetic systems falsifies the hypothesis of admixture of previously isolated branches and the idea that races were pure in the past.

Another test of the hypothesis that equates races to branches on an evolutionary tree arises from phylogenetic reconstructions of the genetic variation found in homologous regions of DNA that show little or no recombination. All the homologous copies of DNA in such a DNA region that are identical at every nucleotide (or, in practice, identical at all scored nucleotide sites) constitute a single haplotype class. In this sense, a haplotype is like an allele, with the main difference being that a haplotype can refer to any segment of DNA and not just a gene. A mutation at any site in this DNA region usually creates a new haplotype that differs initially from its ancestral haplotype by that single mutational change. As time passes, some haplotypes acquire multiple mutational changes at several nucleotides, and, as a result, differ from their ancestral type. All the different copies of a haplotype for each of the haplotypes in a species are subject to mutation, resulting in a diversity of haplotypes in the gene pool that vary in their mutational closeness to one another. If there is little or no recombination in the DNA region (as is the case for human mitochondrial DNA, Y chromosome DNA, and small segments of nuclear DNA), the divergence of haplotypes from one another reflects the order in which mutations occurred in evolutionary history. When mutational accumulation reflects evolutionary history, it is possible to estimate a network that shows how mutational changes transform one haplotype into another from an ancestral haplotype. Such a network is called a haplotype tree. In some circumstances, the ancestral haplotype is known or can be inferred, thereby providing a rooted haplotype tree. In practice, haplotype trees are sometimes difficult to infer from the mutational differences among a set of observed haplotypes, because the same mutation may have occurred more than once or because recombination may have scrambled the DNA region so thoroughly that accumulated mutational differences reflect both evolutionary history and recombination in a confounded fashion. When they can be estimated, haplotype trees directly reflect only the evolutionary history of the genetic diversity being monitored in the DNA region under study. Haplotype trees are *not necessarily* evolutionary trees of species or of populations within species. For example, suppose a species is and always has been mating completely randomly as a single population and, therefore, has no races or subspecies at all: that same randomly mating species will have haplotype trees for all the homologous DNA regions that show little or no recombination.

If a species is truly subdivided into isolates, this should have an effect on the haplotype trees. Therefore, although haplotype trees are not necessarily population trees, they can still contain information about population history. This information in haplotype trees can be used to test the hypothesis that human races are evolutionary lineages whose past purity has been diminished by later admixture. For example, to reconcile the evolutionary tree model with the genetic distance data, it is necessary to regard Europeans as a heavily admixed population (Bowcock et al. 1991; Cavalli-Sforza et al. 1996). When admixture occurs, haplotypes that differ as a result of multiple mutational events should coexist in the admixed population's gene pool, and there should be no intermediate haplotypes (Manderscheid and Rogers 1996; Templeton et al. 1995). The detection of such highly divergent haplotypes requires large sample sizes of the presumed admixed population in order to have statistical power. When large sample surveys have been performed on the presumed admixed European populations, no highly divergent haplotypes or evidence for admixture has been observed for either mtDNA (Manderscheid and Rogers 1996) or Y-DNA (Cooper et al. 1996). In contrast, isolation by distance (the gene flow model) should produce gene pools without strongly divergent haplotypes (i.e., most haplotypes differ by one or at most only a few mutational steps from some other haplotype found in the same population), as has been observed.

The out-of-Africa replacement and trellis hypotheses are models of how genes spread across geographic space and through time, and hence a geographic analysis of haplotype trees provides a direct test of these two models. Statistical techniques exist that separate the influences of historical events (such as population range expansions) from recurrent events (such as gene flow when a population is isolated by distance) when there is adequate sampling in terms of both numbers of individuals and numbers and distribution of sampling sites (Templeton et al. 1995). This statistical approach treats historical and recurrent events as joint possibilities rather than as mutually exclusive alternatives. Moreover, the criteria used to identify range expansions versus gene flow have been empirically validated by analyzing data sets about which strong prior knowledge exists, showing that this approach is accurate and not prone to false positives (Templeton 1998b).

This statistical approach to analyzing human haplotype trees over geographic space reveals that both a mixture of population range expansion events and a recurrent genetic exchange have taken place among the major Old World human populations, as inferred from mitochondrial DNA, Y-DNA, two X-linked DNA regions, and six autosomal DNA regions (Templeton 2002). This multilocus analysis reveals that human evolution from about a million years ago to the last tens of thousands of years has been dominated by two evolutionary forces: (1) population movements and associated range expansions, including an out-of-Africa expansion around one hundred thousand years ago; and (2) gene flow primarily restricted by isolation by distance. Of the ten DNA regions examined, eight of them have an evolutionary history that extends well beyond the one-hundred-thousand-year-old expansion of humans out of Africa. All eight of these DNA regions falsify the hypothesis of an out-of-Africa replacement by providing statistically significant evidence either for earlier range expansions involving Eurasia or for gene flow between Africa and Eurasia prior to the supposed replacement event (if replacement had truly occurred, it would have obliterated all evidence of earlier range expansions and gene flow with Eurasian populations). The fact that all eight informative DNA regions falsify replacement and imply interbreeding is strong evidence that no split occurred between

African and Eurasian populations, and that they were not isolates. The only evidence for any split or fragmentation event in human evolutionary history within this time frame is the one detected with mtDNA involving the colonization of the Americas (Templeton 1997, 1998a). However, this colonization was the outcome of either multiple colonization events or movements by large numbers of peoples (Templeton 1998a), resulting in extensive sharing of genetic polymorphisms of New World with Old World human populations. Moreover, the genetic isolation between the Old and New Worlds was brief (to an evolutionary biologist) and no longer exists. Other than this temporary fragmentation event, the major human populations have been interconnected by gene flow (recurrent at least on a time scale on the order of tens of thousands of years or less) during at least the last six hundred thousand years, with 95 percent statistical confidence (Templeton 2002). Hence, the haplotype analyses of geographical associations strongly reject the existence of multiple evolutionary lineages of humans, reject the idea that Eurasians split from Africans one hundred thousand years ago, and reject the idea that "pure races" existed in the past. Thus, the idea that "races" existed among humans has no biological validity under the evolutionary lineage definition of subspecies.

THE SIGNIFICANCE OF GENETIC DIFFERENCES AMONG HUMAN POPULATIONS

Although human populations do not define races under any of the definitions currently applied to nonhuman organisms, genetic differences do exist among human populations as noted above and quantified by the F_{st} value of about .15—a small value but one greater than zero. These modest genetic differences can still have evolutionary and genetic significance. Therefore, the evolutionary significance of genetic differentiation among human populations (not races, since none exist) is a legitimate issue.

Geographic Differentiation

As indicated earlier (and in more detail in Templeton 1998a), the patterns of genetic diversity found among human populations closely fit an isolation-by-distance model. Under this model, genetic drift is counteracted by genetic exchange (gene flow) among the populations. In many organisms, the amount of gene flow between populations decreases as geographic distance increases. Indeed, populations far from one another geographically may experience no direct genetic interchange at all, but gene flow can still occur as genes are passed on from one adjacent population to the next, a type of gene flow known as "stepping stones" (Wright 1943). Under these models of isolation by geographic distance, genetic differentiation increases with increasing geographic distance. Neutral genetic variants display this expected pattern most clearly (that is, genetic variants that are not subject to natural selection and that, as a result, show the effects of drift and gene flow). When variation is selected, deviations from this pattern can occur. Given the overall excellent fit of human genetic distances to an isolation-by-distance model (figure 12.3), the primary significance of genetic differences among human populations is as selectively neutral indicators of geographic origin. Races are genetically differentiated primarily because they come from different geographic regions, not because the racial classes have any significance per se. For example, Melanesians and Africans exhibit nearly maximal genetic divergence within humanity as a whole with respect to molecular markers (see figure 12.2b). Given that these two populations live on opposite sides of the world, this extreme genetic differentiation is expected under the isolation-by-distance gene flow model. Europeans are genetically closer to both Africans and Melanesians than are Africans to Melanesians, a pattern also expected under the isolation-by-distance gene flow model. However, Melanesians and Africans share dark skin, hair texture, and cranial-facial morphology (Cavalli-Sforza et al. 1996; Nei and Roychoudhury 1993)—the traits typically used to classify people into races. As a result, Europeans were classified into one race and Africans and Melanesians were classified into a single race in the older anthropological literature (e.g., Weidenreich 1946). Such a classification makes no sense in any evolutionary framework and is misleading about the patterns of actual genetic differentiation. Instead, the genetic differentiation among Africans, Europeans, and Melanesians is indicated by their relative geographic positions and not by "racial traits."

Because most genetic differences among humans simply represent the balance of drift versus gene flow in a geographic context, the most extreme genetic differences between human populations are expected between those that are the most geographically distant (e.g., Africans and Melanesians), or between populations that for some reason have had very small population sizes (which accentuates drift) or have had little gene flow with other, even nearby, populations. Indeed, the most dramatic cases of genetic differentiation among human populations are associated with historically small population sizes and not racial categories. For example, the Old Order Amish in North America were established from a relatively small number of founders and have had little subsequent gene flow into their population (McKusick et al. 1964; see Lindee, chapter 2, this volume). The Amish have undergone extreme genetic differentiation from their neighboring populations because of powerful genetic drift. Unlike the differences among races that are primarily neutral indicators of geographic origin, genetic drift is so powerful in populations derived from a small number of founders that selected alleles can drift to highly divergent allele frequencies, even in a non-adaptive direction. As a result, the genetic differentiation found in the Amish and other small human populations has great importance for both clinical and behavioral traits (Ludwig et al. 1997; Pericakvance et al. 1996; Polymeropoulos et al. 1996; Polymeropoulos and Schaffer 1996). In fact, with the examination of molecular markers scattered throughout the genome, such "founder" populations have proved to be extremely valuable resources in mapping and identifying genetic diseases. For example, the first genetic disease gene mapped and ultimately cloned by such positional mapping was the gene for Huntington's disease (Gusella et al. 1983). The key to this original mapping effort was to find an isolated founder population in which this disease was frequent. Such a founder population was discovered in Venezuela. Although such founder populations do have great evolutionary and clinical significance, drift-induced differentiation is generally not used to make "racial" classifications.

Adaptive Differentiation

Genetic drift, then, is sometimes so strong that it can cause human populations to become genetically differentiated even in directions not favored by natural selection. Normally, natural selection tends to keep deleterious alleles rare in all human populations, thereby reducing the differentiation of such genes among human populations below that expected in the case of isolation by distance. However, in some circumstances, natural selection can accentuate differentiation. Adaptive differentiation is expected to occur when a species inhabits different geographic regions that induce divergent selective pressures for some traits. Because humans are a geographically widespread species, humans have indeed adapted to local environmental conditions in a manner that causes genetic differentiation among populations. For example, there is much evidence that the intensity of ultraviolet radiation induces natural selection on the amount of melanin in human skin, with high intensities favoring dark skin, and low intensities light skin (Relethford 1997). As a consequence, human populations have become highly differentiated in skin color in a manner that is adaptive to the area of geographic origin (Relethford 1997). As shown by the comparison of European, Melanesian, and African populations, the pattern of genetic differentiation obtained for this adaptive trait does *not* reflect the overall pattern of genetic differentiation among human populations. Rather, the geographic pattern of the adaptive trait reflects the geographic pattern of the selective conditions that favor the trait in some regions but not others.

One could argue that differentiation for adaptive traits, precisely because these traits are adaptive, does constitute a meaningful method of "racial" classification, that differentiation does lead to special "racial traits." However, using adaptive traits to classify subspecies has no legitimacy in the literature on nonhuman species if there is no concordance across different adaptive traits. Frequently, different adaptive traits show discordant patterns with one another when their selective agents are likewise distributed in a discordant fashion. Human adaptive traits are generally discordant with one another and therefore cannot serve as a basis for racial classification using the standards found in the literature on nonhuman species. For example, malaria has been a major selective influence on human populations. One of the many genetic adaptations to malaria is the allele for sickle-cell anemia at the hemoglobin beta-chain locus (Haldane 1949; Templeton 1982). This *S* allele is selected because it confers resistance to the malarial parasite when heterozygous for the most common *A* allele at this locus. *S* alleles are found in high frequencies only in populations that currently live in, or occupied in the recent past, malarial regions. This includes some, but not all, sub-Saharan African populations, populations in the Mediterranean and Middle East, and populations in India (Oner et al. 1992; El-Hazmi 1990; Schiliro et al. 1990; Boletini et al. 1994; Reddy and Modell 1996). Sickle-cell is not cleanly associated with any "race" but rather is associated with the presence of malaria, its selective agent. Furthermore, its distribution is not concordant with skin color. This example illustrates that it is essential to study each adaptive trait separately and to relate it to its unique selective agents. When populations are interconnected by gene flow, as human populations are, locally adaptive traits are not expected to be concordant with one another or to reflect overall patterns of genetic differentiation. Hence, locally adaptive traits in populations interconnected by gene flow do not define races.

Indeed, the concept of race can be an impediment to a proper understanding of adaptive polymorphism in populations interconnected by gene flow. For example, the water snake *(Nerodia sipedon,* previously *Natrix sipedon)* on islands in Lake Erie has melanic and nonmelanic forms that were used to define two subspecies, *N. s. insularum* (the melanic form), and *N. s. sipedon* (the nonmelanic form that has bands; see Conant and Clay 1937). However, subsequent studies revealed that dispersal patterns and gene flow in these snakes reflected geographic distance rather than skin color category (Camin and Ehrlich 1958), just as in humans. Molecular studies have revealed that the amount of gene flow between the skin-color "races" in these snakes is greater than that of humans (that is, on the geographic scale of Lake Erie the snakes have a smaller F_{st} value than humans on a global scale [King and Lawson 1995]). This result shows that strong patterns of skin color differentiation can be maintained by local selection despite levels of gene flow that are greater than those observed between human "races." Moreover, these studies (Camin and Ehrlich 1958; King and Lawson 1995) indicate that skin color differences are being maintained by selection favoring melanic forms in populations inhabiting cold waters because of the thermal properties of dark skin versus selection favoring banded forms on the mainland for crypsis—that is, coloring that makes it difficult for predators to see them. Hence, these later studies revealed the evolutionary significance of melanism in these snakes. Concerning the older work that had simply placed the melanic snakes into a new subspecies, Joseph H. Camin and Paul R. Ehrlich (1958: 510) wrote, "The subspecies approach has tended to obscure a significant biological problem." Subspecies in this case were actually an impediment to understanding the evolutionary significance of the "racial" variation. The same is certainly true for humans. For example, the fundamental breakthrough in our understanding of sickle-cell anemia occurred when we stopped phrasing its geographic distribution in terms of racial categories and phrased it instead in terms of the distribution of the malarial parasite (Haldane 1949).

The above examples also illustrate that even extreme differentiation for locally adaptive traits is *not* evidence for a lack of gene flow among populations. Local selective forces can cause strong adaptive differentiation even for populations with gene flow levels much higher than those found in humans (e.g., King and Lawson 1995, 1997; Lawson and King 1996; DeSalle et al. 1987; Templeton and Sing 1993; Templeton et al. 1989). However, the human F_{st} value is sufficiently low that it implies the existence of gene flow levels which insure that any trait universally adaptive in all human populations will spread throughout the species (Barton and Rouhani 1993), making humans a single evolutionary lineage. Hence, there is no incompatibility between the idea that humans show local adaptive polymorphisms and the idea that human are a single long-term evolutionary lineage with no races or subspecies.

CONCLUSIONS

The genetic data are consistently and strongly informative about human races. Humans show only modest levels of differentiation among populations when compared to other large-bodied mammals (Templeton 1998a), and this level of differentiation is well below the usual threshold used to identify subspecies (races) in nonhuman species. Hence, human races do not exist under the concept of a subspecies defined by a threshold level of genetic differentiation. A more modern definition of race designates it as a distinct evolutionary lineage within a species. The genetic evidence strongly rejects the existence of distinct evolutionary lineages within humans. The widespread representation of human "races" as branches on an intraspecific population tree is genetically indefensible and biologically misleading, even when the ancestral node is presented as dating to one hundred thousand years ago. Attempts to salvage the idea of human races as evolutionary lineages by presuming that greater racial purity existed in the past and was followed by recent admixture events fail the test. Instead, all the genetic evidence shows that there never was a split, or separation of races, between Africans and Eurasians as postulated by the out-of-Africa replacement hypothesis. Recent human evolution has been characterized by both population range expansions and recurrent genetic interchange among populations. There has been no split between any of the major geographic populations of humanity, with the temporary exception of the split between Native American and Old World populations.

Because of the extensive evidence for genetic interchange through population movements and recurrent gene flow going back hundreds of thousands of years or more, there is only one evolutionary lineage of humanity and there are no subspecies or races under either the threshold or the phylogenetic definition. Human evolution and population structure have been and are characterized by many locally differentiated populations coexisting at any given time, but which have had sufficient genetic contact to make all of humanity a single lineage sharing a common, long-term evolutionary fate. The genetic differences that exist among human populations are explained primarily by geography under an isolation-by-distance model, with some extreme differentiation being due to recent founder events and local adaptations. However, all of humanity shares the vast majority of its molecular genetic variation and the adaptive traits that define it as a single species.

NOTES

I wish to express my deep gratitude to the Wenner-Gren Foundation for the invitation and support to participate in this intellectually stimulating conference. I also thank all the conference participants, whose ideas and comments helped shape the revision of this manuscript, and particularly Alan Goodman for his excellent suggestions and critique of an earlier draft.

REFERENCES

Amato, G., and J. Gatesy. 1994. PCR assays of variable nucleotide sites for identification of conservation units. In *Molecular ecology and evolution: Approaches and applications,* ed. B. Schierwater, B. Streit, G. P. Wagner, and R. DeSalle, 215–26. Basel: Birkhäuser-Verlag.

Avise, J. C. 1994. *Molecular markers, natural history, and evolution.* New York: Chapman and Hall.

Barbujani, G., A. Magagni, E. Minch, and L. L. Cavalli-Sforza. 1997. An apportionment of human DNA diversity. *Proceedings of the National Academy of Sciences, USA* 94:4516–19.

Barton, N. H., and S. Rouhani. 1993. Adaptation and the "shifting balance." *Genetical Research* 61:57–74.

Boletini, E., M. Svobodova, V. Divoky, E. Baysal, M. A. Curuk, A. J. Dimovski, R. Liang, A. D. Adekile, and T. H. Huisman. 1994. Sickle-cell-anemia, sickle-cell beta-thalassemia, and thalassemia major in Albania—characterization of mutations. *Human Genetics* 93:182–87.

Bowcock, A. M., J. R. Kidd, J. L. Mountain, J. M. Hebert, L. Carotenuto, K. K. Kidd, and L. L. Cavalli-Sforza. 1991. Drift, admixture, and selection in human evolution: A study with DNA polymorphisms. *Proceedings of the National Academy of Sciences, USA* 88:839–43.

Brownlow, C. A. 1996. Molecular taxonomy and the conservation of the red wolf and other endangered carnivores. *Conservation Biology* 10:390–96.

Camin, J. H., and P. R. Ehrlich. 1958. Natural selection in water snakes (*Natrix sipedon* L.) on islands in Lake Erie. *Evolution* 12:504–11.

Cavalli-Sforza, L. L. 1997. Genes, peoples, and languages. *Proceedings of the National Academy of Sciences, USA* 94:7719–24.

Cavalli-Sforza, L., P. Menozzi, and A. Piazza. 1996. *The History and geography of human genes.* Princeton, N.J.: Princeton University Press.

Conant, R., and W. M. Clay. 1937. *A new subspecies of water snake from islands in Lake Erie.* Occasional Paper no. 346. Ann Arbor: University of Michigan Museum of Zoology.

Cooper, G., W. Amos, D. Hoffman, and D. C. Rubinsztein. 1996. Network analysis of human Y microsatellite haplotypes. *Human Molecular Genetics* 5:1759–66.

DeSalle, R., A. Templeton, I. Mori, S. Pletscher, and J. S. Johnston. 1987. Temporal and spatial heterogeneity of mtDNA polymorphisms in natural populations of *Drosophila mercatorum. Genetics* 116:215–23.

El-Hazmi, M. 1990. Beta-globin gene haplotypes in the Saudi sickle-cell-anemia patients. *Human Heredity* 40:177–86.

Endler, J. A. 1977. *Geographic variation, speciation, and clines.* Princeton, N.J.: Princeton University Press.

Futuyma, D. J. 1986. *Evolutionary biology.* 2d ed. Sunderland, Mass.: Sinauer Associates.

Gusella, J. F., N. S. Wexler, P. M. Conneally, S. L. Naylor, M. A. Anderson, R. E. Tanzi, P. C. Watkins, K. Ottina, M. R. Wallace, A. Y. Sakaguchi, A. B. Young, I. Shoulson, E. Bonilla, and J. B. Martin. 1983. A polymorphic DNA marker genetically linked to Huntington's disease. *Nature* 306:324–28.

Haldane, J. B. S. 1949. Disease and evolution. *Ricerca Scientifica* 19:3–10.

King, R. B., and R. Lawson. 1995. Color-pattern variation in Lake Erie water snakes—the role of gene flow. *Evolution* 49:885–96.

———. 1997. Microevolution in island water snakes. *Bioscience* 47:279–86.

Lasker, G. W., and D. E. Crews. 1996. Behavioral influences on the evolution of human genetic diversity. *Molecular Phylogenetics and Evolution* 5:232–40.

Lawson, R., and R. B. King. 1996. Gene flow and melanism in Lake Erie garter snake populations. *Biological Journal of the Linnean Society* 59:1–19.

Legge, J. T., R. Roush, R. Desalle, A. P. Vogler, and B. May. 1996. Genetic criteria for establishing evolutionarily significant units in Cryan's buckmoth. *Conservation Biology* 10:85–98.

Lewontin, R. C. 1972. The apportionment of human diversity. *Evolutionary Biology* 6:381–98.

Ludwig, E. H., P. N. Hopkins, A. Allen, L. L. Wu, R. R. Williams, J. L. Anderson, R. H. Ward, J. M. Lalouel, and T. L. Innerarity. 1997. Association of genetic variations in apolipoprotein B with hypercholesterolemia, coronary artery disease, and receptor binding of low density lipoproteins. *Journal of Lipid Research* 38:1361–73.

Manderscheid, E. J., and A. R. Rogers. 1996. Genetic admixture in the late Pleistocene. *American Journal of Physical Anthropology* 100:1–5.

McKusick, V. A., J. A. Hostetler, J. A. Egeland, and R. Eldridge. 1964. The distribution of certain genes in the Old Order Amish. *Cold Spring Harbor Symposium on Quantitative Biology* 29:99–114.

Miththapala, S., J. Seidensticker, and S. J. O'Brien. 1996. Phylogeographic subspecies recognition in leopards *(Panthera pardus)*—molecular genetic variation. *Conservation Biology* 10:1115–32.

Nei, M., and A. K. Roychoudhury. 1974. Genic variation within and between the three major races of man, Caucasoids, Negroids, and Mongoloids. *American Journal of Human Genetics* 26:421–43.

———. 1982. Genetic relationship and evolution of human races. *Evolutionary Biology* 14:1–59.

———. 1993. Evolutionary relationships of human populations on a global scale. *Molecular Biology and Evolution* 10:927–43.

Nei, M., and N. Takezaki. 1996. The root of the phylogenetic tree of human populations. *Molecular Biology and Evolution* 13:170–77.

Oner, C., A. J. Dimovski, N. F. Olivieri, G. Schiliro, J. F. Codrington, S. Fattoum, A. D. Adekile, R. Oner, G. T. Yuregir, C. Altay, A. Gurgey, R. B. Gupta, V. B. Jogessar, M. N. Kitundu, D. Loukopoulos, G. P. Tamagnini, M. Ribeiro, F. Kutlar, L. H. Gu, K. D. Lanclos, and T. Huisman. 1992. Beta-S haplotypes in various world populations. *Human Genetics* 89:99–104.

Pennock, D. S., and W. W. Dimmick. 1997. Critique of the evolutionarily significant unit as a definition for distinct population segments under the U.S. Endangered Species Act. *Conservation Biology* 11:611–19.

Perez-Lezaun, A., F. Calafell, E. Mateu, D. Comas, R. Ruiz-Pacheco, and J. Bertranpetit. 1997. Microsatellite variation and the differentiation of modern humans. *Human Genetics* 99:1–7.

Pericakvance, M. A., C. C. Johnson, J. B. Rimmler, A. M. Saunders, L. C. Robinson, E. G. Dhondt, C. E. Jackson, and J. L. Haines. 1996. Alzheimer's disease and apolipoprotein E-4 allele in an Amish population. *Annals of Neurology* 39:700–704.

Polymeropoulos, M. H., S. E. Ide, M. Wright, J. Goodship, J. Weissenbach, R. E. Pyeritz, E. O. Dasilva, R. I. O. Deluna, and C. A. Francomano. 1996. The gene for the Ellis–van Creveld syndrome is located on chromosome 4p16. *Genomics* 35:1–5.

Polymeropoulos, M. H., and A. A. Schaffer. 1996. Scanning the genome with 1772 microsatellite markers in search of a bipolar disorder susceptibility gene. *Molecular Psychiatry* 1:404–7.

Reddy, P. H., and B. Modell. 1996. Reproductive behaviour and natural selection for the sickle gene in the Baiga tribe of Central India—the role of social parenting. *Annals of Human Genetics* 60:231–36.

Relethford, J. H. 1997. Hemispheric difference in human skin color. *American Journal of Physical Anthropology* 104:449–57.

Santos, E. J. M., J. T. Epplen, and C. Epplen. 1997. Extensive gene flow in human populations as revealed by protein and microsatellite DNA markers. *Human Heredity* 47:165–72.

Schiliro, G., M. Spena, E. Giambelluca, and A. Maggio. 1990. Sickle haemoglobinpathies in Sicily. *American Journal of Hematology* 33:81–85.

Shaffer, H. B., and M. L. McKnight. 1996. The polytypic species revisited—genetic differentiation and molecular phylogenetics of the tiger salamander *Ambystoma tigrinum* (Amphibia, Caudata) complex. *Evolution* 50:417–33.

Slatkin, M. 1991. Inbreeding coefficients and coalescence times. *Genetical Research* 58:167–75.

Smith, H. M., D. Chiszar, and R. R. Montanucci. 1997. Subspecies and classification. *Herpetological Review* 28:13–16.

Templeton, A. R. 1982. Adaptation and the integration of evolutionary forces. In *Perspectives on evolution,* ed. R. Milkman, 15–31. Sunderland, Mass.: Sinauer.

———. 1993. The "Eve" hypothesis: A genetic critique and reanalysis. *American Anthropologist* 95:51–72.

———. 1994. The role of molecular genetics in speciation studies. In *Molecular ecology and evolution: Approaches and applications,* ed. B. Schierwater, B. Streit, G. P. Wagner, and R. DeSalle, 455–477. Basel: Birkhäuser-Verlag.

———. 1997. Out of Africa? What do genes tell us? *Current Opinion in Genetics and Development* 7:841–47.

———. 1998a. Human races: A genetic and evolutionary perspective. *American Anthropologist* 100:632–50.

———. 1998b. Nested clade analyses of phylogeographic data: Testing hypotheses about gene flow and population history. *Molecular Ecology* 7:381–97.

———. 1998c. Species and speciation: Geography, population structure, ecology, and gene trees. In *Endless forms: Species and speciation,* ed. D. J. Howard and S. H. Berlocher, 32–43. Oxford: Oxford University Press.

———. 2002. Out of Africa again and again. *Nature* 416:45–51.

Templeton, A. R., H. Hollocher, and J. S. Johnston. 1993. The molecular through ecological genetics of abnormal abdomen in *Drosophila mercatorum.* V. Female phenotypic expression on natural genetic backgrounds and in natural environments. *Genetics* 134:475–85.

Templeton, A. R., H. Hollocher, S. Lawler, and J. S. Johnston. 1989. Natural selection and ribosomal DNA in *Drosophila. Genome* 31:296–303.

Templeton, A. R., E. Routman, and C. Phillips. 1995. Separating population structure from population history: A cladistic analysis of the geographical distribution of mitochondrial DNA haplotypes in the tiger salamander, *Ambystoma tigrinum. Genetics* 140:767–82.

Templeton, A. R., and C. F. Sing. 1993. A cladistic analysis of phenotypic associations with haplotypes inferred from restriction endonuclease mapping. IV. Nested analyses with cladogram uncertainty and recombination. *Genetics* 134:659–69.

Terrell, J. E., and P. J. Stewart. 1996. The paradox of human population genetics at the end of the twentieth century. *Reviews in Anthropology* 25:13–33.

Vogler, A. P. 1994. Extinction and the formation of phylogenetic lineages: Diagnosing units of conservation management in the tiger beetle *Cicindela dorsalis.* In *Molecular ecology and evolution: Approaches and applications,* ed. B. Schierwater, B. Streit, G. P. Wagner, and R. DeSalle, 261–73. Basel: Birkhäuser-Verlag.

Weidenreich, F. 1946. *Apes, giants, and man.* Chicago: University of Chicago Press.
Wilson, E. O., and W. L. Brown. 1953. The subspecies concept and its taxonomic applications. *Systematic Zoology* 2:97–111.
Wolpoff, M., and R. Caspari. 1997. *Race and human evolution.* New York: Simon and Schuster.
Wright, S. 1943. Isolation by distance. *Genetics* 28:114–38.
———. 1969. *Evolution and the genetics of populations.* Vol. 2, *The Theory of Gene Frequencies.* Chicago: University of Chicago Press.

Chapter 13

Buried Alive

The Concept of Race in Science

Troy Duster

FLUIDITY IN THE SCIENTIFIC STATUS OF THE CONCEPT OF RACE

A consortium of leading scientists across the disciplines from biology to physical anthropology issued a "Revised UNESCO Statement on Race" in 1995—a definitive declaration that summarizes eleven central issues and concludes that, in terms of scientific discourse, the concept of race has no scientific utility: "The same scientific groups that developed the biological concept over the last century have now concluded that its use for characterizing human populations is so flawed that it is no longer a scientifically valid concept. In fact, the statement makes clear that the biological concept of race as applied to humans has no legitimate place in biological science" (Katz 1995: 4, 5).

Note that the statement is not only about the utility of the concept of race for biological science but also about its legitimacy. For more than two centuries, the intermingling of scientific and commonsense thinking about race has produced remarkable exchanges between scientists and the laity about the salience of race as a stratifying practice (itself worthy of scientific investigation) and as a socially decontextualized, biologically accurate, and meaningful taxonomy. The current decade is no exception. In the rush to purge commonsense thinking of groundless beliefs about the biological basis of racial classifications, scientists have overstated the simplicity of very complex interactive feedback loops between biology and culture and social stratification.

In this essay I demonstrate how and why purging science of race—when race and ethnic classifications are embedded in the routine collection and analysis of data (from oncology to epidemiology, from hematology to social anthropology, from genetics to sociology)—is not practicable, possible, or even desirable. Rather, we should recognize, engage, and clarify the complexity of the interaction between *any* taxonomies of race and biological, neurophysiological, social, and health outcomes. Whether race is a legitimate concept for scientific inquiry depends on the criteria for defining race, and will in turn be related to the analytic purposes for which the concept is deployed. This may seem heretical at the outset. However, that a concept is variable in its meaning does not mean it has no important analytic use. Scientific inquiry abounds in such concepts, including a range that extends from "genetic disorders" to "economic markets." On close examination, these apply to widely divergent empirical sites. I propose a framework that makes sense of how scientific studies deploy the concept of race: not a radical surgical removal, not an uncritical acceptance of old taxonomies, but an acknowledgement and recognition of complex feedback loops.

My strategy is threefold. First, I summarize an emerging problem in clinical genetics that is forcing scientific medicine to reconsider the practical or efficacious meaning of race with reference to blood transfusions. Second, I examine recent attempts to identify individuals from ethnic and racial populations using molecular genetics. Third, I address the possible, even likely, interaction among racial or ethnic identity, nutritional intake, and disease, notably cancer and heart disease. I conclude with some remarks about how anthropologists (and others working on aggregate data on selected populations designated by race) should try to advance our understanding of *how* race is always going to be a complex interplay of social and biological realities with ideology and myth.

CONTEXT AND CONTENT FOR FEEDBACK LOOPS: THE EMPIRICAL PROBLEM

By the mid-1970s, it was abundantly clear that there is more genetic variation within the most current common socially used categories of race than between them (Polednak 1989; Bittles and Roberts 1992; Chapman 1993; Shipman 1994). This consensus, however, was a recent development. In the early part of this century, scientists in several countries tried to connect the major blood groups in the ABO system to racial and ethnic groups.[1] They had learned that blood type B was more common in certain ethnic and racial groups, which some believed to be more inclined to criminality and mental illness (Gundel 1926; Schusterov 1927). Their attempts were fruitless, because there was nothing in the ABO system that could predict behavior. Although that effort ended a half century ago, hematology (the study of blood) is now once again being linked with race.

It has been observed that blood from Americans of European ancestry (i.e., mainly whites) tends to contain a greater number of antigens than blood from Americans of African or Asian ancestry (Vichinsky et al. 1990). This means there is a greater chance of serious transfusion reactions for blacks and Asians receiving blood from whites, but a lower risk for whites receiving blood from Asians or blacks. In the United States, whites make up approximately 80 percent of the population, and proportionally fewer blacks and Asian Americans donate blood than do whites. This social fact has biological consequences, which in turn has social consequences.

Approximately four hundred red-blood-cell-group antigens have been identified. The antigens have been classed into a number of fairly well-defined systems: the best-known are the ABO and Rhesus (Rh) systems, but there are other systems such as P, Lewis, MNSs, and Kell (standard hematology texts note ten systems, including ABO and Rh). The clinical significance of blood groups is that, in the case of a blood transfusion, individuals who lack a particular blood-group antigen may produce antibodies reacting with that antigen in the transfused blood. This immune response to alloantigens (nonself antigens) may produce hemolytic reactions, the most serious being complete hemolysis (destruction of all red blood cells), which can be life threatening. Once generated, the capacity to respond to a particular antigen is more or less permanent because the immune system generates "memory cells" that can be activated by future exposures to the antigen. For those with chronic conditions requiring routine blood transfusion, this aspect of the immune response increases the likelihood of future transfusion incompatibility. The clinical goal is to minimize immune responses to antigens in transfused blood, in part because a crisis (such as trauma surgery) may require transfusion of whatever blood is available, regardless of its antigen composition.

Most blood banks test only for ABO and Rh. Testing for the other systems is considered inefficient and increases the cost of blood. It is essential to minimize the production of antibodies against blood group antigens in all recipients. However, the current method of blood typing puts members of racial and ethnic minorities at greater risk for adverse reactions to frequent transfusions. "Phenotypic matching" of blood basically means using the superficial appearances of race as a rough screening process to minimize the production of antibodies (along with ABO and Rh).

Transfusion therapy for sickle-cell anemia is limited by the tendency of recipients to develop antibodies to foreign red cells (Vichinsky et al. 1990). In one important study, researchers evaluated the frequency and risk factors associated with such alloimmunization. They obtained the transfusion history, red-cell phenotype, and information on the development of alloantibodies in 107 black patients with sickle-cell anemia who received transfusions. They then compared the results with those from similar studies of 51 black patients with sickle-cell disease who did not receive transfusions and of 19 nonblack patients who received transfusions for other forms of chronic anemia:

> We assessed the effect that racial differences might have in the frequency of alloimmunization by comparing the red-cell phenotypes of patients and blood-bank donors ($N = 200$, 90 percent white). Although they received transfusions less frequently, 30 percent of the patients with sickle cell anemia became alloimmunized, in contrast to 5 percent of the comparison-group patients with other forms of anemia ($p < 0.001$). Of the 32 alloimmunized patients with sickle cell anemia, 17 had multiple antibodies and 14 had delayed transfusion reactions. Antibodies against the K, E, C, and Jkb antigens accounted for 82 percent of the alloantibodies.

They conclude: "These differences are most likely racial. We conclude that alloimmunization is a common, clinically serious problem in sickle-cell anemia and that it is partly due to racial differences between the blood-donor and recipient populations" (Vichinsky et al. 1990).

True enough, this may not be race in any essentialist conception, but that is precisely the point. A full eighty years ago, Hirschfeld and Hirschfeld (1919: 675) posited that, when the blood of one species is introduced into that of the same species, "those antigen properties which are common to the giver and receiver of blood can not give rise to antibodies, since they are not felt as foreign by the immunized animal." Although the Hirschfelds were talking about dogs, they were drawing a straight line to humans, human classification, and racial taxonomy: "If we inject into dogs the blood of other dogs it is in many cases possible to produce antibodies. By means of these antibodies we have been able to show that there are in dogs two antigen types. These antigen types, which we recognize by means of the iso-antibodies, may designate two biochemical races" (Hirschfeld and Hirschfeld 1919: 675–76).

Using this hypothesis, they went on to perform the first systematic and comprehensive serological study of a variety of ethnic and racial groups. Their classification system did not survive the test of time, but it produced "a way of thinking" that persists (Marks 1995). Moreover, with the data reported in the Vichinsky study (given that increasing the number of blood donations from blacks is a key policy goal intended to help those suffering from a relatively common genetic disease—sickle-cell anemia), the resuscitation of race through blood antigen theorizing and empirical research means that the concept is very much still with us in clinical medicine. That persons who are phenotypically white can and do have sickle-cell anemia complicates any essentialized racial theorizing, to be sure—but for the purposes of further action (blood donation requests and transfusion direction), racial phenotyping as a shorthand for racial differentiation still exists.

This provides a remarkably interesting intersection. While the full range of analysts, commentators and scientists—from postmodern essayists to molecular geneticists to social anthropologists—have been busily pronouncing "the death of race," for practical clinical purposes the concept is resurrected in the conflation of blood donation frequencies, by race. I am not trying to resurrect race here as a social construct (with no biological meaning), any more than I am trying to resurrect race as a biological construct with no social meaning. Rather, when race is used as a stratifying practice (which can be apprehended empirically and systematically), there is often a reciprocal interplay of biological outcomes that makes it impossible to disentangle the biological from the social. That may be obvious to some, but it is alien to some of the key players in current debates about the biology of race.

In late September 1996, Tuskegee University hosted a conference on the Human Genome Project and addressed specifically the project's relevance to the subject of race (Smith and Sapp 1997). In attendance was Luca Cavalli-Sforza, a preeminent population geneticist from Stanford University and perhaps the leading figure behind the Human Genetic Diversity Project.[2] Cavalli-Sforza had appeared on the cover of *Time* a few years earlier, a hero to the forces attacking the genetic determinism in Richard Herrnstein and Charles Murray's *The Bell Curve* (1994). At this conference, he repeated what he had said in the *Time* article: "One important conclusion of population genetics is that races do not exist":

> If you take differences between two random individuals of the same population, they are about 85% of the differences you would find if you take two individuals at random from the whole world. This means two things: (1) The differences between individuals are the bulk of the variation; (2) the differences among populations, races, continents are very small—the latter are only the rest, 15%, about six times less than that between two random individuals of one perhaps very small population (85%). Between you and your town grocer there is on average a variation which is almost as large as that between you and a random individual of the whole world. This person could be from Africa, China, or [could be] an Australian aborigine. (Smith and Sapp 1997: 53, 55)

Cavalli-Sforza speaks here as a population geneticist. In that limited frame of what is important and different about us as humans, he may be empirically correct. But humans give meaning to differences. At a particular historical moment, to make this same statement to an Albanian in Kosovo, a Hutu among the Tutsi, a Zulu among the Boers, or a German Jew among the Nazis may be as convincing for the purposes of further action as telling it to an audience of mainly African Americans at Tuskegee University.[3] Indeed, David Botstein, delivering the keynote address, said about *The Bell Curve:*

> So from a scientific point of view, this whole business of *The Bell Curve,* atrocious though the claims may be, is nonsense and is not to be taken seriously. People keep asking me why I do not rebut *The Bell Curve.* The answer is

> because it is so stupid that it is not rebuttable. You have to remember that the Nazis who exterminated most of my immediate family did that on a genetic basis, but it was false. Geneticists in Germany knew that it was false. The danger is not from the truth, the danger is from the falsehood. (Smith and Sapp 1997: 212)

David Botstein is also a preeminent molecular geneticist. In this statement, he takes the position that if people just understood the genetic truth, that would be sufficient and would even correct racist thinking and action.[4] Even though people may someday come to understand that they are basically similar at the level of DNA, RNA, immunology, and kind of blood systems, it is language group, kinship, religion, region, and race that are still far more likely to generate their pledge of allegiance.

COMPLEMENTARY STATEMENTS ON RACE BY THE AMERICAN ANTHROPOLOGICAL ASSOCIATION STATEMENT AND THE AMERICAN SOCIOLOGICAL ASSOCIATION

In May 1998 the American Anthropological Association issued its own statement on race.[5] In attempting to address myths and misconceptions, it takes a corrective stance toward folk beliefs about race. The statement declares that "physical variations in the human species have no meaning except the social ones that humans put on them."[6] But in casting the problem in this fashion, it gives the impression that the biological meanings that scientists attribute to any attempt at phenotypical variation by race are refutable by biological facts, while the social meanings that lay persons give to race are either errors or mere artificial social constructions, and, moreover, that these errors are mere social constructions not themselves capable of affecting biochemical, neurophysiological, and cellular aspects of our bodies that, in turn, can be studied scientifically. The statement of the Anthropological Association is consistent with the UNESCO statement on race. However, defining race as consisting *only* in the social meanings that humans provide implies that mere lay notions of race furnish a rationale for domination but have no other utility.

There is profound misunderstanding of the implications of a social constructivist notion of social phenomena. How humans identify themselves—whether in religious or ethnic or racial or aesthetic terms—matters to their subsequent behavior. Places of worship are socially constructed with human variations of meaning, interpretation, and use very much in mind. Whether a cathedral or mosque, a synagogue or Shinto temple, those "constructions" are no less real because one has accounted for and documented the social forces at play that resulted in such a wide variety of socially constructed places of worship. Race as social construction can and does have a substantial effect on how people behave.

In response to the growing controversy about the use of race as a meaningful category in scientific and policy research, the American Sociological Association issued its own "Statement on Race" at the ninety-fifth annual meetings, in August 2002.[7] This is most profitably viewed as a complementary statement that fills in some of the gaps left by the anthropology statement, most notably the policy implications. The primary concern was to provide the social scientific rationale for the continuing study of race in societies that use the concept as a means of stratifying members' access to resources. The statement explains why it is important to continue to collect and analyze data on such matters as racial differences in health outcomes, educational achievement, and contacts with the criminal justice system, even after we know there are no discrete genetic or biological categories for racial classification.

One important arena for further scientific exploration and investigation is the feedback between that behavior and the biological functioning of the body. To restate the well-known social analytic aphorism of W. I. Thomas, but to refocus it on human taxonomies of other humans: If humans define situations as real, they can and often do have real biological and social consequences.

Explicating the Conflation of Crime, Genetics, and Race

If race is a concept with no scientific utility, what are we to make of a series of articles that have appeared in the scientific literature over the last several years, and that have detailed a search for genetic markers of population groups that coincide with commonsense, lay renditions of ethnic and racial phenotypes? The forensic applications have generated much of this interest. Bernard Devlin and Neil Risch (1992a) published an article titled "Ethnic Differentiation at VNTR Loci, with Specific Reference to Forensic Applications," a research report that appeared prominently in the *American Journal of Human Genetics*. In it, they state:

> The presence of null alleles leads to a large excess of single-band phenotypes for blacks at D17S79 (Devlin and Risch 1992b), as Budowle et al. predicted. This phenomenon is less important for the Caucasian and Hispanic populations, which have fewer alleles with a small number of repeats (figs. 2–4). . . .
>
> It appears that the FBI's data base is representative of the Caucasian population. Results for the Hispanic ethnic groups, for the D17S79 locus, again suggest that the data bases are derived from nearly identical populations, when both similarities and expected biases are considered (for approximate biases, see fig. 9). For the allele frequency distributions derived from the black population, there may be small differences in the populations from which the data bases are derived, as the expected bias is .05. (540, 546)

When researchers try to make probabilistic statements about which group a person belongs to, they look at variations at several different locations in the DNA—usually from three to seven loci.[8] For any particular locus, they examine the frequency of the allele at that locus, and for that population. In other words, what is being assessed is the frequency of genetic variation at a particular spot in the DNA in each population.

Occasionally, these researchers find a locus where one of the populations being observed and measured has, for example, what I will call alleles H, I, and J, and another population has alleles H, I, and K. We know, for example, that certain alleles are found primarily among subpopulations of North American Indians. When comparing a group of North American Indians with a group of Finnish people, one might find a single allele that is present in some Indians but in no Finns (or it occurs at such a low frequency in the Finns that it is rarely, if ever, seen). However, it is important to stress that this does *not* mean that all North American Indians, even in this subpopulation, will have that allele.[9] Indeed, it is inevitable that some will have a different set of alleles, and that many of them will have the same alleles found in some of the Finns. Also, if comparing North American Indians from Arizona to North American Caucasians from Arizona, we would probably find a low level of the "Indian allele" in the Caucasians, because there has been interbreeding. Which leads to the next point:

It is possible to make arbitrary groupings of populations (geographic, linguistic, self-identified by faith, identified by others by physiognomy, etc.) and still find statistically significant allelic variations between those groupings. For example, we could examine all the people in Chicago, and all those in Los Angeles, and find statistically significant differences in allele frequency at *some* loci. Of course, at many loci, even most loci, we would not find statistically significant differences. When researchers claim to be able to assign people to groups based on allele frequency at a certain number of loci, they have chosen loci that show differences between the groups they are trying to distinguish. The work of Devlin and Risch (1992a,b), Ian Evett and colleagues (1993, 1996), and others suggests that only about 10 percent of the sites in DNA are useful for making distinctions. At the other 90 percent, the allele frequencies do not vary between groups such as "Afro-Caribbean people in England" and "Scottish people in England." Nonetheless, even though we cannot find a single site where allele frequency matches some phenotype that we wish to identify, there may be several sites that *will* be effective for the purposes of aiding the FBI, Scotland Yard, or other criminal justice systems around the globe in making highly probabilistic statements about suspects and the likely ethnic, racial, or cultural populations to which they can be assigned statistically.

While Devlin and Risch expressed skepticism about the utility of this approach, Mark Shriver and his associates (1997) were far more enthusiastic in suggesting that forensic science would be able to deploy a strategy of combining several points along DNA as useful markers "for ethnic-affiliation estimation." But although ethnicity was in the title of the Shriver article, the content pointed decidedly to the old racial taxonomies of black and white and Asian. A few years later, the skepticism was absent in a piece titled "Inferring Ethnic Origin by Means of an STR Profile" in *Forensic Science International Journal* (Lowe et al. 2001).

In the July 8, 1995, issue of the *New Scientist*, in an article titled, "Genes in Black and White," some extraordinary claims are made about what one can learn about socially defined categories of race from reviewing information gathered by means of the new molecular genetic technology. In 1993, a British forensic scientist published what is perhaps the first DNA test explicitly acknowledged as providing intelligence information about ethnic makeup for "investigators of unsolved crimes." Ian Evett, of the Home Office's forensic science laboratory in Birmingham, England, and his colleagues in the Metropolitan Police, claimed that their DNA test can distinguish between Caucasians and Afro-Caribbeans in nearly 85 percent of the cases.

Evett's work, published in the *Journal of the Forensic Science Society*, draws on apparent genetic differences in three sections of human DNA. Like most stretches of human DNA used for forensic typing, each of these three regions differs widely from person to person, irrespective of race. But by looking at all three, say the researchers, it is possible to estimate the probability that someone belongs to a particular racial group. The implications of this for determining, for legal purposes, who is and who is not "officially" a member of some racial or ethnic category are profound.

Two years after the publication of the UNESCO statement intended to bury the concept of race for the purposes of scientific inquiry and analysis, and during the period when the American Anthropological Association was generating a parallel statement, an article appeared in the *American Journal of Human Genetics*, authored by Evett and his associates, summarized thus:

> Before the introduction of a four-locus multiplex short-tandem-repeat (STR) system into casework, an extensive series of tests [was] carried out to determine robust procedures for assessing the evidential value of a match between crime and suspect samples. Twelve databases were analyzed from the three main ethnic groups encountered in casework in the United Kingdom: Caucasians, Afro-Caribbeans, and Asians from the Indian subcontinent. Independent tests resulted in a number of significant results, and the impact that these might have on forensic casework was investigated. It is demonstrated that previously published methods provide a similar procedure for correcting allele frequencies—and that this leads to conservative casework estimates of evidential value. (1996: 398)

These new technologies have the potential to be used in developing and "authenticating" typologies of human ethnicity and race. The old question of how to decide on a person's "degree of whiteness" or "degree of Indianness" may be recast in the terms of molecular genetics. The U.S. Congress passed the Allotment Act of 1887, denying land rights to Native Americans who were "less than half-blood." The government still requires American Indians to produce Certificates with Degree of Indian Blood in order to qualify for a number of entitlements. The Indian Arts and Crafts Act of 1990 makes it a crime to for an artist to identify herself as a Native American when selling artwork unless she has federal certification demonstrating that she is authentic American Indian (that is, has "one-quarter blood"). However, it is not art but law and forensics that ultimately will prompt the use of genetic technologies to identify who is "authentically" in one category or another. Geneticists in Ottawa have been trying to set up a system "to distinguish between 'Caucasian Americans' and 'Native Americans' on the basis of a variable DNA region used in DNA fingerprinting" (Genes in black and white 1995: 37). For practical purposes, authenticating a person's membership in a group (racial, ethnic, or cultural) can be brought to the level of DNA analysis. The efficaciousness of testing and screening for genetic disorders in at-risk populations that are ethnically and racially designated poses a related set of vexing concerns for the separation of the biological and cultural taxonomies of race.

Here, we must be alert to the social tendency to regard the DNA analysis as real and social relations as more problematic. There is a parallel in according parent status to the biological parents rather than the social parents who have bonded with a child over time. Many Native American tribes and nations wrestle with this problem—and some try to resolve it by distinguishing between those who identify with and live among their people and those who have long since left the community. This suggests that DNA analysis for group membership (inclusion or exclusion) will never be as definitive as DNA analysis for forensic purposes. Nonetheless, we can expect to see more uses for these new technologies in determining allele frequencies, because they serve an institutional need to certify and make legitimate individuals' claims to membership in specific groups.

Genetic Testing and Genetic Screening

When members of social groups with a strong endogamous tradition (such as ethnic or racial groups) intermarry for centuries, they are at higher risk for pairing recessive genes and passing on a genetic disorder. In the United States, the best known of these clustered autosomal recessive disorders are Tay-Sachs disease, beta-thalassemia, sickle-cell anemia, and cystic fibrosis (see tables 13.1 and 13.2). Tay-Sachs, which occurs primarily among Ashkenazi

TABLE 13.1 Selected High Incidence of Genetic Disorders

Condition	*Estimated Incidence*[a]
Cleft lip or palate	1 in 675 individuals
Clubfoot	1 in 350 individuals
Cystic fibrosis	1 in 2,500 Caucasians
Diabetes	1 in 80 individuals
Down syndrome	1 in 1,050 individuals
Hemophilia	1 in 10,000 males
Huntington's disease	1 in 2,500 individuals
Duchenne muscular dystrophy	1 in 7,000 males
Phenylketonuria	1 in 12,000 Caucasians
Rh incompatibility	1–2 in 100 individuals
Sickle-cell anemia	1 in 625 African Americans
Tay-Sachs disease	1 in 3,000 Ashkenazi Jews
Beta-thalassemia (Cooley's anemia)	1 in 2,500 Mediterranean peoples

[a]Incidence refers to the number of cases occurring among live births. As in the original table, the term *genetic disorders* corresponds to typical usage; however, many of these are really multifactorial disorders with some known genetic component. All figures refer to the U.S. population, except where noted otherwise. These figures vary according to ethnic background. For example, Rh incompatibility is much lower among those with Asian ancestry. Phenylketonuria is rare in those of black or Ashkenazi Jewish ancestry (most Jewish people in the United States are of Ashkenazi descent).

SOURCE: Burhansstipanov, Giarratano, Koser, and Mongoven 1987: 6–7.

TABLE 13.2 Ethnicities or Groups Primarily Affected by Disorders (United States)

Condition	*Ethnicities Primarily Affected*
Duchenne muscular dystrophy	Northeastern British
Adult lactase deficiency	African Americans, Chinese
Cleft lip or palate	North American Indians, Japanese
Cystic fibrosis	Northern Europeans
Familial Mediterranean fever	Armenians
Phenylketonuria	Caucasians (especially Irish)
Sickle-cell anemia	African Americans
Alpha-thalassemia	Chinese, Southeast Asians
Beta-thalassemia	Mediterraneans
Spina bifida or anencephaly	Caucasians (especially Welsh and Irish)
Tay-Sachs disease	Ashkenazi Jews (with origins in central and eastern Europe)

SOURCE: Burhansstipanov, Giarratano, Koser, and Mongoven 1987: 6–7.

Jews of northern and eastern European ancestry, is carried by about 1 in 30 with this ancestry, and approximately 1 in every 3,000 newborns in this ethnic group will have the disorder. About 1 in 30 Americans of European descent is a carrier of cystic fibrosis, with a similar incidence rate.[10] In contrast, approximately 1 in 12 American blacks is a carrier for sickle-cell disease, and 1 in 625 black newborns has the disorder. Irish and northern Europeans are at greater risk for phenylketonuria. In the United States, 1 in 60 whites is a carrier, and about 1 in 12,000 newborns is affected.

When both members of a mating couple are carriers of the autosomal recessive gene, the probability that each live birth will be affected by the disorder is 25 percent. However, being a carrier, or passing on the gene so that one's offspring is also a carrier, typically poses no more of a health threat than carrying a recessive gene for a different eye color. That is, carrier status typically poses no health threat at all. The health rationale behind carrier screening is to inform prospective parents about their chances of having a child with a genetic disorder.

In the United States, the two most widespread genetic-screening programs for carriers have screened Jews of northern European descent (Tay-Sachs) and blacks of West African descent (sickle-cell disease). From 1972 to 1985, there was extensive prenatal screening for both disorders, and by 1988, newborn screening for sickle-cell disease had become common (Duster 1990). Autosomal recessive disorders such as these, located in risk populations that coincide with ethnicity and race, are of special interest as we turn to address genetic screening for populations at greatest risk for a disorder.

It is important to distinguish between a genetic screen and a genetic test. The latter is done when there is reason to believe that a particular individual is at high risk for having a genetic disorder or for being a carrier of a gene for a disorder. For example, a sibling of someone diagnosed with Huntington's disease (a late-onset neurological disorder) would be a candidate for a genetic test for that disorder. A genetic screen, on the other hand, is used for a *population* at higher risk for a genetic disorder. For example, Ashkenazi Jews have been the subjects of genetic screening for Tay-Sachs.

As with most of the genetic disorders mentioned above, the incidence of cystic fibrosis varies remarkably among different groups. For the purposes of my argument, what is most striking is the fact that the sensitivity of the current genetic test for cystic fibrosis differs considerably according to group. For example, the test is 97 percent sensitive for Ashkenazi Jews but only 30 percent for Asian Americans. The National Institutes of Health convened a Consensus Conference on Genetic Testing for Cystic Fibrosis in 1997 and, as a result of that meeting, issued a statement declaring that cystic fibrosis testing should be made available to all couples planning a pregnancy. Yet, with such low sensitivity in the cases of some groups, the test will result in frequent

TABLE 13.3 Ethnic or Group Variation with Incidence of Cystic Fibrosis, with Sensitivity to Delta F508 Test (United States)

Group	*Incidence*	*Carrier Frequency*	*% Delta F508*	*% Sensitivity*
Caucasians	1:3,300	1:29	70	90
Ashkenazi Jews	Not tested or unknown	1:29	30	97
Native Americans	1:3,950	Not tested or unknown	0	94
Hispanics	1:8,500	1:46	46	57
African Americans	1:15,300	1:63	48	75
Asian Americans	1:32,100	1:90	30	30

false negatives.[11] For example, notice that table 13.3 reports that sensitivity among Latinos is only 57 percent, despite a relatively high incidence of the disorder.

The incidence of cystic fibrosis among all Native Americans is only 1 in 11,200—but for the Zuni the rate is 1 in 1,580, or seven times higher. Table 13.1 shows the incidence rate among Caucasians as being approximately 1 in 2,500—yet the test developed for cystic fibrosis is much more accurate for whites. Moreover, we now know that there is a particular allele for cystic fibrosis peculiar to Zuni Indians. This finding is an important harbinger of the next period of research on allele frequencies for group categories. For the past decade, there has been a strong and persistent hunt for the genetic basis of alcoholism.[12] Given the fact that some Native American tribes have extraordinarily high rates of alcoholism, and given the nature of the quest for "the gene for alcoholism," it is likely that research scientists will find some kind of allele "associated with" certain Native American tribes. Once that happens, it is inevitable that an allele "associated with" the drinking patterns of select social groups will be marked—even if it turns out to be a spurious relationship.[13]

This brings us to an interesting intersection of genetic markings and the social and political power of those marked, and the capacity to accept or reject the science of ethnicity and race. Let us consider Ashkenazi Jews and Zuni Indians and their respective capacities to direct or divert genetic testing based on ethnicity, race, or relative social power in the medical profession. This is not a mere speculative exercise. In the late 1990s, Ashkenazi Jewish women began to protest strongly their identification as being at higher risk for breast cancer. The Zuni have yet to be heard from in a parallel manner regarding their identification with cystic fibrosis.

The Interaction between Race as Identity, Nutrient Consumption, and Health

The scientific literature on cancer rates for different diseases in racially and ethnically designated populations is fairly well developed. For example, Ashkenazi women were initially reported, clinically, to have higher rates of breast cancer than other groups (Richards et al. 1997: 1096).[14] Among men under age sixty-five, African Americans have almost double the rate of prostate cancer in Caucasians, according to reports released by the National Cancer Institute.[15] How can this be explained in a way that uses race not simply as an outcome but as a factor that helps produce the outcome? Certain forms of cancer may be a function of nutrition and diet. Groups with certain dietary patterns or restrictions might then be systematically at greater risk for cancer. If members of a certain group who identify themselves as, say, Ashkenazi, have a diet that follows a certain pattern, they may routinely have rates of certain groups of cancers—both lower and higher—than groups with different dietary habits. African American males, for example, may, by identifying as African Americans, be more likely to eat a category of food ("soul food") that systematically puts them at higher risk for prostate cancer (Braun 2002; Frank 2001; Lee et al. 2001). With this formulation, I am "bringing the systematic study of race" back into the scientific inquiry—even though I am not going to the DNA level to attempt a reductionist account of race as determined by DNA difference.

Even with sturdy epidemiological evidence that heart disease and hypertension among African Americans are strongly associated with social factors such as poverty, there has been a persistent attempt to pursue the scientific study of hypertension through a link to the genetics of race. Dark pigmentation is indeed associated with hypertension in America. Michael Klag and colleagues (1991) report the results of a carefully controlled study looking at the relationship between skin color and high blood pressure. The researchers found that darker skin color is a good predictor of hypertension among blacks of low socioeconomic status, but not among blacks of any shade who are "well employed or better educated." The study further suggested that poor blacks with darker skin color experience greater hypertension not for genetic reasons but because darker skin color subjects them to greater discrimination, with consequently greater stress and psychological and medical consequences. Of course, from another perspective, darker skin color is dark *mainly* for genetic reasons, so it is a matter of how one chooses to direct theorizing about the location of causal arrows. When practicing physicians see darker skin color, their diagnostic interpretation and their therapeutic recommendations are systematically affected. K. A. Schulman and colleagues (1999) indicate that, in clinical practice, physicians are likely to make systematically different recommendations for treatment of heart disorders according to race, even when patients have the same presenting symptoms. Thus, when researchers analyze outcome data such as cause of death by race and find that blacks have a higher incidence of death from heart failure, it would be easy to incorrectly infer causation and direction of the relationship between the variables.

When African Americans show up with higher rates of prostate cancer, and the prevailing paradigm for oncology is molecular genetics, there will be a tendency to search for oncogenes (Fujimura 1996). Taking a different tack, Nancy Press and colleagues (1998) held an anthropological lens to the report, mentioned earlier, that Ashkenazi Jewish women are at higher risk for certain breast cancers (Richards et al. 1997). When these researchers looked at a complex interplay of factors, the higher incidence rate disappeared. A parallel development can be found in the study of the relationship between ethnicity, race, and violence.

When the National Institutes of Health convened a panel to review its full portfolio of research on violence in the early 1990s, the panel concluded that the great bulk of studies focused on the individual, or smaller, units of analysis (i.e., processes or units inside the body). Nonetheless, blacks commit over 60 percent of the homicides yet represent only 12 percent of the population. When queried about their focus of study, research scientists working on neurological or biochemical aspects stated, "We are engaged in basic science." They regarded study of violence among groups as not science but a policy matter. This is the justification for assigning the bulk of science funding to "basic processes" (Greenberg 1999). But how can the molecular geneticists and neurologists have it both ways? If they scientifically legitimate explanations of the basic processes of violence behind the high rate of homicide in the African American community, all the while saying there is nothing to the *scientific* classification of race, then by their own admission the study of violence at the molecular level, or at the level of neurotransmission (existence of a decreased serotonin level remains the most popular theory), will have nothing to do with race as phenotypically reported in the FBI crime statistics.

By heading toward an unnecessarily binary, socially constructed fork in the road, by forcing ourselves to think that we must choose between "race as biological" (now out of favor) and "race as *merely* a social construction," we fall into an avoidable trap. A refurbished and updated insight from W. I. Thomas can help us. It is not an either/or proposition. In some cases, we must conduct systematic investigations, guided by a body of theory, into the role of race (or ethnicity or religion—they are all socially constructed) as an organizing force in social relations and as a stratifying practice (Oliver and Shapiro 1995). In other cases, we must conduct systematic investigations, guided by theory, into the role of the interaction of race (or ethnicity or religion)—however flawed as a biologically discrete and coherent taxonomic system—with feedback loops into the biological functioning of the human body, and then again in relation to medical practice. For example, hypertension, a biological or medical health condition, is influenced by social treatment (Klag et al. 1991) but then is sometimes diagnosed as genetic. Such studies might include examination of the effects of the systematic administration until the 1960s of higher doses of X rays to African Americans;[16] of the creation of genetic tests with high rates of sensitivity to some ethnic and racial groups and low sensitivity to others; and of the systematic treatment, or lack of it, using diagnostic and therapeutic interventions for heart and cancer patients according to race. To throw out the concept of race is to take the official approach to race and ethnicity pioneered and celebrated by the French government: "We do not collect data on that topic. Therefore, it does not exist!"[17] It is as much a convenient bureaucratic posture as it is a political fiction in France that—because the French choose not to count the number of Algerians and sub-Sahara Africans in Paris who are unemployed, who are subject to housing discrimination, and who have limited access to health care—no such problems exist. Understanding what remains problematic about the concept of race—the complex interrelationships between sociopolitical processes and scientific knowledge claims or biomedical practices—is also fundamental to understanding clearly the natural-cultural relations that shape our lives (Xie et al. 2001; Risch et al. 2002; Evans and Relling 1999).

NOTES

1. For the discussion in this paragraph, and for my references to the German literature, I am indebted to William H. Schneider (1996).

2. The Human Genetic Diversity Project is not to be confused with the Human Genome Project. The latter is a $3-billion effort jointly funded in the United States by the National Institutes of Health and the Department of Energy. From the project's outset approximately a decade ago, its goal has been to map and sequence the entire human genome. The major rationale for the project has been to provide information that would assist medical genetics in decoding genetic disorders, better understanding them, and eventually, producing gene therapeutic interventions for them. In contrast, the Human Genetic Diversity Project has been concerned with tracing human populations through an evolutionary history of many centuries. Its primary goal has been to better understand human evolution.

3. Tuskegee, after all, was the site of the infamous syphilis experiments on black males, in which the U.S. Public Health Service studied the racial effects of how the disease ravages blacks in contrast to whites (Jones 1981).

4. Botstein's assertion that the German geneticists knew that Nazi claims about Aryan racial purity were false is highly contestable, and the work by Robert Proctor (1988, 1998) has abundant counterevidence.

5. This statement, approved by the executive board on May 17, 1998, can be

retrieved at http://www.aaanet.org/stmts/racepp.htm (accessed on 11 December 2002).

6. The statement of the American Anthropological Association is admirable in its attempt to thwart biologically based scientific claims that race should be used to abrogate human rights. However, the accumulation of wealth, political power, and social privilege based on a biologically erroneous classification system is no less real and consequential, and no less amenable to systematic empirical investigation.

7. Although I served as chair of the task force that issued the American Sociological Association's statement on race, this was a consensus document approved by the eighteen-member task force and by the Council of the Association. The statement can be found at http://www.asanet.org/media/racestmt02.pdf (accessed on 1 December 2002).

8. Each location is called a locus, and the plural, or several in combination, is called loci.

9. This major point is made by the two statements about race by UNESCO and the American Anthropological Association, and it cannot be repeated too often.

10. Note that the estimated incidence rates of cystic fibrosis are different for Caucasians in tables 13.1 and 13.3. This is in part a result of the lack of any general population screen, and estimates are continually reformulated on the basis of clinical data, diagnostic skills at different locations and at different periods, and so on. Moreover, collapsing the category of Caucasian ignores the fact that individuals of northern European ancestry are at much greater risk for cystic fibrosis than are those of southern European ancestry.

11. Among any population in which there is a low incidence, a low carrier rate, and low sensitivity for the test, there will be a substantial number of false negatives.

12. In 1987, Robert Cloninger published an article in *Science* in which he denoted a class of alcoholics, what he called type II, who started their alcohol abuse early in life and had a strong urge to seek alcohol throughout their lives. Using studies of twins reared apart, Cloninger said that the children of type II alcoholics were four times more likely to abuse alcohol, whether or not they were raised in the home of an alcoholic (410–11). Shortly thereafter, full attention turned to the dopamine receptor mutation (DRD2). In 1990, Blum and colleagues reported that a particular variant of the DRD2 receptor, the A1 allele, was present in the DNA of 69 percent of their alcoholic subjects, compared to 20 percent of the controls. In the last decade, researchers have had little success in replicating this study, but the search continues.

13. We already know that Asians are more likely to have the allele that produces skin flushing from alcohol, but it is empirically demonstrable that the social norms and practices associated with alcohol consumption have far more analytic power to explain the lower rate of alcoholism among Asian Americans. See Kim 1995.

14. That finding was subsequently challenged. See Press et al. 1998.

15. See "Prostate cancer (invasive) trends in SEER incidence and mortality, by race," http://www.seer.ims.nci.nih.gov (accessed in May 1999).

16. In its 1963 edition of *How to Prepare an X-ray Technic Chart,* the General Electric Company suggested that "Negroid patients" be given an increased dosage 40 to 60 percent higher than that given to whites. The reference was to a standard work of the period by C. A. Jacobi and D. Q. Paris. The third edition of this publication, in 1964, still contained this suggestion (102).

17. Perhaps this is an internally consistent emanation from a society that gave the world the Cartesian formulation about thought and existence, as well as subject/object dualities.

REFERENCES

Bittles, A. H., and D. F. Roberts, eds. 1992. *Minority populations: Genetics, demography, and health.* London: Macmillan.

Blum, K., E. P. Noble, P. J. Sheridan, A. Montgomery, T. Richie, P. Jagadeeswaran, H. Nogami, A. H. Briggs, and J. B. Cohn. 1990. Allelic association of human dopamine D2 receptor gene in alcoholism. *Journal of the American Medical Association* 263, no. 15:2055–60.

Braun, L. 2002. Race, ethnicity, and health: Can genetics explain disparities? *Perspectives in Biology and Medicine* 45, no. 2 (spring): 159–74.

Burhansstipanov, L., S. Giarratano, K. Koser, and J. Mongoven. 1987. *Prevention of genetic and birth disorders,* 6–7. Sacramento: California State Department of Education.

Chapman, Malcolm, ed. 1993. *Social and biological aspects of ethnicity.* New York: Oxford University Press.

Cloninger, R. 1987. Neurologic adaptive mechanisms in alcoholism. *Science* 236 (24 April): 410–16.

Committee on Human Genetic Diversity. 1997. *Scientific and medical value of research on human genetic variation.* Washington, D.C.: National Research Council and National Academy Press.

Devlin, B., and N. Risch. 1992a. Ethnic differentiation at VNTR loci, with specific reference to forensic applications. *American Journal of Human Genetics* 51:534–48.

———. 1992b. A note on the Hardy-Weinberg equilibrium of VNTR data by using the Federal Bureau of Investigation's fixed-bin method. *American Journal of Human Genetics* 51:549–53.

Duster, T. 1990. *Backdoor to eugenics.* New York: Routledge.

Evans, W., and M. Relling. 1999. Pharmacogenomics: Translating functional genomics into rational therapeutics. *Science* 286 (15 October): 487–91.

Evett, I. W. 1993. Criminalistics: The future of expertise. *Journal of the Forensic Science Society* 33, no. 3:173–78.

Evett, I. W., I. S. Buckleton, A. Raymond, and H. Roberts. 1993. The evidential value of DNA profiles. *Journal of the Forensic Science Society* 33, no. 4:243–44.

Evett, I. W., P. D. Gill, J. K. Scranage, and B. S. Wier. 1996. Establishing the robustness of short-tandem-repeat statistics for forensic application. *American Journal of Human Genetics* 58:398–407.

Frank, R. 2001. A reconceptualization of the role of biology in contributing to race/ethnic disparities in health outcomes. *Population Research and Policy Review* 20:441–55.

Fujimura, J. H. 1996. *Crafting science: A sociohistory of the quest for the genetics of cancer.* Cambridge: Harvard University Press.

Genes in black and white. 1995. *New Scientist* (July 8): 34–37.

Greenberg, D. S. 1999. Hardly an ounce for prevention. *Washington Post,* 22 March, A19.

Gundel, M. 1926. Einige Beobachtungen bei der rassenbiologischen Durchforschung Schleswig-Holsteins. *Klinische Wochenschrift* 5:1186.

Herrnstein, R. J., and C. Murray. 1994. *The bell curve: Intelligence and class structure in American life.* New York: Free Press.

Hirschfeld, L., and H. Hirschfeld. 1919. Serological difference between the blood of different races. *Lancet* (October 18): 675–79.

Jacobi, C. A., and D. Q. Paris. 1964. *X-ray technology.* 3d ed. St. Louis, Mo.: C. V. Mosby.

Jones, J. H. 1981. *Bad blood: The Tuskegee syphilis experiment—a tragedy of race and medicine.* New York: Free Press.

Katz, S. H. 1995. *Is race a legitimate concept for science? The AAPA revised statement on race: A brief analysis and commentary.* Philadelphia: University of Pennsylvania, February.

Kim, S-O. 1995. Cultural influences on drinking practices among people of Japanese descent. Ph.D. diss., University of California, Berkeley.

Klag, M., P. K. Whelton, J. Coresh, C. E. Grim, and L. H. Kuller. 1991. The association of skin color with blood pressure in U.S. blacks with low socioeconomic status. *Journal of the American Medical Association* 265, no. 5:599–602.

Lee, S. S. J., J. Mountain, and B. A. Koenig. 2001. The meanings of "race" in the new genomics: Implications for health disparities research. *Yale Journal of Health Policy, Law, and Ethics* 12 (3 May): 15, 33–75.

Lowe, A. L., A. Urquhart, L. A. Foreman, I. Evett. 2001. Inferring ethnic origin by means of an STR profile. *Forensic Science International* 119:17–22.

Marks, J. 1995. *Human biodiversity: Genes, race, and history.* New York: Aldine de Gruyter.

Oliver, M., and T. Shapiro. 1995. *Black wealth/white wealth: A new perspective on racial inequality.* New York: Routledge.

Polednak, A. P. 1989. *Racial and ethnic differences in disease.* New York: Oxford University Press.

Press, N., W. Burke, and S. Durfy. 1998. How are Jewish women different from all other women? *Journal of Law-Medicine* 7, no. 1 (winter): 1–17.

Proctor, R. 1988. *Racial hygiene: Medicine under the Nazis.* Cambridge: Harvard University Press.

———. 1998. *Cancer wars: How politics shapes what we know and don't know about cancer.* New York: Basic Books.

Richards, C. S., P. A. Ward, B. R. Roa, L. C. Friedman, A. Boyd, G. Kuenzli, J. Dunn, and S. Plon. 1997. Screening for 185delAG in the Ashkenazim. *American Journal of Human Genetics* 60:1085–98.

Risch, N., E. Burchard, E. Ziv, and H. Tang. 2002. Categorizations of humans in biomedical research: Genes, race, and disease. *Genome Biology* 3, no. 7:2007.1–2007.12. Also available at http://genomebiology.com/2002/3/7/comment/2007.1 (accessed on 1 December 2002).

Schneider, W. H. 1996. The history of research on blood group genetics: Initial discovery and diffusion. *History and Philosophy of the Life Sciences* 18 3:277–303.

Schulman, K. A., J. A. Berlin, W. Harless, J. F. Kenner, S. Sistrunk, B. J. Gersh, R. Dube, C. K. Taleghani, J. E. Burke, S. Williams, J. M. Eisenberg, and J. J. Escarce.

1999. The effect of race and sex on physicians' recommendations for cardiac catheterization. *New England Journal of Medicine* 34, no. 8:618–26.

Schusterov, G. A. 1927. Isohaemoagglutinierenden Eigenschaften des menschlichen Blutes nach den Ergebnissen einer Untersuchung an Straflingen des Reformatoriums (Arbeitshauses) zu Omsk. *Moskovskii Meditsinksii Jurnal* 1:1–6.

Shipman, P. 1994. *The evolution of racism: Human differences and the use and abuse of science.* New York: Simon and Schuster.

Shriver, M. D., M. W. Smith, L. Jin, A. Marcini, J. M. Akey, R. Deka, and R. E. Ferrell. 1997. Ethnic-affiliation estimation by use of population-specific DNA markers. *American Journal of Human Genetics* 60:957–64.

Smith, E., and W. Sapp, eds. 1997. *Plain talk about the Human Genome Project.* Tuskegee, Ala.: Tuskegee University.

Vichinsky, E. P., A. Earles, R. A. Johnson, M. S. Hoag, A. Williams, and B. Lubin. 1990. Alloimmunization in sickle cell anemia and transfusion of racially unmatched blood. *New England Journal of Medicine* 322, no. 23:1617–21.

Xie, H. G., R. B. Kim, A. J. Wood, and C. M. Stein. 2001. Molecular basis of ethnic differences in drug disposition and response. *American Review of Pharmacology and Toxicology* 41:815–50.

Zerjal, T., B. Dashnyam, A. Pandya, M. Kayser, L. Roewer, F. R. Santos, W. Schiefenhovel, N. Fretwell, M. A. Jobling, S. Harihara, K. Shimizu, D. Semjidmaa, A. Sajantila, P. Salo, M. H. Crawford, E. K. Ginter, O. V. Evgrafov, and C. Tyler-Smith. 1997. Genetic relationships of Asians and northern Europeans revealed by Y-chromosomal DNA analysis. *American Journal of Human Genetics* 60:1174–83.

Chapter 14

The Good, the Bad, and the Ugly

Promise and Problems of Ancient DNA for Anthropology

Frederika A. Kaestle

Recent technological advances, including the development of the automated polymerase chain reaction (PCR), improved DNA detection systems, and new DNA extraction protocols, many of them developed for forensic applications, now allow us to recover genetic information from ancient individuals and groups (see Herrmann and Hummel 1993). This permits researchers to apply to ancient samples many of the techniques and analyses previously available only for modern samples. Hypotheses regarding the genetics of these ancient peoples can be explored directly, rather than through the indirect methods of archaeology, linguistics, and similar disciplines. Aided by these tools, we can use ancient DNA (aDNA) to help answer many questions of interest to a wide range of fields, including anthropology, archaeology, linguistics, genetics, ecology, conservation, history, and forensics, among others. My own work has concentrated on utilizing ancient mitochondrial DNA (mtDNA) extracted from prehistoric skeletal remains to test hypotheses, based on archaeological and linguistic data, of population movement and ancestor-descendant relationships in the Americas.

However, when applied to the study of humans, DNA analyses raise certain ethical dilemmas, which are often overlooked or minimized. These considerations, although subject to increasing debate, are not amenable to simple solutions. In this essay I discuss promising new avenues of anthropological research using aDNA, and raise both methodological and ethical questions involving the application of aDNA techniques to human remains, utilizing some specific examples from my own work and that of other researchers.

APPLICATIONS

The possible applications of aDNA techniques to anthropological questions are numerous (see table 14.1). These range from testing hypotheses about single individuals to testing those concerning large populations and the human species as a whole, and include not only studies of human aDNA but also of aDNA derived from animals and plants. In many cases, aDNA studies present the only way to directly test hypotheses generated by other anthropological or archaeological methods and data.

A significant portion of aDNA research in anthropology concerns nonhuman DNA. It is often difficult or impossible to identify the genus or species of animal and plant remains found at archaeological sites using their morphology alone. Ancient DNA offers the opportunity to make these distinctions using genetic markers and has created a minor revolution in our understanding of prehistoric environments and human exploitation of them. For example, R. L. Reese and colleagues (1996) have used aDNA extracted from pictographs from Texas to determine what organic components were present in the paint. They concluded that the paint was bound with material derived from an ungulate, most likely a deer or bison (but see Mawk et al. 2002). A few studies have also detected the DNA from diseases that still infect us today, and have allowed us to trace the history and patterns of these diseases. For example, Mark Braun and colleagues (1998) have detected the presence of *Mycobacterium tuberculosis* DNA in the remains of two pre-Columbian Native American individuals, confirming the presence of TB in the Americas before European contact and contradicting the hypothesis that it was carried here by the first European explorers.

However, at the moment, the majority of anthropologists using aDNA methods are studying ancient humans themselves. These studies, at the finest level, concern attributes of the individual. For example, we are now able to determine the genetic sexes of ancient individuals, including juveniles, infants, and those from whom we have only partial remains, none of whom are generally good subjects for conventional sexing techniques. In one case, genetic sex determination of ancient individuals was used to explore infanticide in the Late Roman Era (Faerman et al. 1998). The authors have shown that the majority of the remains of neonates found below a Roman bathhouse were male, and they speculate that there was selection in favor of female infants by the courtesans or prostitutes employed in the bathhouse.

Comparisons between individuals are also possible using aDNA techniques. In most cases these involve the investigation of kinship between individuals buried together or presumed to be related for other reasons. For example, Tobias Schultes and colleagues (2000) found that they could identify paternal and maternal lineages, and indeed could reconstruct indi-

TABLE 14.1 Anthropological Applications of Ancient DNA Techniques

Application	*Implications*	*Examples*
Animal and plant mtDNA	Understand hunting and dietary patterns, ecology, domestication of animals and plants, environmental reconstruction	Loreille et al. 1997; Bailey et al. 1996; Reese et al. 1996; Brown 1999; Brown et al. 1994
Disease mtDNA	Trace history and patterns of prehistoric and historic diseases	Braun et al. 1998; Drancourt et al. 1998; Kolata 1999; Taubenberger et al. 1997
Genetic sexing	Understand marriage and burial patterns, differential patterns by gender of disease, diet, status and material possessions, forensics	Faerman et al. 1998; Lin et al. 1995; Stone et al. 1996
Maternal and paternal kinship	Understand social structure, status, marriage patterns, burial customs, migration, demography, forensics	Kaestle and Smith 2001; Gill et al. 1994; Schafer 1998; Schultes et al. 2000
Population continuity and replacement	Trace prehistoric population movement, ancestor-descendant relationships between groups, and relationships among ancient groups connected by similar morphology or material culture	Kaestle 1997; Kaestle and Smith 2001; Hagelberg 1997; Wang et al. 2000; Parr et al. 1996; O'Rourke et al. 1996; Stone and Stoneking 1999; Handt, Hoss, Krings, and Paabo 1994; Krings et al. 1997

vidual families, from the remains of thirty-six individuals recovered from the Late Bronze Age collective burial in Liechtenstein Cave, Germany. These results are not consistent with the previous interpretation of the site as a sacrifice cave.

On a broader level, comparisons of groups of prehistoric people to each other and to living populations can help us understand events of prehistoric population movement and issues of population continuity or replacement, especially when a large number of ancient individuals are available for testing. Because genetic variation is inherited from a group's ancestors, modern groups are expected to have frequencies of genetic markers similar to those of their ancestors, while ancient and modern groups with very different frequencies are not likely to be closely related. In addition, certain genetic markers have limited distribution and can be used as indicators of relationship. Using this knowledge we can approach questions of ancestor-descendant relationships at many scales.

My own research has involved testing hypotheses of prehistoric population movement in the American Great Basin (Kaestle and Smith 2001 and references therein). Both linguistic and archaeological evidence has been used to suggest that the current inhabitants of the Great Basin, speakers of Numic languages, are recent immigrants into the area (within the last one thousand years) who replaced the previous inhabitants (see Madsen and Rhode 1994). However, others have interpreted this same evidence as a sign of local adaptation to a changing environment (see Madsen and Rhode 1994). As part of a larger project to study the prehistory of the western Great Basin, begun by the Nevada State Museum with permission from local Native American tribal groups, I analyzed the mitochondrial DNA (mtDNA) from approximately forty ancient individuals from western Nevada.

Mitochondrial DNA is inherited only through the maternal line, and maternal relatives therefore share mitochondrial mutations with each other. These mutations, or genetic markers, can be used to divide the variation found in human mtDNA into family trees in which closely related individuals share more mutations with each other than do distantly related individuals. In addition, because each cell has hundreds of copies of mtDNA, it is more likely that at least one of these copies will survive the ravages of time than will occur in the case of the two copies of nuclear DNA present in each cell. For these reasons, many aDNA studies focus on mitochondrial variation. Modern Native Americans possess genetic markers in their mtDNA that divide them into at least five matrilineages, or haplogroups, called A, B, C, D, and X (Schurr et al. 1990; M. Brown et al. 1998; Smith et al. 1999). These haplogroups represent a subset of those currently found in Asia. Recent studies of ancient mtDNA from the prehistoric inhabitants of the Americas have confirmed that the majority of ancient Native Americans also fall into these five haplogroups (Stone and Stoneking 1993; Parr et al. 1996; O'Rourke et al. 1996; Kaestle and Smith 2001). However, the frequencies of these haplogroups vary significantly among both modern and ancient Native American groups, often following linguistic or geographic boundaries (Merriwether et al. 1994; Lorenz and Smith 1996; Kaestle and Smith 2001).

I found that the frequencies of these haplogroups in the ancient western Nevadans were statistically different from those of the modern inhabitants, and in fact from all modern Native Americans from the western United States studied, except for some groups in California. This dissimilarity in mtDNA haplogroup frequencies supports the hypothesis that the Numic presence in the Great Basin is quite recent, and suggests that the previous inhabitants are most closely related to the modern Native Americans of central California (with whom they appear to have had cultural ties; see Hattori 1982; Moratto 1984). However, phylogenetic analysis of these data also suggests that there was a limited amount of admixture between the expanding Numic group and the previous inhabitants of the western Great Basin (Kaestle and Smith 2001 and references therein).

Many of the studies mentioned in the above examples involved samples from multiple ancient individuals. However, some studies have gleaned information from analyses of the aDNA of a single individual. The study of one individual that has generated the most media coverage to date is the study by Mattias Krings and colleagues (1997) reporting the recovery of mtDNA from a Neanderthal type-specimen more than thirty thousand years old, found near Düsseldorf in the mid–nineteenth century. The authors found that the DNA sequence of a short region of the mtDNA from this Neanderthal was highly divergent from that of a sample of modern humans. The authors concluded that the large average difference in sequence from modern humans, and an equidistant relationship to all modern humans rather than a closer relationship to those in Europe, suggests that "Neandertals did not contribute mtDNA to modern humans," but rather diverged from the modern human lineage more than five hundred thousand years ago (Krings et al. 1997: 27). Although subsequent studies have shown that mtDNA sequences from two other Neanderthals are consistent with this conclusion (Krings et al. 2000; Ovchinnikov et al. 2000), others have suggested alternative explanations that do not require the complete replacement of archaic humans by modern humans (e.g., Relethford 2001).

METHODOLOGICAL DIFFICULTIES

Although the above discussion includes numerous examples of how ancient DNA research can inform many aspects of anthropological inquiry, the study of aDNA is still a young field hampered by many methodological and interpretive difficulties. These include the large obstacle presented by the necessity to detect and eliminate contamination of the samples by exogenous DNA; other impediments are the small sample sizes, problematic methods of group definition, and the modification of analytical methods originally developed for modern samples. In many cases, analyses of aDNA may not be the most appropriate method of inquiry.

Contamination is the most common problem (see Handt, Hoss, Krings, and Paabo 1994; Cooper and Wayne 1998). In general, ancient samples arrive with unknown genetic provenance. That is, we do not know whose DNA has been deposited on the surface of the remains. The sample might include DNA from archaeologists and anthropologists involved in the excavation and subsequent study of the remains. Once the sample reaches the genetic laboratory, lab workers might contaminate the bone, or it might become contaminated by DNA extracted from other samples (modern or ancient) or by the DNA deposited on supposedly sterile lab disposables (tubes, pipette tips, reagents) by the manufacturer.

For these reasons it is vital to any study of aDNA to decontaminate the surface of the sample (by means of ultraviolet irradiation, washing with bleach or hydrochloric acid, mechanical removal of the surface layers, or a combination of these precautions) and to prevent subsequent contamination. All extractions should take place in a sterile room isolated from both modern DNA work and work with amplified aDNA, using sterile equipment that is decontaminated regularly; workers should change protective clothing frequently. Detection of any contamination present is essential and can be accomplished by means of negative controls at every step of the extraction and amplification, by multiple extractions per sample, by multiple amplifications per extraction, and by the further replication of anomalous results. In addition, another laboratory should attempt to replicate at least a subset of the results (see Kelman and Kelman 1999). Instances of contamination could be reduced or eliminated, and probably will be, with modifications of archaeological collection and curation methods. Ideally, the selection of samples for DNA analysis could be performed on-site during the excavation, and an excavator wearing latex gloves could remove those samples to a sterile container immediately. These samples would then be exempt from the usual cleaning or preservation procedures, and would be unavailable for any additional study other than aDNA extraction. Some forms of study presumed to be nondestructive, such as x-ray analysis, should not be applied to samples intended for aDNA analysis, as these methods can destroy intact DNA (Gotherstrom et al. 1995). Given that most aDNA studies require only a few grams of bone per sample, this would not represent a large loss.

Because aDNA studies are expensive and deal mostly with archaeological remains, sample sizes are often limited. Most of the studies published to date include samples from fewer than ten individuals, sometimes from a wide geographic range (e.g., Handt, Hoss, Krings, and Paabo 1994; Hanni et al. 1995; Colson et al. 1997; Krings et al. 1997). This obviously severely limits the statistical analyses that can be applied, reduces the power of those statistical methods that are used, and often precludes the use of this type of data in the evaluation of hypotheses dependent on large sample size. In addition, researchers are often stymied by the small sample sizes of the modern groups with whom the ancient individuals are to be compared. For example, I have found that a survey of the literature on mitochondrial DNA variation in modern Native Americans shows that 14 out of 31 South American groups studied are each represented by 15 or fewer individuals; the same is true of 2 out of 8 Central American groups studied, and 17 out of 46 North American groups studied (Merriwether et al. 1994: fig. 2, my calculations). Only 28 (33 percent) of the 85 Native American groups reported by Andrew Merriwether and colleagues (1994) are represented by more than 30 individuals. Although this problem could be somewhat ameliorated by the large number of nucleotides hypothetically available for study (Takahata and Nei 1985), the linkage of mtDNA sites, the difficulty of amplifying and sequencing additional aDNA fragments, and the lack of survival of intact nuclear DNA in ancient samples negate this advantage to a large degree (Cummings et al. 1995). In addition, population-level differences are more often a matter of difference in frequency of genetic markers, rather than presence or absence of such markers, and thus the small sample sizes of modern and ancient populations remain problematic.

Defining a prehistoric group or population is also more complicated than defining modern groups. Samples often span large time periods, and therefore it is sometimes unclear where the temporal boundaries should be drawn. In addition, membership in modern groups is often determined using linguistic, cultural, and self-identification information, while group identity of ancient individuals is often inferred from limited material, geographic, and morphological evidence. Therefore, the assignment of ancient samples to populations is considerably more problematic. These difficulties result in a higher probability of error in defining ancient populations.

Furthermore, the extended temporal factors associated with aDNA might introduce large errors in conventional population genetic analyses. The majority of population genetic techniques were developed under assumptions that generally hold true for modern populations—including sampling across geographic space but not across much temporal space (i.e., from one or a few generations). New models incorporating the time factor must be found, including simulation studies of the effects of this type of sampling; the latter are currently being developed (e.g., Cabana 2002).

The more conventional morphological studies of bones and teeth can avoid some of these problems. Such studies often are nondestructive, are not subject to contamination concerns, are in some cases cheaper, and can utilize much larger sample sizes when available. Because morphological traits are the product of the interaction of genes and environment, they often can be used as a proxy for genetic data. However, for the same reason, morphological traits are subject to the effects of climate, nutrition, human intervention, and other environmental factors, which can confound their genetic information content. By utilizing noncoding DNA regions, the effects of selection (natural and human) often can be minimized. For these reasons, whereas in general both methods can yield important insights into anthropological questions, and in many cases both data sets will be relevant, it is important to determine which approach is more appropriate to the hypotheses being tested. Neither should be considered inherently superior.

CONTEMPORARY IMPLICATIONS

Most reviews of the utility of aDNA analyses for anthropology, or science in general, reach their conclusion at this point, addressing no other questions (see Hagelberg et al. 1991; Sykes 1993; Brown and Brown 1994; Lister 1994; Merriwether et al. 1994; Stoneking 1995; Cooper and Wayne 1998). However, qualitatively different questions must be addressed when studying the remains of ancient human groups, questions that have as yet barely been considered. These larger concerns stem from the fact that when we study ancient humans we may be studying the ancestors of people alive today who have widely different social, political, religious, and legal beliefs and interests. These studies, therefore, can affect the lives of living humans in unforeseen ways, and their results, especially given today's atmosphere of genetic essentialism, can have serious social consequences and policy implications.

An obvious example of the conflict between scientific study and the rights of the descendants of the ancient people in question is the case of Kennewick Man, which has been discussed in newspapers, scientific journals, television news reports (e.g., *Sixty Minutes,* 25 October 1998), and even a science fiction magazine (Silverberg 1998). In July 1996 the nearly complete skeletal remains of an individual were found near the banks of the Columbia River in Washington. Initial study of these remains suggested that he might be a Euro-American from the historic period, but the stone spear point embedded in his hip suggested otherwise (Chatters 2001; Preston 1997). Given the unusual combination of morphology and material remains, the local coroner authorized additional study of the individual, including carbon 14 dating and aDNA analysis (Chatters 2001). This information was important because Kennewick Man's age and ethnic identity would determine whether further investigation was up to the coroner, or whether the Archaeological Resource Protection Act or the Native American Graves Protection and Repatriation Act (NAGPRA) might apply instead.

After initial study, the morphology of the individual (which most agree is not typical of modern Native Americans), combined with preliminary carbon dating results suggesting an ancient date (ninety-two hundred years before the present), supported the view that the colonization of the Americas was more complicated than many had previously thought—a viewpoint with a growing number of proponents (see Jantz and Owsley 1997; Chatters 2001 and references therein). It was suggested that the identity of the ancestors (and perhaps the descendants) of this ancient individual might be informed by the analysis of his mtDNA (Chatters 2001), because modern and ancient Native Americans possess certain mitochondrial DNA mutations unique to Asians and their descendant populations (as discussed above). I agreed to perform the analyses while a graduate student in David G. Smith's laboratory at the University of California, Davis. That same year, I was forced to abandon it without results when we received a cease-and-desist order from the Army Corps of Engineers (COE), who controlled the land on which the remains were found and who planned to repatriate the remains to local Native American tribal groups without further study. A few years later, at the request of the Department of the Interior, two other aDNA researchers and I completed these genetic studies, only to find that we were unable to amplify aDNA from the samples provided (Kaestle 2000; Merriwether and Cabana 2000; Smith et al. 2000).

Recently, many anthropologists have proposed that the presence of ancient individuals like Kennewick Man may indicate that Native American populations originated in regions of Asia farther west than previously thought, and possessed morphology and genes characteristic of current European and Near Eastern populations (see Jantz and Owsley 1997; Thomas 2000 and references therein). Scientists have pointed to some of the other ancient individuals discovered in the Americas, who also seem to have uncharacteristic morphology, as support for this view. Additional supporting evidence had been seen in the discovery of mitochondrial variants among modern and ancient Native Americans that previously had been found only among modern Europeans and Middle Easterners (M. Brown et al. 1998; note that these variants have more recently been discovered among North Asians [Derenko et al. 2001]). Other anthropologists have disagreed, pointing out that modern Native Americans are quite variable morphologically, as were their ancestors, and that studies of both genes and morphology have shown that racial categories are not, and have never been, valid biological subdivisions of the human species, but rather are an enforcement of typological social categories onto the continuous range of human variation (e.g., Goodman 1997; Marks 1998; Thomas 2000). Further exploration of these hypotheses has obvious scientific interest for a wide range of academic fields, especially anthropology, and some might argue (and have argued) that this imperative overwhelms any Native American objections on religious grounds to the destructive analysis of an individual they consider to be their ancestor.

In fact, protesting what they felt to be a premature identification of the remains as legally Native American and a denial of their rights to study the remains, a group of prominent scientists in the field of Native American prehistory filed suit against the COE to prevent repatriation, and the remains have been held pending a final decision on the case (Chatters 2001; Preston 1997; McDonald 1998). As a result of the suit, scientists chosen by the Department of the Interior (DOI, who took over responsibilities for the case in March 1998), beginning in March 1999, were allowed to examine the remains in an attempt to determine whether they were Native American and, if so, to what modern Native American group they are most closely related (to whom they should be repatriated). Accelerator mass spectrometer dating was authorized, and the results ultimately were consistent with the initial dating of the remains to about ninety-two hundred years before the present (McManamon 2000); therefore, Kennewick Man was Native American by NAGPRA standards.

Although studies of morphology, linguistics, and archaeology suggested, if anything, cultural discontinuity between Kennewick Man and the living Native American tribes in the area (American Association of Physical Anthropologists 2000; Ames 2000; Hunn 2000; Hackenberger 2000; Society for American Archaeology 2000), Interior Secretary Bruce Babbit decided that cultural continuity was indicated by the traditional historical and ethnographic data (Boxberger 2000; Department of the Interior 2000), as well as the geographic proximity, and ordered the remains repatriated (Babbit 2000). The suit against the COE to prevent repatriation, which had been on hold during the government's investigation, went forward. In August 2002, the presiding judge, John Jelderks, found in favor of the plaintiffs. The case is currently on appeal to the Ninth U.S. Circuit Court (Lee 2002).

However, the Native American Graves Protection and Repatriation Act, as many have claimed (see Thomas 2000), was not written as science legislation but as human rights legislation, and the federal government has made it clear that "claimants do not have to establish cultural affiliation with scientific certainty" (*Federal Register* 58, no. 102 [28 May 1993]: 31132). The results of the scientific study of Kennewick Man and other individuals of similar antiquity might have enormous implications for Native Americans. If these individuals are shown to possess mitochondrial lineages that are not Asian-derived, and their "nontypical" morphology is confirmed, this might fuel arguments against Native American rights to sovereignty, land rights, oil rights, and so on. In fact, the Kennewick case has already prompted Richard Hastings, representative from the Washington State district in which the remains were found, to introduce legislation (H.R. 2893 and H.R. 2643) to amend NAGPRA to "ensure that[,] when a federal agency repatriates remains, we can be reasonably confident that the remains are affiliated with that particular tribe or group" (Hastings 1998: 18). In particular, this legislation would have required cultural affiliation to be determined through scientific study, the results of which would be published in the *Federal Register.* The National Congress of American Indians (Abrams 1998) and the Clinton administration condemned this legislation (Clinton administration opposes bill to study Kennewick man 1998), and it died in committee. However, the ongoing legal battle over Kennewick Man continues.

In addition to having legal and political implications, aDNA evidence could be interpreted as contradicting the creation myths of the tribes living in the region, many of whom believe they were created de novo in the area where they currently live (Minthorn 1996; Preston 1997; Lee 1998b). However, a further issue, raised by Nicholas Nicastro, is the question of whose human rights are being protected: "If Kennewick Man is repatriated to an unrelated group or even ancient adversaries (recall he was found with an archaic spear point lodged in his pelvis), have his human rights been respected, or violated?" (1998: 4). Complicating these issues, the Asatru Folk Assembly, a neo-Viking religious group, also filed for repatriation of the Kennewick remains on the grounds that his Caucasoid features and the recently proposed hypotheses of circumpolar migration into the Americas allow for the possibility that Kennewick Man is more closely related to that group than to Native Americans (Lee 1997, 1998a). Although the Asatru Folk Assembly has since dropped its claim due to lack of legal funds (Chatters 2001), this does not change the fact that NAGPRA legislation makes no provision for mediating between two opposing religious views should this issue arise again.

Additional examples of the social and legal implications of my own studies of ancient DNA , some of them unforeseen, include the impact of the results of my study of the ancient inhabitants of the Great Basin. *In combination* with the archaeological and linguistic evidence suggesting a biological *and* cultural discontinuity, the aDNA evidence provides a powerful argument in favor of the hypothesis that the current Native American inhabitants of the Great Basin are relative newcomers to the area (however, it also suggests that a low level of genetic admixture with the previous inhabitants took place during this expansion). Although some Native American inhabitants of the Great Basin find that these results conflict with their own understanding of their creation (M. Brewster pers. comm. 1999; A. Moyle pers. comm. 1999), the results may be seen as consistent with other accounts of the Northern Paiute creation myth and their arrival in the Great Basin (Stewart 1939; Heizer 1970; N. Wright pers. comm. 1999). The oldest mummy discovered to date in the Americas is the Spirit Cave Man, found in 1940 in the western Nevada desert and recently dated to 9,415 years ago (Tuohy and Dansie 1997). These remains resemble those of Kennewick Man in some measurements and do not appear to have close morphological affinities with modern Native American groups (Dansie 1997; Jantz and Owsley 1997, 1998). However, the remains were discovered on land currently associated with the modern Fallon Paiute–Shoshone tribe. In negotiations with the tribe regarding repatriation of these remains, the Nevada state office of the Bureau of Land Management included a consideration of my genetic studies on the ancient inhabitants of the western Great Basin when they suggested that the Spirit Cave remains are not affiliated with the modern Fallon Paiute–Shoshone tribe (Barker et al. 2000; Rose 2000).

These mtDNA results also suggest that the closest living relatives of the ancient inhabitants of the Great Basin are to be found in California. This study was undertaken, under NAGPRA, with the permission of certain Native American groups from the Great Basin. However, NAGPRA states that remains should be repatriated to the group or individual who "can show cultural affiliation by a preponderance of the evidence based upon geographical, kinship, biological, archaeological, anthropological, linguistic, folkloric, oral traditional, historical, or other relevant information or expert opinion" (NAGPRA, Section 7a[4]). Do the results of this study suggest that further study should take place only under the auspices of Native Americans in California? What about the ancient remains that have already been repatriated to local Great Basin groups and reburied by them according to local religious practices? Ironically, anthropological studies have also shown that in many cases genes, and biology in general, may have little to do with *cultural* affiliation, despite the fact that they are listed as valid sources of evidence by NAGPRA.

Contemporary attitudes about the potential for genetic research to solve numerous human problems, coupled with increased media attention to genetic research such as the successful extraction of Neanderthal DNA, appear to have resulted in a rather blind faith that genetic research is the key to studying the remains of ancient individuals. This bias toward scientific research is evidenced in a 1997 *New Yorker* article discussing the recent history of Kennewick Man. The author, Douglas Preston, directly quotes at least thirteen scientists (anthropologists, archaeologists, geologists, geneticists). In contrast, he quotes directly (and for that matter mentions at all) only one Native American, and he does not appear to have discussed the case with the Army Corps of Engineers, the federal agency with jurisdiction over the bones who made the initial decision to repatriate.

This prevailing Western (or perhaps American) attitude that scientific study takes precedence over all else, and that it can be performed in a vacuum with no social, legal, political, or spiritual repercussions or influences, has begun to be challenged within anthropology, medical genetics, and related fields (Greely 1997; Dukepoo 1998; Foster et al. 1998; Foster and Freeman 1998; Juengst 1998). The results of studies on human genetic variation have many implications, including the possibility of social, legal, medical, and economic discrimination against certain groups; the use of the data for racist purposes; the contradiction of oral history and deeply held beliefs about one's origins; the patenting of particular genetic variants; and the rejection of legal claims to membership in a specific group or land rights, to name a few. In addition, the physical act of genetic sampling may be morally or spiritually repugnant to groups or individuals. For these reasons, many people have called for changes in the laws governing the sampling of individuals for genetic and medical studies, and for more stringent informed consent requirements. However, this process is problematic. How do you explain your study in terms and contexts that enable a broad variety of people with widely divergent worldviews to make an informed decision? How do you define a group? Whom do you ask for group consent? These are all questions that apply equally to studies of modern genetic variation, and that have been discussed extensively in the literature, although with varied results (see Marks 1995; Greely 1997; Foster et al. 1998; Juengst 1998; Mittman 1998; Foster and Freeman 1998).

Research involving ancient remains entails additional problems. Obviously, the individual concerned cannot be consulted directly. We are left with two options—consult nobody, or consult people living today. If we choose (or are required by law) to consult people living today, the obvious choice would be the modern descendant(s) of the ancient individual. Unfortunately, this is easier said than done. We often are not sure who the most likely modern descendants might be. In fact, in many cases the genetic study is proposed to answer precisely that question. What evidence should be examined to help us make a decision? If different pieces of evidence are contradictory, how do we decide? What if we gain permission from one group but preliminary data suggest that they are not, in fact, the modern descendants? The above questions point to the inadequacies of the current legislation for dealing with the situation (and perhaps the impossibility of any legislation ever dealing satisfactorily with such complex issues), as well as the common conflicts inherent in the juxtaposition or attempted union of scientific and cultural or religious knowledge. In my experience, the answers to these questions must be determined on a case-by-case basis and reevaluated throughout the research.

In general, consultation with the local inhabitants, regardless of the probability that they are descendants of the people under study, can be mutually beneficial. Researchers may gain insights into the local history and prehistory, as well as gain the modern inhabitants' cooperation and interest in the research project. Local inhabitants sometimes find the results of such studies valuable pieces of knowledge and, in some cases, decide to continue the study of their regional history independently. In addition, the possible implications of the results of such studies for nonlocal groups should be assessed to determine if consultation with other individuals or groups is indicated.

CONCLUSIONS

As with any new field, the potential uses of ancient DNA analysis have only begun to be realized. Perhaps most important among these uses is aDNA's utility for directly testing hypotheses for which we previously had only indirect evidence. The pitfalls are also slowly revealing themselves. Studies of aDNA cannot be used to address all questions with equal utility and must be undertaken with much more care than most analyses of modern human genetics. In addition, the application of these new techniques to the study of humans inherently involves ethical questions, which often are examined only cursorily, if at all. New proposals for aDNA research should be evaluated not only for their scientific merit and promise but also for adequate consideration of the possible social, legal, and political implications of the proposed study.

NOTES

I thank David G. Smith, Dennis O'Rourke, Andrew Merriwether, Jonathan Marks, Keith Hunley, Graciela Cabana, Debra Harry, Nila Wright, and Alvin Moyle for their insightful comments on this essay and ancient DNA research in general. I also thank the Native American individuals and groups who donated modern genetic samples and granted permission for the ancient DNA studies discussed in this paper. My study of ancient DNA from the Great Basin populations was supported by a predoctoral graduate fellowship and a dissertation improvement grant from the National Science Foundation. This paper was written with the support of an Alfred P. Sloan/NSF postdoctoral fellowship in molecular evolution.

REFERENCES

Abrams, G. H. J. 1998. Finding NAGPRA's middle ground. *Anthro. Newsletter* (April): 5.

American Association of Physical Anthropologists. 2000. Statement by the American Association of Physical Anthropologists on the secretary of interior's letter of 21 September 2000 regarding cultural affiliation of Kennewick Man. Http://www.physanth.org/positions/kennewick.html (accessed on 1 December 2002).

Ames, K. M. 2000. Review of the archaeological data. In *Cultural Affiliation Report,* chap. 2. National Parks Service, http://www.cr.nps.gov/aad/kennewick/index.htm (accessed on 1 December 2002).

Babbit, B. 2000. Letter from Secretary of the Interior Bruce Babbitt to Secretary of the Army Louis Caldera regarding disposition of the Kennewick human remains. National Parks Service, http://www.cr.nps.gov/aad/kennewick/index.htm (accessed on 1 December 2002).

Bailey, J. F., M. B. Richards, V. A. Macaulay, I. B. Colson, I. T. James, D. G. Bradley, R. E. M. Hedges, and B. C. Sykes. 1996. Ancient DNA suggests a recent expansion of European cattle from a diverse wild progenitor species. *Proc. R. Soc. Lond.,* ser. B, 263:1467–73.

Barker, P., C. Ellis, and S. Damadio. 2000. Determination of cultural affiliation of ancient human remains from Spirit Cave, Nevada. Reno, Nevada: Bureau of Land Management, Nevada State Office. Http://www.nv.blm.gov/cultural/spirit_cave_man/SC_final_July26.pdf (accessed on 1 December 2002).

Boxberger, D. L. 2000. Review of traditional historical and ethnographic information. In *Cultural Affiliation Report,* chap. 3. National Parks Service, http://www.cr.nps.gov/aad/kennewick/index.htm (accessed on 1 December 2002).

Braun, M., D. Collins Cook, and S. Pfeiffer. 1998. DNA from *Mycobacterium tuberculosis* complex identified in North American, pre-Columbian human skeletal remains. *J. Arch. Sci.* 25:271–77.

Brown, M. D., S. H. Hosseini, A. Torroni, H. J. Bandelt, J. C. Allen, T. G. Schurr, R.

Scozzari, F. Cruciani, and D. C. Wallace. 1998. MtDNA haplogroup X: An ancient link between Europe, western Asia, and North America? *Am. J. Hum. Genet.* 63, no. 6:1852–61.

Brown, T. A. 1999. How ancient DNA may help in understanding the origin and spread of agriculture. *Phil. Trans. R. Soc. Lond.*, ser. B, 354, no. 1379:89–97.

Brown, T. A., R. G. Allaby, K. A. Brown, K. O'Donoghue, and R. Sallares. 1994. DNA in wheat seeds from European archaeological sites. *Experientia* 50:571–75.

Brown, T. A., and K. A. Brown. 1994. Ancient DNA: Using molecular biology to explore the past. *BioEssays* 16, no. 10:719–26.

Cabana, G. 2002. A demographic simulation model to assess prehistoric migrations. Ph.D. diss., Department of Anthropology, University of Michigan, Ann Arbor.

Chatters, J. 2001. *Ancient encounters: Kennewick Man and the first Americans.* New York: Simon and Schuster.

Clinton administration opposes bill to study Kennewick Man. 1998. *Tri-City Herald*, Kennewick/Pasco/Richland, Wash., 10 June 1998, http://www.kennewick-man.com/news/ (accessed on 1 December 2002).

Colson, I. B., M. B. Richards, J. F. Bailey, and B. C. Sykes. 1997. DNA analysis of seven human skeletons excavated from the Terp of Wijnaldum. *J. Arch. Sci.* 24:911–17.

Cooper, A., and R. Wayne. 1998. New uses for old DNA. *Curr. Biol.* 9:49–53.

Cummings, M. P., S. P. Otto, and J. Wakeley. 1995. Sampling properties of DNA sequence data in phylogenetic analysis. *Mol. Biol. Evol.* 12:814–23.

Dansie, A. 1997. Early Holocene burials in Nevada: Overview of localities, research, and legal issues. *Nevada Hist. Soc. Qtly.* 40, no. 1:4–14.

Department of the Interior. 2000. Human culture in the southeastern Columbia plateau, 9500–9000 BP, and cultural affiliation with present-day tribes. Enclosure 3 in Letter from Secretary of the Interior Bruce Babbitt to Secretary of the Army Louis Caldera regarding disposition of the Kennewick human remains. National Parks Service, http://www.cr.nps.gov/aad/kennewick/index.htm (accessed on 1 December 2002).

Derenko, M. V., T. Grzybowski, B. A. Malyarchuk, J. Czarny, D. Miscicka-Sliwka, and I. A. Zakharov. 2001. The presence of mitochondrial haplogroup X in Altains from south Siberia. *Am. J. Hum. Genet.* 69, no. 1:237–41.

Drancourt, N., G. Aboudharam, M. Signoli, O. Dutour, and D. Raoult. 1998. Detection of 400-year-old *Yersinia pestis* DNA in human dental pulp: An approach to the diagnosis of ancient septicemia. *Proc. Natl. Acad. Sci., USA* 95, no. 21:12637–40.

Dukepoo, F. C. 1998. The trouble with the Human Genome Diversity Project. *Molec. Med. Today* (June): 242–43.

Faerman, M., G. Kahila Bar-Gal, D. Filon, C. L. Greenblatt, L. Stager, A. Oppenheim, and P. Smith. 1998. Determining the sex of infanticide victims from the Late Roman Era through ancient DNA analysis. *J. Arch. Sci.* 25:861–65.

Foster, M. W., D. Bernsten, and T. H. Carter. 1998. A model agreement for genetic research in socially identifiable populations. *Am. J. Hum. Genet.* 63:696–702.

Foster, M. W., and W. L. Freeman. 1998. Naming names in human genetic variation research. *Genome Res.* 8:755–57.

Gill, P., P. L. Ivanov, C. Kimpton, R. Piercy, N. Benson, G. Tully, I. Evett, E. Hagelberg, and K. Sullivan. 1994. Identification of the remains of the Romanov family by DNA analysis. *Nature Genetics* 6:130–35.

Goodman, A. 1997. Racializing Kennewick Man. *Anthro. Newsletter* (October): 3–5.

Gotherstrom, A., C. Fischer, K. Linden, and K. Liden. 1995. X-raying ancient bone: A destructive method in connection with DNA analysis. *Laborativ Arkeologi: J. of Nordic Archeol. Sci.* 8:26–28.

Greely, H. T. 1997. The control of genetic research: Involving the "groups between." *Houston Law Rev.* 33, no. 5:1397–1430.

Hackenberger, S. 2000. Cultural affiliation study of the Kennewick human remains: Review of bio-archaeological information. In *Cultural Affiliation Report,* chap. 5. National Parks Service, http://www.cr.nps.gov/aad/kennewick/index.htm (accessed on 1 December 2002).

Hagelberg, E. 1997. Ancient and modern mitochondrial DNA sequences and the colonization of the Pacific. *Electrophoresis* 18:1529–33.

Hagelberg, E., L. S. Bell, T. Allen, A. Boyde, S. J. Jones, and J. B. Clegg. 1991. Analysis of ancient bone DNA: Techniques and applications. *Phil. Trans. R. Soc. Lond.,* ser. B, 333:339–407.

Handt, O., M. Hoss, M. Krings, and S. Paabo. 1994. Ancient DNA: Methodological challenges. *Experientia* 50:524–29.

Handt, O., M. Richards, M. Trommsdorff, C. Kilger, J. Simanainen, O. Georgiev, K. Bauer, A. Stone, R. Hedges, W. Schaffner, G. Utermann, B. Sykes, and S. Paabo. 1994. Molecular genetic analyses of the Tyrolean ice man. *Science* 264:1775–78.

Hanni, C., A. Begue, V. Laudet, and D. Stehelin. 1995. Molecular typing of neolithic human bones. *J. Arch. Sci.* 22:649–58.

Hastings, R. 1998. Kennewick Man inspires NAGPRA amendments. *Anthro. Newsletter* (January): 18–19.

Hattori, E. M. 1982. The archaeology of Falcon Hill, Winnemucca Lake, Washoe County, Nevada. Ph.D. diss., Department of Anthropology, Washington State University, Pullman.

Heizer, R. 1970. Ethnographic notes on the Northern Paiute of the Humboldt Sink, west central Nevada. In *Languages and cultures of western North America: Essays in honor of Sven S. Liljeblad,* ed. E. Swanson Jr. Pocatello: Idaho State University Press.

Herrmann, B., and S. Hummel. 1993. *Ancient DNA: Recovery and analysis of genetic material from paleontological, archaeological, museum, medical, and forensic specimens.* New York: Springer-Verlag.

Hunn, E. S. 2000. Review of linguistic information. In *Cultural Affiliation Report,* chap. 4. National Parks Service, http://www.cr.nps.gov/aad/kennewick/index.htm (accessed on 1 December 2002).

Jantz, R. L., and D. W. Owsley. 1997. Pathology, taphonomy, and cranial morphometrics of the Spirit Cave mummy. *Nev. Hist. Soc. Qtly.* 40, no. 1:62–84.

———. 1998. How many populations of early North Americans were there? *Am. J. Phys. Anthro.* 26, suppl.:128.

Juengst, E. T. 1998. Group identity and human diversity: Keeping biology straight from culture. *Am. J. Hum. Genet.* 63:673–77.

Kaestle, F. A. 1997. Molecular analysis of ancient Native American DNA from western Nevada. *Nev. Hist. Soc. Qtrly.* 40, no. 1:85–96.

———. 2000. Report on DNA analyses of the remains of "Kennewick Man" from Columbia Park, Washington. In *Report on the DNA testing results of the Kennewick human remains from Columbia Park, Kennewick, Washington,* chap. 2. National Parks

Service, http://www.cr.nps.gov/aad/kennewick/index.htm (accessed on 1 December 2002).

Kaestle, F. A., and D. G. Smith. 2001. Ancient mtDNA evidence for prehistoric population movement: The Numic expansion. *Am. J. Phys. Anthro.* 115:1–12.

Kelman, L. M., and Z. Kelman. 1999. The use of ancient DNA in paleontological studies. *J. Vert. Paleontol.* 19:8–20.

Kolata, G. 1999. *Flu: The story of the great influenza pandemic of 1918 and the search for the virus that caused it.* New York: Farrar, Straus, and Giroux.

Krings, M., C. Capelli, F. Tschentscher, H. Geisert, S. Meyer, A. von Haeseler, K. Grosschmidt, G. Possnert, M. Paunovic, and S. Paabo. 2000. A view of Neanderthal genetic diversity. *Nat. Genet.* 26:144–46.

Krings, M., A. Stone, R. W. Schmitz, H. Krainitzki, M. Stoneking, and S. Paabo. 1997. Neanderthal DNA sequences and the origin of modern humans. *Cell* 90:19–30.

Lee, M. 1997. Ancient ritual pays tribute to Kennewick Man. *Tri-City Herald,* Kennewick/Pasco/Richland, Wash., 28 August, http://www.kennewick-man.com/news/ (accessed on 1 December 2002).

———. 1998a. Tribes, Asatru pay respect to old bones before move to Seattle museum. *Tri-City Herald,* Kennewick/Pasco/Richland, Wash., 30 October, http://www.kennewick-man.com/news/ (accessed on 1 December 2002).

———. 1998b. Ancient one belongs to land, tribal leader says. *Tri-City Herald,* Kennewick/Pasco/Richland, Wash., 1 May, http://www.kennewick-man.com/news/ (accessed on 1 December 2002).

———. 2002. NW tribes announce Kennewick Man appeal. *Tri-City Herald,* Kennewick/Pasco/Richland, Wash., 23 October, http://www.kennewick-man.com/news/ (accessed on 1 December 2002).

Lin, Z., T. Kondo, T. Minamino, M. Ohtsuji, J. Ishigami, T. Takayasu, R. Sun, and T. Ohshima. 1995. Sex determination by polymerase chain reaction on mummies discovered at Taklamakan Desert in 1912. *Forensic Sci. Int.* 75, no. 2–3:197–205.

Lister, A. M. 1994. Ancient DNA: Not quite *Jurassic Park. Trends Ecol. Evol.* 9, no. 3:82–84.

Loreille, O., J-D. Vigne, C. Hardy, C. Callou, F. Treinen-Claustre, N. Dennebouy, and M. Monnerot. 1997. First distinction of sheep and goat archaeological bones by the means of their fossil mtDNA. *J. Arch. Sci.* 24:33–37.

Lorenz, J. G., and D. G. Smith. 1996. MtDNA variation among native North Americans: Geographic and ethnic distribution of the four founding mtDNA haplogroups. *Am. J. Phys. Anthro.* 101:307–23.

Madsen, D. B., and D. Rhode. 1994. *Across the West: Human populations movement and the expansion of the Numa.* Salt Lake City: University of Utah Press.

Marks, J. 1995. The Human Genome Diversity Project. *Anthro. Newsletter* (April): 72.

———. 1998. Replaying the race card. *Anthro. Newsletter* (May): 1–5.

Mawk, E. J., M. Hyman, and M. W. Rowe. 2002. Re-examination of ancient DNA in Texas rock paintings. *J. Arch. Sci.* 29:301–6.

McDonald, K. A. 1998. Researchers battle for access to a 9,300-year-old skeleton. *Chron. Higher Ed.* (22 May): A18–22.

McManamon, F. P. 2000. Memorandum: Results of radiocarbon dating the Kennewick human skeletal remains. National Parks Service, http://www.cr.nps.gov/aad/kennewick/index.htm (accessed on 1 December 2002).

Merriwether, D. A., and G. S. Cabana. 2000. Kennewick Man ancient DNA analysis: Final report submitted to the Department of the Interior, National Park Service. In *Report on the DNA testing results of the Kennewick human remains from Columbia Park, Kennewick, Washington,* chap. 3. National Parks Service, http://www.cr.nps.gov/aad/kennewick/index.htm (accessed on 1 December 2002).

Merriwether, D. A., F. Rothhammer, and R. E. Ferrell. 1994. Genetic variation in the New World: Ancient teeth, bone, and tissue as sources of DNA. *Experientia* 50:592–601.

Minthorn, A. 1996. Human remains should be reburied. http://www.umatilla.nsn.us/kennman.html (accessed on 1 December 2002).

Mittman, L. S. 1998. Genetic education to diverse communities: Employing a community empowerment model. *Community Genetics* 1, no. 3:160–65.

Moratto, M. J. 1984. *California archaeology.* Orlando, Fla.: Academic Press.

Nicastro, N. 1998. Repatriation law re-buries the past. *Anthro. Newsletter* (January): 4.

O'Rourke, D. H., S. W. Carlyle, and R. L. Parr. 1996. Ancient DNA: A review of methods, progress, and perspectives. *Am. J. Hum. Biol.* 8, no. 5:557–71.

Ovchinnikov, I. V., A. Gotherstrom, G. P. Romanova, V. M. Kharitonov, K. Liden, and W. Goodwin. 2000. Molecular analysis of Neanderthal DNA from the northern Caucasus. *Nature* 404:490–93.

Parr, R. L., S. W. Carlyle, and D. H. O'Rourke. 1996. Ancient DNA analysis of Fremont Amerindians of the Great Salt Lake wetlands. *Am. J. Phys. Anthro.* 99:507–18.

Preston, D. 1997. The lost man: Why are scientists suing the government over the nine-thousand-year-old Kennewick man? *New Yorker* (16 June): 70–81.

Reese, R. L., M. Hyman, M. W. Rowe, J. N. Derr, and S. K. Davis. 1996. Ancient DNA from Texas pictographs. *J. Arch. Sci.* 23:269–77.

Relethford, J. H. 2001. Absence of regional affinities of Neanderthal DNA with living humans does not reject multiregional evolution. *Am. J. Phys. Anthro.* 115:95–98.

Rose, M. 2000. Spirit Cave and Kennewick. *Archaeology* (27 September), http://www.archaeology.org/magazine.php?page=online/news/kennewick5 (accessed on 1 December 2002).

Schafer, S. M. 1998. Pentagon IDs "unknown soldier." Associated Press, 29 June 1998.

Schultes, T., S. Hummel, and B. Herrmann. 2000. Reconstruction of kinship based on Bronze Age materials from the Liechtenstein Grotto. *Homo* 51:S118.

Schurr, T. G., S. W. Ballinger, Y. Gan, J. A. Hodge, D. A. Merriwether, D. N. Lawrence, W. C. Knowler, K. M. Weiss, and D. C. Wallace. 1990. Amerindian mitochondrial DNAs have rare Asian mutations at high frequencies, suggesting they derived from four primary maternal lineages. *Am. J. Hum. Genet.* 46:613–23.

Silverberg, R. 1998. Dem bones, dem bones. *Asimov's Science Fiction* (October-November): 4–7.

Smith, D. G., R. S. Malhi, J. A. Eshleman, and F. A. Kaestle. 2000. Report on DNA analyses of the remains of "Kennewick Man" from Columbia Park, Washington. In *Report on the DNA testing results of the Kennewick human remains from Columbia Park, Kennewick, Washington,* chap. 4. National Parks Service, http://www.cr.nps.gov/aad/kennewick/index.htm (accessed on 1 December 2002).

Smith, D. G., R. S. Malhi, J. Eshleman, J. G. Lorenz, and F. A. Kaestle. 1999. Distribution of mtDNA haplogroup X among native North Americans. *Am. J. Phys. Anthro.* 110, no. 3:271–84.

Society for American Archaeology. 2000. *Society for American Archaeology position paper,* SAA Web site, http://www.saa.org/repatriation/lobby/kennewickc8.html (accessed on 1 December 2002).

Stewart, O. 1939. The Northern Paiute bands. *Univ. of Cal. Anthro. Records* 2, no. 3:127–49.

Stone, A. C., G. R. Milner, S. Paabo, and M. Stoneking. 1996. Sex determination of ancient human skeletons using DNA. *Am. J. Phys. Anthro.* 99:231–38.

Stone, A. C., and M. Stoneking. 1993. Ancient DNA from a pre-Columbian Amerindian population. *Am. J. Phys. Anthro.* 92:463–71.

———. 1999. Analysis of ancient DNA from a prehistoric Amerindian cemetery. *Phil. Trans. R. Soc. Lond.,* ser. B, 354:153–59.

Stoneking, M. 1995. Ancient DNA: How do you know when you have it, and what can you do with it? *Am. J. Hum. Genet.* 57:1259–62.

Sykes, B. 1993. Less cause for grave concern. *Nature* 366:513.

Takahata, N., and M. Nei. 1985. Gene genealogy and variance of interpopulational nucleotide differences. *Genetics* 110:325–44.

Taubenberger, J. K., A. H. Reid, A. E. Krafft, K. E. Bijwaard, and T. G. Fanning. 1997. Initial genetic characterization of the 1918 "Spanish" influenza virus. *Science* 275:1793–95.

Thomas, D. H. 2000. *Skull wars: Kennewick Man, archaeology, and the battle for Native American identity.* New York: Nevraumont Publishing.

Tuohy, D. R., and A. Dansie. 1997. Papers on Holocene burial localities presented at the Twenty-fifth Great Basin Anthropology Conference, 10–12 October 1996. *Nev. Hist. Soc. Qtrly.* 40, no. 1:1–3.

Wang, L., H. Oota, N. Saitou, F. Jin, T. Matsushita, and S. Ueda. 2000. Genetic structure of a 2,500-year-old human population in China and its spatiotemporal changes. *Mol. Biol. Evol.* 17, no. 9:1396–1400.

CONTRIBUTORS

Troy Duster, Institute for the History of the Production of Knowledge, New York University, and Institute for the Study of Social Change, University of California, Berkeley

Arturo Escobar, Department of Anthropology, University of North Carolina, Chapel Hill

Sarah Franklin, Department of Sociology, Cartmel College, Lancaster University, Lancaster, United Kingdom

Joan H. Fujimura, Department of Sociology and the Program in Science and Technology Studies, University of Wisconsin, Madison

Alan H. Goodman, School of Natural Science, Hampshire College, Amherst, Massachusetts

Donna Haraway, Department of History of Consciousness, University of California, Santa Cruz

Deborah Heath, Department of Sociology and Anthropology, Lewis & Clark College, Portland, Oregon

Chaia Heller, Department of Anthropology, University of Massachusetts, Amherst, and Institute of Social Ecology, Plainfield, Vermont

Frederika A. Kaestle, Department of Anthropology, Indiana University, Bloomington

Rick Kittles, National Human Genome Center at Howard University, Washington, D.C

M. Susan Lindee, Department of the History and Sociology of Science, University of Pennsylvania, Philadelphia

Jonathan Marks, Department of Sociology and Anthropology, University of North Carolina, Charlotte

Rayna Rapp, Department of Anthropology, New York University

Hilary Rose, Department of Sociology, City University, London

Charmaine Royal, National Human Genome Center at Howard University, Washington, D.C

Ricardo Ventura Santos, Escola Nacional de Saúde Pública/Fiocruz, Rio de Janeiro

Himla Soodyall, Division of Human Genetics, National Health Laboratory Service, and University of the Witwatersrand, Johannesburg

Karen-Sue Taussig, Department of Anthropology, University of Minnesota, Minneapolis

Alan R. Templeton, Department of Biology, Washington University, St. Louis, Missouri

INDEX

Page numbers in italics indicate illustrative material.

Indexer: Kevin Millham
Compositor: Binghamton Valley Composition, Inc.
Text: 10/12 Baskerville
Display: Baskerville
Printer and Binder: Malloy Lithographing, Inc.